Recent Advances in Motion Planning and Control of Autonomous Vehicles

Recent Advances in Motion Planning and Control of Autonomous Vehicles

Editors

Bai Li
Youmin Zhang
Xiaohui Li
Tankut Acarman

Basel • Beijing • Wuhan • Barcelona • Belgrade • Novi Sad • Cluj • Manchester

Editors

Bai Li
College of Mechanical and
Vehicle Engineering,
Hunan University
Changsha, China

Youmin Zhang
Department of Mechanical,
Industrial and Aerospace
Engineering,
Concordia University
Montreal, QC, Canada

Xiaohui Li
College of Intelligence
Science and Technology,
National University of
Defense Technology
Changsha, China

Tankut Acarman
Department of Computer
Engineering, Galatasaray
University
Istanbul, Turkey

Editorial Office
MDPI
St. Alban-Anlage 66
4052 Basel, Switzerland

This is a reprint of articles from the Special Issue published online in the open access journal *Electronics* (ISSN 2079-9292) (available at: https://www.mdpi.com/journal/electronics/special_issues/auto_vehicles_motion).

For citation purposes, cite each article independently as indicated on the article page online and as indicated below:

Lastname, A.A.; Lastname, B.B. Article Title. *Journal Name* **Year**, *Volume Number*, Page Range.

ISBN 978-3-0365-9766-9 (Hbk)
ISBN 978-3-0365-9767-6 (PDF)
doi.org/10.3390/books978-3-0365-9767-6

© 2023 by the authors. Articles in this book are Open Access and distributed under the Creative Commons Attribution (CC BY) license. The book as a whole is distributed by MDPI under the terms and conditions of the Creative Commons Attribution-NonCommercial-NoDerivs (CC BY-NC-ND) license.

Contents

Bai Li, Xiaoming Chen, Tankut Acarman, Xiaohui Li and Youmin Zhang
Recent Advances in Motion Planning and Control of Autonomous Vehicles
Reprinted from: *Electronics* **2023**, *12*, 4881, doi:10.3390/electronics12234881 1

Bai Li, Shiqi Tang, Youmin Zhang and Xiang Zhong
Occlusion-Aware Path Planning to Promote Infrared Positioning Accuracy for Autonomous Driving in a Warehouse
Reprinted from: *Electronics* **2021**, *10*, 3093, doi:10.3390/electronics10243093 7

Hao Wang, Guoqing Li, Jie Hou, Lianyun Chen and Nailian Hu
A Path Planning Method for Underground Intelligent Vehicles Based on an Improved RRT* Algorithm
Reprinted from: *Electronics* **2022**, *11*, 294, doi:10.3390/electronics11030294 23

Qi Zhang, Yukai Song, Peng Jiao and Yue Hu
A Hybrid and Hierarchical Approach for Spatial Exploration in Dynamic Environments
Reprinted from: *Electronics* **2022**, *11*, 574, doi:10.3390/electronics11040574 41

Taekgyu Lee, Dongyoon Seo, Jinyoung Lee and Yeonsik Kang
Real-Time Drift-Driving Control for an Autonomous Vehicle: Learning from Nonlinear Model Predictive Control via a Deep Neural Network
Reprinted from: *Electronics* **2022**, *11*, 2651, doi:10.3390/electronics11172651 63

Jianxiang Sun and Yadong Liu
A Hybrid Asynchronous Brain–Computer Interface Based on SSVEP and Eye-Tracking for Threatening Pedestrian Identification in Driving
Reprinted from: *Electronics* **2022**, *11*, 3171, doi:10.3390/electronics11193171 81

Biao Xu, Shijie Yuan, Xuerong Lin, Manjiang Hu, Yougang Bian and Zhaobo Qin
Space Discretization-Based Optimal Trajectory Planning for Automated Vehicles in Narrow Corridor Scenes
Reprinted from: *Electronics* **2022**, *11*, 4239, doi:10.3390/electronics11244239 93

Kangle Hu and Kai Cheng
Trajectory Planning for an Articulated Tracked Vehicle and Tracking the Trajectory via an Adaptive Model Predictive Control
Reprinted from: *Electronics* **2023**, *12*, 1988, doi:10.3390/electronics12091988 113

Xiao Hu, Kai Hu, Datian Tao, Yi Zhong and Yi Han
GIS-Data-Driven Efficient and Safe Path Planning for Autonomous Ships in Maritime Transportation
Reprinted from: *Electronics* **2023**, *12*, 2206, doi:10.3390/electronics12102206 135

Yong Fang and Xiaoyan Peng
Micro-Factors-Aware Scheduling of Multiple Autonomous Trucks in Open-Pit Mining via Enhanced Metaheuristics
Reprinted from: *Electronics* **2023**, *12*, 3793, doi:10.3390/electronics12183793 155

Tantan Zhang, Hu Li, Yong Fang, Man Luo and Kai Cao
Joint Dispatching and Cooperative Trajectory Planning for Multiple Autonomous Forklifts in a Warehouse: A Search-and-Learning-Based Approach
Reprinted from: *Electronics* **2023**, *12*, 3820, doi:10.3390/electronics12183820 171

Yuheng Chen and Yougang Bian
Tube-Based Event-Triggered Path Tracking for AUV against Disturbances and Parametric Uncertainties
Reprinted from: *Electronics* **2023**, *12*, 4248, doi:10.3390/electronics12204248 . . . . **197**

Editorial

Recent Advances in Motion Planning and Control of Autonomous Vehicles

Bai Li [1], Xiaoming Chen [1], Tankut Acarman [2], Xiaohui Li [3] and Youmin Zhang [4,*]

1. College of Mechanical and Vehicle Engineering, Hunan University, Changsha 410082, China; libai@zju.edu.cn (B.L.); xiaomingchen@hnu.edu.cn (X.C.)
2. Department of Computer Engineering, Galatasaray University, Istanbul 34349, Turkey; tacarman@gsu.edu.tr
3. College of Intelligence Science, National University of Defense Technology, Changsha 410073, China; xiaohui_lee@nudt.edu.cn
4. Department of Mechanical, Industrial and Aerospace Engineering, Concordia University, Montreal, QC H3G 1M8, Canada
* Correspondence: ymzhang@encs.concordia.ca

1. Introduction

An autonomous vehicle operates without human intervention, marking advancements in navigating structured urban roads and unstructured environments. Central to its operation are two pivotal components: planning and control. The planning module creates an open-loop trajectory, which lays out a spatiotemporal path for the ego vehicle to follow. The control module's role is to accurately follow this planned trajectory in a closed-loop manner and adeptly handle a spectrum of conditions ranging from varying road and weather situations to disruptive driving scenarios and even extending to abnormal circumstances, such as physical malfunctions and cyber threats. These modules are crucial because they embody the core intelligence of an autonomous system. This Special Issue aims to showcase the latest developments in planning and control strategies, which play critical roles in the evolution of autonomous vehicle technology.

Qualified submitted papers should focus on how the proposed planning and/or control method solves real-world problems. The editorial board would maintain a high standard in prescreening submitted drafts that are methodology oriented instead of task oriented. Notably, this Special Issue also welcomes papers that discuss methods indirectly related to planning and control as long as they facilitate the planning or control module. Topics of interest include but are not limited to the following:

- Path/trajectory/motion planning and replanning;
- Path/trajectory/motion control;
- On-road/off-road planning and control;
- Modeling and simulation methods for planning and/or control;
- Testing and validation methods related to planning and/or control;
- Safety-related issues with planning and control;
- Security-related issues with planning and control;
- Human–machine interaction related to planning and/or control;
- Intelligent techniques/methods for planning and/or control;
- Integration of planning and control;
- Reviews of planning or control methodologies;
- Data-driven/model-based planning or control;
- Comparisons among different types of planning or control methods;
- Fault-tolerant planning and control;
- Cooperative planning and control;
- Real-world applications of planning and control.

The rest of this editorial paper is organized as follows. Section 2 briefly reviews the 11 papers published in this Special Issue. Section 3 presents our perspective, and Section 4 concludes the article.

2. Overview of Contributions

The first paper published in this Special Issue is titled Occlusion-aware Path Planning to Promote Infrared Positioning Accuracy for Autonomous Driving in a Warehouse. The paper proposes an occlusion-aware path planning method for autonomous vehicles in indoor environments and specifically addresses challenges in infrared positioning systems. In these systems, the vehicle, equipped with an infrared emitter, often faces signal occlusion issues due to obstacles, which can lead to inaccurate positioning if fewer than three receivers detect the signal. To tackle this issue, the study introduces a four-layered planner to enhance the accuracy of indoor infrared positioning by ensuring that the vehicle navigates areas with minimal occlusion while adhering to collision avoidance and kinematic constraints. The planner's effectiveness in reducing positioning errors and maintaining trackability is validated through simulations.

The second paper is titled A Path Planning Method for Underground Intelligent Vehicles based on an Improved RRT* Algorithm. This paper addresses the challenge of path planning for underground intelligent vehicles, with a particular focus on environments with complex drifts and vehicles with articulated structures (i.e., tractor–trailer vehicles). Recognizing the unique demands/features of underground scenarios, such as narrow, curved spaces without GPS, and the need for precise control due to the vehicles' complex kinematics, the study proposes an enhanced path planning method. This method is based on an improved version of the rapidly exploring random tree star (RRT*) algorithm and is tailored to meet the specific constraints of underground intelligent vehicles. The improvements to the RRT* algorithm include dynamic step size adjustment, steering angle constraints, and optimal tree reconnection strategies. These modifications aim to ensure that the path planning is efficient and compatible with the articulated structure of the vehicles and the challenging underground environment. The effectiveness of this improved algorithm is demonstrated through simulation case studies to showcase the improved algorithm's ability to generate paths with short lengths, few explored nodes, and high steering angle efficiency.

The third paper is titled A Hybrid and Hierarchical Approach for Spatial Exploration in Dynamic Environments. This paper introduces a novel, three-tiered hierarchical model called IRHE-SFVO for spatial exploration in dynamic environments and for addressing the challenge in AI system tasks, such as search and rescue. The model combines deep reinforcement learning for high-level exploration with a rule-based, real-time, obstacle-avoidance approach for local movement. It uses a global exploration module to identify distant, reachable targets for exploration and a local movement module that employs an optimistic A* algorithm and an improved finite-time velocity obstacle method for safe, efficient navigation. This approach results in smooth, natural movements and improved exploration efficiency, as demonstrated by tests on various 2D dynamic maps. The model's hierarchical structure simplifies training and enhances movement quality, marking an improvement over current spatial exploration techniques.

The fourth paper is titled Real-time Drift-driving Control for an Autonomous Vehicle: Learning from Nonlinear Model Predictive Control via a Deep Neural Network. This study focuses on developing a drift control method for autonomous vehicles and specifically addresses the challenge of managing oversteers in hazardous conditions, such as sharp curves or slippery roads. Initially, a nonlinear model predictive control (NMPC) method was designed to enable an autonomous vehicle to perform drift maneuvers. However, NMPC's reliance on real-time numerical optimization posed limitations in computational efficiency. To address this issue, the study introduces a deep neural network (DNN)-based controller trained using datasets generated from the NMPC method. This DNN-based controller effectively replaces NMPC; it reduces the computational load while maintaining similar control performance. The study's innovation lies in its successful integration of a data-driven approach into drift control. It demonstrates the DNN-based controller's capability to accurately track predefined trajectories under realistic simulation scenarios. This approach enhances real-time performance and has potential for broad applications in autonomous vehicle control, particularly in safety-critical situations.

The fifth paper is titled A Hybrid Asynchronous Brain–computer Interface Based on SSVEP and Eye-tracking for Threatening Pedestrian Identification in Driving. This paper introduces a multimodal hybrid brain–computer interface (BCI) system that integrates eye-tracking and electroencephalogram (EEG) signals to identify potentially threatening pedestrians in a driving context. Traditional steady-state visual evoked potential (SSVEP) BCIs in automatic driving can cause driver fatigue, so this study aims to improve interaction efficiency and reduce fatigue. The system works by superimposing stimulus arrows of different frequencies and directions on pedestrian targets, and subjects scan these arrows until they identify a threatening pedestrian. The hybrid BCI system distinguishes between active and idle states by using thresholds established in offline experiments, and subjects select pedestrians on the basis of their judgment in online experiments. This approach enhances selection accuracy by filtering low-confidence results. The system's effectiveness is demonstrated through experiments with six subjects. Its performance is superior to that of a single SSVEP-BCI system. It has an average selection time of 1.33 s, 95.83% accuracy, and an information transfer rate of 67.50 bits/min. This hybrid BCI system that combines eye tracking and SSVEP offers a promising solution for dynamic pedestrian identification in driving while enhancing safety and comfort.

The sixth paper is titled Space Discretization-based Optimal Trajectory Planning for Automated Vehicles in Narrow Corridor Scenes. The study addresses the challenge of optimal trajectory planning for automated vehicles in narrow corridor scenarios, a task characterized by limited space and the need for safe, feasible, and smooth navigation. The study introduces a novel space discretization-based optimal trajectory planning method that focuses on minimizing travel time and avoiding boundary collisions. The approach involves creating a mathematically described driving corridor model and a trajectory optimization model that incorporates various constraints, such as vehicle kinematics, collision avoidance, and actuator limitations, to ensure feasibility and comfort. The method's effectiveness is demonstrated through simulations and field tests, which show its superiority over baseline methods in terms of smoothness, computational efficiency, and reducing tracking errors.

The seventh paper is titled Trajectory Planning for an Articulated Tracked Vehicle and Tracking the Trajectory via an Adaptive Model Predictive Control. This study addresses the challenge of trajectory planning and tracking control for articulated tracked vehicles (ATVs), which are complex because of their unique steering mechanisms and nonlinear dynamics. The study introduces a two-step approach. First, the hybrid A-star method combined with minimum snap smoothing is employed for feasible kinematic trajectory planning. Second, an adaptive model predictive controller (AMPC) is implemented for precise trajectory tracking. The hybrid A-star method is selected for its effectiveness in generating smooth paths suitable for ATVs' nonholonomic nature, and the AMPC is designed to manage ATVs' nonlinearity through a linear-parameter-varying kinematic error-tracking model. This approach improves tracking accuracy and reduces computational load compared with standard model predictive controllers.

The eighth paper is titled GIS-data-driven Efficient and Safe Path Planning for Autonomous Ships in Maritime Transportation. This paper introduces a novel path planning method for autonomous ships and addresses the limitations of existing approaches that often disregard the dynamic constraints of ships, leading to impractical and inefficient trajectories. The proposed method, efficient and safe path planning (ESP), integrates ship dynamics into the planning process to generate real-time optimal trajectories that are fuel-efficient and smooth. ESP is distinct because of its threefold approach. First, it employs a modified A* search algorithm called A-turning to find an optimal path with minimal sharp turns by using geographic data from a geographic information system. Second, it formulates a minimum-snap trajectory optimization problem by incorporating dynamic ship constraints to ensure a smooth, collision-free trajectory with minimal fuel consumption. Last, ESP includes a local trajectory replanner based on B-spline for real-time avoidance of unexpected obstacles. The method's effectiveness is demonstrated through data-driven simulations, which show ESP's ability to plan safe global trajectories, minimize turning points, and reduce fuel consumption while swiftly adapting to avoid unforeseen obstacles.

The ninth paper is titled Micro-Factors-Aware Scheduling of Multiple Autonomous Trucks in Open-pit Mining via Enhanced Metaheuristics. The paper addresses the task of scheduling autonomous trucks in open-pit mines, which is complex because of the need for efficient coordination in dynamic environments. The study introduces a high-precision scheduling model that considers micro-level temporal and spatial factors to optimize the energy consumption, time, and output in the transportation system. Unique to this model is the inclusion of charging requirements for autonomous trucks, a critical aspect that is often overlooked in traditional scheduling. The methodological core of this research is the integration of a Voronoi diagram for accurately estimating the traverse average speed of each autonomous truck.

The tenth paper is titled Joint Dispatching and Cooperative Trajectory Planning for Multiple Autonomous Forklifts in a Warehouse: A Search-and-learning-based Approach. The paper explores the task of dispatching and cooperative trajectory planning for multiple autonomous forklifts in warehouse environments. Traditionally, dispatching and planning were treated as separate processes, leading to suboptimal motion quality in forklift teams. The study introduces a novel joint dispatching and planning method that simultaneously addresses these issues and aims to optimize cooperative trajectories. This method stands out because of its rapid execution, minimal computational demands, and high-quality solutions. It involves enumerating potential goals for each forklift and evaluating dispatch solutions by using an improved hybrid A* search algorithm enhanced with an artificial neural network for improved cost assessment. This approach ensures computational efficiency, kinematic feasibility, and collision avoidance and can rapidly find optimal solutions. The integration of neural networks into the dispatching process reduces the warehouse mission completion time by 2% compared with that in strategies without machine learning. The study highlights the importance of balancing shelf-filling states to prevent end-mission deadlocks and demonstrates that the proposed method allows forklifts to cooperatively find feasible trajectories quickly while maintaining efficiency even when priorities shift during tasks.

The eleventh paper is titled Tube-based Event-triggered Path Tracking for AUV against Disturbances and Parametric Uncertainties. The paper introduces a novel tube-based event-triggered path-tracking strategy to improve disturbance rejection in autonomous underwater vehicle (AUV) path tracking. The strategy combines a speed control law that uses linear model predictive control (LMPC) and a tube model predictive control (tube MPC) scheme. The LMPC controller is designed to reduce path-tracking deviation, and the tube MPC scheme calculates optimal control inputs, thus enhancing disturbance rejection. This approach considers AUVs' nonlinear hydrodynamic characteristics. It uses linear matrix inequality to establish tight constraints and a feedback matrix, both of which are calculated offline for real-time efficiency. An event-triggering mechanism adjusts these constraints based on surge speed changes, allowing for adaptive control. The strategy's effectiveness is demonstrated through numerical simulations, which show improved path-tracking performance and real-time capability. The paper addresses the challenge of robust control in AUVs in consideration of the complexities of underwater dynamics and the need for efficient, real-time solutions.

3. Emerging Trends and Perspectives

Based on the insights obtained from these papers, we anticipate several future trends in the field of autonomous driving planning and control.

Enhancing Efficiency in Planners for Tailored Autonomous Driving Scenarios

In autonomous driving, motion planning methods are broadly categorized into sampling-based [1], search-based [2], optimization-based [3], simulation-based [4], and learning-based [5] techniques. Each of these methodologies has its strengths and limitations. However, every real-world autonomous driving scenario presents distinct characteristics, necessitating specific task-related constraint design or trajectory preference design in motion planning formulation. Although various planner types are available for general applications, they often struggle to meet the specific needs of real-world scenarios. The autonomous driving research area, which has grown in the past decade, no longer gains substantial benefits from generic planners that do not effectively tackle real-world challenges. In the future, the development of motion planners is expected to increasingly focus on distinct and tangible scenarios. The emphasis is likely to shift toward devising solutions that are not only theoretically robust but also practically adept at addressing real-world issues in autonomous driving [6].

Leveraging Multimodal and Hierarchical Control Strategies

Future developments in autonomous driving are expected to increasingly focus on complex multimodal and hierarchical control strategies that can handle the dynamic and intricate nature of driving environments. The integration of deep learning's adaptability and advanced processing with the consistent reliability of rule-based methods is a promising direction. This integration can boost the adaptability and efficiency of autonomous driving systems, enabling them to effectively manage and adapt to changing conditions and complex tasks.

Enhancing Trajectory Planning Intelligence via Human–Machine Interaction

By integrating brain–computer interfaces and eye-tracking technologies, autonomous systems can now incorporate human input in various forms, such as human languages. This integration taps into the potential of cutting-edge fields, such as large language models, allowing for a dynamic and interactive driving experience. Such advancements mean that human drivers can assist autonomous systems through verbal commands or receive guidance through machine-interpreted human language, thus creating a collaborative and intuitive driving environment. This approach not only adds an extra layer of information to the decision-making process but also provides an additional safety net, making autonomous driving safe and aligned with human instincts and responses.

Re-evaluating the Role of Machine Learning in Autonomous Driving Planning

The integration of machine learning, particularly deep and reinforcement learning, is a growing trend in the field of autonomous driving motion planning. These methods, known for their ability to process and learn from large datasets, promise advancements in understanding and navigating complex and interactive driving environments. However, their current applications remain confined to simulations, with limited real-world deployment on real vehicles. This situation is partly due to the methods' generalization challenges; they sometimes act as supplementary tools relying on rule-based systems for core decision-making [7]. The true suitability of machine learning approaches for autonomous driving planning is a subject of ongoing research. A balance must exist between the innovative potential of these learning-based methods and the reliability of traditional rule-based approaches. This balance is crucial to developing practical, safe, efficient autonomous driving systems that can operate effectively in diverse and unpredictable real-world scenarios.

4. Conclusions

This Special Issue has successfully showcased a range of innovative approaches in the planning and control of autonomous vehicles by drawing insights from 11 groundbreaking papers. Our editorial has revisited these contributions, emphasizing their role in tackling real-world challenges and setting the stage for future advancements in autonomous driving. The collected papers highlight the shift toward highly specialized planning and control methods, the integration of complex control strategies, the enhancement of human–machine interaction, and the evolving application of machine learning in this dynamic field.

The field of autonomous driving is poised for considerable advancements. The focus on the development of planners and controllers that are effective in real-world scenarios marks a major progression. The incorporation of human input through advanced interfaces and the strategic use of machine learning techniques are redefining the capabilities of autonomous vehicles. These developments are not just technological achievements; they represent a shift toward highly intuitive, safe, and efficient transportation solutions.

Autonomous driving technology is filled with promise and potential. As we continue to innovate, the vision of safe, efficient, accessible transportation becomes increasingly realistic. This Special Issue not only captures the current state of the art but also serves as a guidepost for the future, where autonomous vehicles are expected to become an integral part of our daily lives by reshaping our approach to mobility and connectivity. The path forward is filled with excitement and endless possibilities. We are at the threshold of a new era in transportation driven by intelligence, adaptability, and a commitment to enhancing the human experience.

Author Contributions: Conceptualization, B.L.; investigation, X.C.; writing—original draft preparation, B.L., X.L. and X.C.; writing—review and editing, T.A. and Y.Z.; funding acquisition, Y.Z. All authors have read and agreed to the published version of the manuscript.

Funding: This research was funded by the Lotus Youth Talent Program of Hunan Province, China under grant number 2023RC3115.

Conflicts of Interest: The authors declare no conflict of interest.

References

1. Ma, L.; Xue, J.; Kawabata, K.; Zhu, J.; Ma, C.; Zheng, N. Efficient sampling-based motion planning for on-road autonomous driving. *IEEE Trans. Intell. Transp. Syst.* **2015**, *16*, 1961–1976. [CrossRef]
2. Ajanovic, Z.; Lacevic, B.; Shyrokau, B.; Stolz, M.; Horn, M. Search-based optimal motion planning for automated driving. In Proceedings of the 2018 IEEE/RSJ International Conference on Intelligent Robots and Systems (IROS), Madrid, Spain, 1–5 October 2018; pp. 4523–4530.
3. Li, B.; Ouyang, Y.; Li, X.; Cao, D.; Zhang, T.; Wang, Y. Mixed-integer and conditional trajectory planning for an autonomous mining truck in loading/dumping scenarios: A global optimization approach. *IEEE Trans. Intell. Veh.* **2023**, *8*, 1512–1522. [CrossRef]
4. Xie, S.; Hu, J.; Bhowmick, P.; Ding, Z.; Arvin, F. Distributed motion planning for safe autonomous vehicle overtaking via artificial potential field. *IEEE Trans. Intell. Transp. Syst.* **2022**, *23*, 21531–21547. [CrossRef]
5. Qi, Z.; Wang, T.; Chen, J.; Narang, D.; Wang, Y.; Yang, H. Learning-based Path Planning and Predictive Control for Autonomous Vehicles with Low-Cost Positioning. *IEEE Trans. Intell. Veh.* **2023**, *8*, 1093–1104. [CrossRef]
6. Chen, L.; Li, Y.; Huang, C.; Xing, Y.; Tian, D.; Li, L.; Hu, Z.; Teng, S.; Lv, C.; Wang, J.; et al. Milestones in Autonomous Driving and Intelligent Vehicles—Part I: Control, Computing System Design, Communication, HD Map, Testing, and Human Behaviors. *IEEE Trans. Syst. Man Cybern. Syst.* **2023**, *53*, 5831–5847. [CrossRef]
7. Dauner, D.; Hallgarten, M.; Geiger, A.; Chitta, K. Parting with Misconceptions about Learning-based Vehicle Motion Planning. *arXiv* **2023**, arXiv:2306.07962.

Disclaimer/Publisher's Note: The statements, opinions and data contained in all publications are solely those of the individual author(s) and contributor(s) and not of MDPI and/or the editor(s). MDPI and/or the editor(s) disclaim responsibility for any injury to people or property resulting from any ideas, methods, instructions or products referred to in the content.

Article

Occlusion-Aware Path Planning to Promote Infrared Positioning Accuracy for Autonomous Driving in a Warehouse

Bai Li [1], Shiqi Tang [1], Youmin Zhang [2] and Xiang Zhong [1,*]

1. College of Mechanical and Vehicle Engineering, Hunan University, Changsha 410082, China; libai@zju.edu.cn (B.L.); tangshiqi@hnu.edu.cn (S.T.)
2. Department of Mechanical, Industrial and Aerospace Engineering, Concordia University, Montreal, QC H3G 1M8, Canada; ymzhang@encs.concordia.ca
* Correspondence: zx5587@126.com

Abstract: Infrared positioning is a critical module in an indoor autonomous vehicle platform. In an infrared positioning system, the ego vehicle is equipped with an infrared emitter while the infrared receivers are fixed onto the ceiling. The infrared positioning result is accurate only when the number of valid infrared receivers is more than three. An infrared receiver easily becomes invalid if it does not receive light from the infrared emitter due to indoor occlusions. This study proposes an occlusion-aware path planner that enables an autonomous vehicle to navigate toward the occlusion-free part of the drivable area. The planner consists of four layers. In layer one, a homotopic A* path is searched for in the 2D grid map to roughly connect the initial and goal points. In layer two, a curvature-continuous reference line is planned close to the A* path using numerical optimal control. In layer three, a Frenet frame is constructed along the reference line, followed by a search for an occlusion-aware path within that frame via dynamic programming. In layer four, a curvature-continuous path is optimized via quadratic programming within the Frenet frame. A path planned within the Frenet frame may violate the curvature bounds in a real-world Cartesian frame; thus, layer four is implemented through trial and error. Simulation results in CarSim software show that the derived paths reduce the poor positioning risk and are easily tracked by a controller.

Keywords: autonomous vehicle; infrared positioning; occlusion-aware path planning; numerical optimal control; dynamic programming; quadratic program

1. Introduction

Indoor positioning is an important area of development with wide applications in surveillance, human motion analysis, logistics, and entertainment [1–5]. As one of the most well-known indoor positioning approaches, infrared positioning is characterized by low energy consumption and high precision [6–8]. In an infrared positioning system, an infrared emitter is installed on a movable target that is required to be localized, and infrared receivers are fixed onto the ceiling. The infrared positioning result is accurate only when the number of valid infrared receivers is more than three. An infrared receiver becomes invalid if it does not receive the light originating from the infrared emitter due to indoor occlusions.

Automated guided vehicles (AGVs) are commonly used in warehouses for cargo delivery [9,10]. However, AGV positioning results easily become inaccurate when cargoes in a warehouse occlude the light beams originating from the emitter installed on an AGV. Instead of improving the positioning of the sensors, the current work considers the planning of occlusion-aware paths that would enable an AGV to drive in the occlusion-free part of the drivable area in a warehouse.

If the cargoes in a warehouse are permanently fixed, then the poor positioning regions may be estimated a priori and regarded as static obstacles in an AGV path planning scheme. In most cases, however, cargo placement is always changing, resulting in unfixed poor

positioning regions. Thus, the concerned occlusion-aware path planner must work fast while guaranteeing its outputs are collision-free and kinematically feasible [11].

1.1. Related Work

The most prevalent path planners in robotics may be classified into sampling-, search-, optimization-, and learning-based methods.

Sampling-based planners generate candidate path/trajectory primitives and then connect selected ones to form a complete solution. Such primitives are typically formed using polynomials [12–14], state lattices [15–17], and closed-loop tracking [18].

Search-based methods divide a continuous solution space into nodes in a graph and then search for a link between the nodes in the graph. Typical searchers include dynamic programming (DP) [19,20], A* [16,21], and rapidly exploring random tree (RRT) [18,22].

An optimization-based planner formulates the concerned planning task as an optimal control problem (OCP) before solving the OCP numerically. Herein, solving an OCP numerically refers to discretizing it into a mathematical programming problem and solving it using a gradient-based optimizer. Typical mathematical programming problems include quadratic programming (QP) [20,23,24], quadratically constrained QP (QCQP) [25], nonlinear programming (NLP) [26–28], and unconstrained optimization problems [12,29]. Most gradient-based optimizers suitable for path planning only exhibit local optimization capabilities because global convergence requires a much longer runtime than one can afford. Given this property, the finally derived optimal path is close to the initial guess [30,31]. Therefore, identifying a good initial guess with global optimality contributes considerably toward finding a high-quality optimum. A sampling-/search-based planner is commonly used to provide a good initial guess [32,33].

A learning-based method generates vehicle motions based on trained data. Artificial neural networks [34], spline-constrained policy networks [35], and double deep Q-networks [36] are used in offline training. Reinforcement learning-based methods [37,38] are applied to online training.

The aforementioned four types of planners have their respective strengths and limitations. Sampling- and search-based methods do not handle the configuration space in its continuous form. Thus, they find suboptimal rather than optimal solutions. They even fail to find any feasible solution if the complexity of the planning scheme is beyond the sampling/search resolution level [23]. Increasing the resolution level is inapplicable when using a sampling-/search-based planner due to the curse of dimensionality. Despite their drawbacks, sampling-/search-based methods can efficiently describe the non-differentiable occlusion-related cost [39,40]. An optimization-based planner finds optimal paths in the continuous solution space; however, it has two typical limitations: (1) occlusion-related constraints/costs cannot be handled due to non-differentiability, and (2) the runtime is considerably longer than that of a sampling-/search-based method. Learning-based methods work fast online after a long offline training process, but they lack interpretability, and thus are rarely tested on real-world vehicles [41,42]. According to the above analysis, concluding that one type of planner fully outperforms another is difficult; instead, maximizing the advantages of each planner type while combining multiple types has been a common practice in this community.

Combining a sampling-/search-based planner with an optimization-based one renders a hierarchical planning framework; however, the optimization layer is time-consuming if the planning scheme is described in the Cartesian frame. At this point, the majority of path planners for on-road cruising scenarios describe the movements of an autonomous vehicle in the Frenet frame to enhance real-time performance. The Frenet frame, also known as the curvilinear frame, has been widely used to standardize the irregular trend of a road [43]. When using the Frenet frame, the ego vehicle is regarded as driving in a straight tunnel with left and right bounds. The Frenet frame helps convert an NLP problem into an easier form, making an optimization-based planner faster. However, the paths optimized in the Frenet frame may violate vehicle kinematic constraints because the conversions between the Cartesian and Frenet frames are neither unique nor uniform [44]. The situation worsens when the road is curvy, as it might be in our warehouse scenario. Therefore, a perfect solution that simultaneously considers kinematic feasibility, optimality, runtime efficiency, and occlusion-avoidance performance is not yet available.

1.2. Contributions

The current study aims to develop an occlusion-aware path planner for enhancing the indoor infrared positioning accuracy of an autonomous vehicle system. The planner is expected to be optimal and fast without violating fundamental restrictions, such as collision-avoidance and kinematic constraints. In particular, we adopt the first-search-then-optimize framework to combine search-based and optimization-based planners to find the global optimum. Both planners work within the Frenet frame, and thus time efficiency is enhanced. The optimizer is designed through trial and error; hence, the finally derived path is kinematically feasible within a real-world Cartesian frame.

1.3. Organization

The remainder of this paper is organized as follows. Section 2 briefly presents the concerned path planning task. Section 3 introduces our occlusion-aware path planner. Section 4 provides the simulation results and discusses them. Finally, Section 5 concludes the study.

2. Problem Statement

The purpose of this work is to generate a kinematically feasible and collision-free path for a car-like robot in a warehouse, such that the impact of infrared positioning inaccuracy can be reduced. The path planning task is described as the following OCP:

$$
\begin{aligned}
&\min J(\mathbf{z}(s), \mathbf{u}(s)), \\
&\text{s.t., } f(\mathbf{z}(s), \mathbf{u}(s)) = 0, \ s \in [0, S_{\max}]; \\
&\underline{\mathbf{z}} \leq \mathbf{z}(s) \leq \overline{\mathbf{z}}, \ \underline{\mathbf{u}} \leq \mathbf{u}(s) \leq \overline{\mathbf{u}}, \ s \in [0, S_{\max}]; \\
&\mathbf{z}(0) = \mathbf{z}_{\text{init}}, \ \mathbf{u}(0) = \mathbf{u}_{\text{init}}; \\
&\mathbf{z}(S_{\max}) = \mathbf{z}_{\text{end}}, \ \mathbf{u}(S_{\max}) = \mathbf{u}_{\text{end}}; \\
&fp(\mathbf{z}(s)) \subset \chi_{\text{free}}, \ s \in [0, S_{\max}].
\end{aligned}
\quad (1)
$$

Herein, variable s stands for the distance that the ego vehicle has to travel; variable $S_{\max}$ denotes the traveled distance when the ego vehicle reaches the destination, thus $S_{\max}$ may not be known *a priori*; $J(\mathbf{z}(s), \mathbf{u}(s))$ denotes the cost function w.r.t. travel efficiency, path smoothness, and positioning quality; $\mathbf{z}$ denotes the state profiles; $\mathbf{u}$ represents the control profiles; $f(\mathbf{z}(s), \mathbf{u}(s)) = 0$ denotes the kinematic constraints written in the form of ordinary differential equations; $\underline{\mathbf{z}} \leq \mathbf{z}(s) \leq \overline{\mathbf{z}}$ and $\underline{\mathbf{u}} \leq \mathbf{u}(s) \leq \overline{\mathbf{u}}$ denote the kinematic constraints written as algebraic box constraints; $\mathbf{z}(0) = \mathbf{z}_{\text{init}}$, $\mathbf{u}(0) = \mathbf{u}_{\text{init}}$, $\mathbf{z}(S_{\max}) = \mathbf{z}_{\text{end}}$, and $\mathbf{u}(S_{\max}) = \mathbf{u}_{\text{end}}$ denote the two-point boundary constraints; χ_{obs} denotes the partial workspace occupied by obstacles; suppose that χ denotes the entire workspace, then $\chi_{\text{free}} \equiv \chi/\chi_{\text{obs}}$ denotes the free space drivable for the ego vehicle; $fp(\cdot)$ is a mapping from the vehicle state $\mathbf{z}$ to its footprint, thus $fp(\mathbf{z}(s)) \subset \chi_{\text{free}}$ represents the collision-avoidance constraints [45].

In addition, one may define the poor positioning regions in the workspace as $\chi_{\text{occlusion}}$, wherein the ego vehicle cannot touch enough infrared receivers. Ideally, one would expect the ego vehicle to always keep free from poor positioning and collisions, i.e., $fp(\mathbf{z}(s)) \subset \chi/(\chi_{\text{obs}} \cup \chi_{\text{occlusion}})$ for any $s \in [0, S_{\max}]$. However, this condition is too harsh when the infrared beams are seriously blocked by the cargoes. As depicted in Figure 1, no kinematically feasible paths exist in $\chi/(\chi_{\text{obs}} \cup \chi_{\text{occlusion}})$. Empirically, the temporary loss of positioning accuracy is not a serious problem because the inertial measurement units (IMUs) equipped onboard can accurately estimate the vehicle's configuration within a short period. Therefore, (1) allows the ego vehicle to travel in $\chi_{\text{occlusion}}$, but penalizes traveling in $\chi_{\text{occlusion}}$ for a long distance using the cost function J.

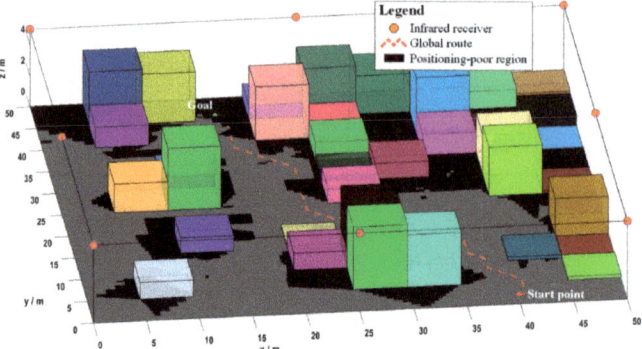

Figure 1. A warehouse workspace with poor positioning regions. Note that there are no kinematically feasible paths if the ego vehicle follows the global route because the goal lies in the poor positioning region.

3. A Four-Layer Path Planning Method

Our proposed path planner consists of four layers. In layer one, an A* path is searched for in the 2D grid map to coarsely connect the initial and goal points. The A* path is deployed to determine the homotopy class. In layer two, a curvature-continuous reference line is planned close to the A* path via numerical optima control. The reference line is deployed to construct a Frenet frame. In layer three, a coarse occlusion-aware path is searched for within the Frenet frame using DP. In layer four, a curvature-continuous path is optimized using QP. The overall architecture is shown in Figure 2.

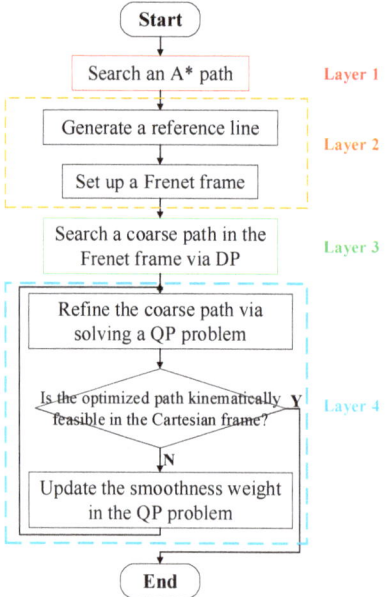

Figure 2. Overall framework of occlusion-aware path planning method.

3.1. Layer One: Search an A* Path

As a preliminary step, the homotopy class needs to be identified, which decides how the ego vehicle bypasses each of the obstacles (i.e., cargoes) from the start point to the goal. This work uses the A* algorithm to find a coarse path, given that it explicitly reflects the determined homotopy class. Concretely, an occupancy grid map is formed based on the warehouse layout and cargo locations. Dilating the occupancy grid map by L renders a dilated map. In this work, L is set to the half-width of the ego vehicle. A 2D path is searched for in the dilated map via the A* algorithm [46], which coarsely connects the assigned starting and goal points. The output of layer one is a path presented in the form of N waypoints stored in a set, i.e., $W_{gr} = \left\{ (x_i^{gr}, y_i^{gr}) | i = 1, \ldots, N \right\}$.

3.2. Layer Two: Generate a Reference Line

The preceding layer identifies a global route from the start point to the goal. This layer is focused on generating a curvature-continuous path that is close to the global route. The curvature-continuous path is denoted as a reference line, which is used to construct a Frenet frame for future usage. The principle for generating a reference line is as follows.

The global route $W_{gr} = \left\{ (x_i^{gr}, y_i^{gr}) | i = 1, \ldots, N \right\}$ derived in layer one is resampled as N_{FE} equidistant waypoints $W_{grrs} = \left\{ (x_i^{grrs}, y_i^{grrs}) | i = 1, \ldots, N_{FE} \right\}$. A reference line is generated by driving a virtual vehicle to track the waypoints. This process can be described as a trajectory planning-oriented OCP:

$$\begin{aligned} &\min J(\mathbf{z}(t), \mathbf{u}(t)), \\ &\text{s.t., } f(\mathbf{z}(t), \mathbf{u}(t)) = 0 \\ &\underline{\mathbf{z}} \leq \mathbf{z}(t) \leq \overline{\mathbf{z}}, \\ &\underline{\mathbf{u}} \leq \mathbf{u}(t) \leq \overline{\mathbf{u}}, \ t \in [0, t_{max}]. \end{aligned} \quad (2)$$

In (2), t is the time index, $\mathbf{z}(t)$ denotes the ego vehicle's state profiles in the Cartesian frame, i.e., $[x(t), y(t), \theta(t), v(t), a(t), \phi(t)]$. Furthermore, $(x(t), y(t))$ refers to the location of the rear-axle midpoint of the ego vehicle, $\theta(t)$ is the orientation angle, $v(t)$ is the longitudinal velocity, $a(t)$ is the corresponding acceleration, and $\phi(t)$ is the steering angle. $\mathbf{u}(t)$ denotes the control profiles $[jerk(t), \omega(t)]$, wherein $jerk(t)$ is the derivative of $a(t)$, and $\omega(t)$ is the angular velocity of $\phi(t)$. All of the constraints in (2) are kinematic constraints, which are presented by the well-known bicycle model [47]:

$$\begin{aligned} \frac{dx(t)}{dt} &= v(t) \cdot \cos \theta(t) \\ \frac{dy(t)}{dt} &= v(t) \cdot \sin \theta(t) \\ \frac{d\theta(t)}{dt} &= \frac{v(t) \cdot \tan \phi(t)}{L_W} \\ \frac{dv(t)}{dt} &= a(t) \\ \frac{d\phi(t)}{dt} &= \omega(t) \\ \frac{da(t)}{dt} &= jerk(t) \end{aligned}, \ t \in [0, t_{max}]. \quad (3)$$

Herein, L_W denotes the vehicle wheelbase (Figure 3). The boundary constraints $\underline{\mathbf{z}} \leq \mathbf{z}(t) \leq \overline{\mathbf{z}}$ and $\underline{\mathbf{u}} \leq \mathbf{u}(t) \leq \overline{\mathbf{u}}$ are defined as

$$\begin{aligned} &-jerk_{max} \leq jerk(t) \leq jerk_{max}, \ a_{min} \leq a(t) \leq a_{max}, \ 0 \leq v(t) \leq v_{max}, \\ &-\Omega_{max} \leq \omega(t) \leq \Omega_{max}, -\Phi_{max} \leq \phi(t) \leq \Phi_{max}, \ t \in [0, t_{max}]. \end{aligned} \quad (4)$$

The cost function $J(\mathbf{z}(t), \mathbf{u}(t))$ is defined as:

$$J = \int_{\tau=0}^{t_{max}} \left\{ [x(\tau) - x_{grrs}(\tau)]^2 + [y(\tau) - y_{grrs}(\tau)]^2 \right\} \cdot d\tau + w_u \cdot \int_{\tau=0}^{t_{max}} \left[jerk^2(\tau) + \omega^2(\tau) \right] \cdot d\tau, \quad (5)$$

wherein $w_u > 0$ is a weighting parameter, and $(x_{grrs}(t), y_{grrs}(t))$ is a parametric trajectory formed by linearly connecting the N_{FE} waypoints $\left\{ (x_i^{grrs}, y_i^{grrs}) | i = 1, \ldots, N_{FE} \right\}$ in a sequence.

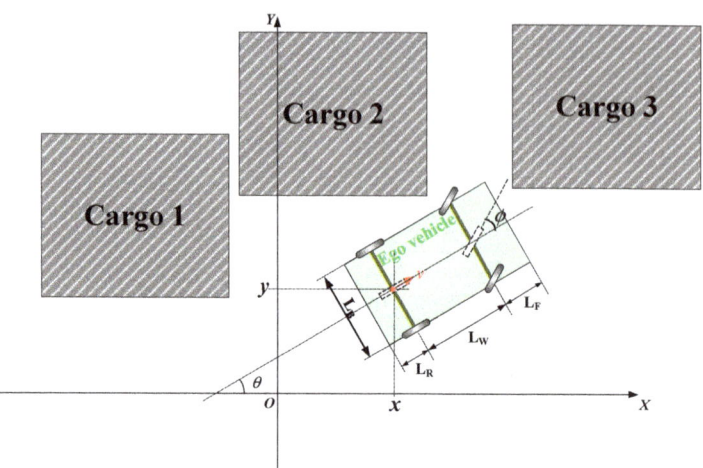

Figure 3. Schematics for vehicle kinematics and geometrics.

Compared with (1), OCP (2) does not contain two-point constraints or collision-avoidance constraints. Thus, the resultant trajectory may not connect the start point to the goal and may not guarantee collision avoidance despite being kinematically feasible. At this point, it should be clarified that the resultant reference line is only used to construct a Frenet frame; the concerns about collision avoidance, occlusion avoidance, etc., will be handled in a subsequent layer based on the constructed Frenet frame.

OCP (2) is solved numerically, which involves discretizing the OCP into an NLP problem and then solving it via a gradient-based NLP solver, such as the interior-point method (IPM) [48,49]. The solution to the NLP problem is a vector of waypoints together with the corresponding state/control profiles in their discretized forms. A reference line is formed by connecting the resultant waypoints smoothly via spline interpolation.

3.3. Layer Three: Search for a DP Path in the Frenet Frame

This layer is focused on generating a coarse path within the Frenet frame with collision avoidance, occlusion awareness, travel efficiency, and path smoothness considered.

The first step is to map the start point and goal to the reference line so that their Frenet coordinate values are identified as (s_{start}, l_{start}) and (s_{end}, l_{end}), respectively (Figure 4). The second step is to generate $(N_S + 1)$ equidistant skeleton points along the reference line ranging from $(s_{start}, 0)$ to $(s_{end}, 0)$. A normal line is drawn along each skeleton point, which is orthogonal to the reference line. Along each normal line, N_L candidate grids are sampled (Figure 4), which range in an interval around the skeleton point. For the nominal line passing through the last skeleton point located at $(s_{end}, 0)$, N_L is set to 1 and the only candidate grid left is specified as the goal (s_{end}, l_{end}).

The planning task in layer three is to select one and only one candidate grid along each of the normal lines such that the sequentially connected candidate grids are collision free, occlusion minimized, short, and smooth. DP is adopted to find the optimal choice for the candidate grid along each normal line [50]. Compared to enumeration, DP reduces the search complexity from $O(N_L^{N_S})$ to $O(N_S \cdot N_L^2)$, and thus promises to find the optimum in a graph consisting of candidate grids. The brief principle of the DP search-based path planning method in layer three is presented as follows.

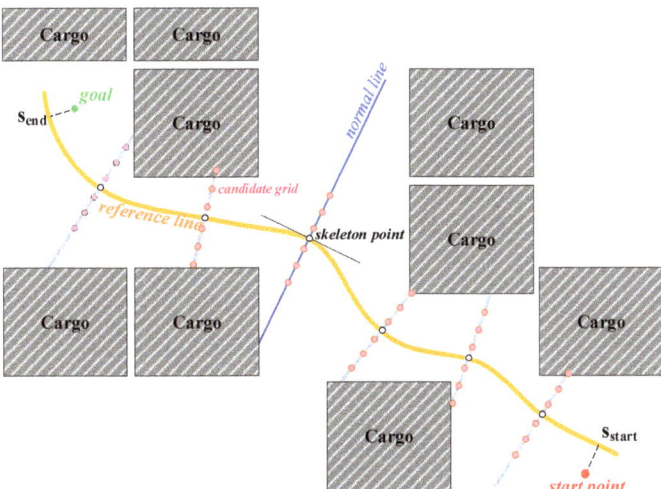

Figure 4. Schematics for a graph of sampled grids in DP search.

In Algorithm 1, each candidate grid is regarded as a node. *InitializeNodes*() is used to initialize the cost of each node as $+\infty$. *TraceBack*(opti_id) is used to identify a sequence of nodes from Node(N_S, 1), Node($N_S - 1$, opti_id), to its ancestors until Node(0, 1). The node sequence, if output in inverse order, forms a coarse path within the Frenet frame. *MeasureCost*(Node1, Node2) measures the cost of the path segment from Node1 to Node2. We define the cost function as a weighted sum of collision cost $J_{\text{collision}}$, travel efficiency cost $J_{\text{efficiency}}$, smoothness cost $J_{\text{smoothness}}$, and positioning-related cost $J_{\text{positioning}}$. The collision cost penalizes the case that the ego vehicle collides with the surrounding cargoes when driving along the concerned path segment. $J_{\text{collision}}$ is set to a large value (e.g., 10^{20}) if a collision occurs, otherwise, $J_{\text{collision}}$ is set to 0. The efficiency cost $J_{\text{efficiency}}$ is written as the length of the concerned path segment because this term encourages the ego vehicle to travel across a short distance. Suppose the parent of Node1 is Node0, $J_{\text{smoothness}}$ is defined as $|(\text{Node0.config} - \text{Node1.config}) \times (\text{Node1.config} - \text{Node2.config})|$. Intuitively speaking, the smoothness cost $J_{\text{smoothness}}$ penalizes the case that the heading direction changes from Node0, Node1, to Node2. $J_{\text{positioning}}$ penalizes the case that the ego vehicle stays in the positioning-poor regions for a long distance. Furthermore, N_{sample} waypoints $\left\{(s_i^{\text{wp}}, l_i^{\text{wp}}) \mid i = 1, \ldots, N_{\text{sample}}\right\}$ are evenly sampled along the concerned line segment from Node1 to Node2. Regarding the ith sampled waypoint $(s_i^{\text{wp}}, l_i^{\text{wp}})$, the corresponding coordinate value in the Cartesian frame is identified as $(x_i^{\text{wp}}, y_i^{\text{wp}})$ via frame conversion. Suppose the infrared emitter is installed at the height of h onto the ego vehicle, one may draw a line from the 3D point $(x_i^{\text{wp}}, y_i^{\text{wp}}, h)$ to each infrared receiver and then check if the line is occluded by cargoes in the warehouse. If there is no occlusion, then the receiver is regarded as valid (Figure 5). If the total number of valid infrared receivers is larger than three, then the concerned waypoint $(s_i^{\text{wp}}, l_i^{\text{wp}})$ is regarded as valid. $J_{\text{positioning}}$ measures the rate of valid waypoints along the concerned line segment from Node1 to Node2.

3.4. Layer Four: Optimize a Curvature-Continuous Path

The preceding layer helps to identify a collision-free and occlusion-aware path, which consists of (N_S + 1) waypoints. Since N_S is not large, the derived path is quite coarse. This section is focused on how to refine the coarse path via numerical optimization within the Frenet frame. In this work, path refinement is performed via Baidu Apollo EM planner [20], which involves implementing path-velocity decomposition in an iterative loop before an optimum (rather than sub-optimum) is finally derived. Since there are only static obstacles, the EM planner is degraded as a run-once path planning method, the details of which are given as follows.

Algorithm 1. Path planning via DP search.

Input: Reference line, scenario layout, location of cargoes;
Output: A path $\Lambda = \left\{ (s_i^{dp}, l_i^{dp}) | i = 0, \ldots, N_S \right\}$;

1. *InitializeNodes()*;
2. Set $Node(0,1).config = (x_{start}, y_{start})$;
3. **For each** $j \in \{1, \ldots, N_L\}$, **do**
4.   Set $Node(1,j).parent = Node(0,1)$;
5.   Identify $Node(1,j).config$;
6.   Set $Node(1,j).cost = MeasureCost(Node(0,1), Node(1,j))$;
7. **End for**
8. **For each** $i \in \{1, \ldots, N_S - 2\}$, **do**
9.   **For each** $j \in \{1, \ldots, N_L\}$, **do**
10.   **For each** $k \in \{1, \ldots, N_L\}$, **do**
11.   Identify $Node(i+1,k).config$;
12.   cost_candidate = $MeasureCost(Node(i,j), Node(i+1,k))$;
13.   **If** $Node(i+1,k).cost > Node(i,j).cost + cost\_candidate$, **then**
14. $Node(i+1,k).parent = Node(i,j)$;
15. $Node(i+1,k).cost = cost\_candidate$;
16.   **End if**
17.   **End for**
18.   **End for**
19. **End for**
20. opti_id = $\arg\min_{j=1,\ldots,N_L}(Node(N_S - 1, j).cost + MeasureCost(Node(N_S - 1, j), Node(N_S, 1)))$;
21. $\Lambda = TraceBack(opti\_id)$;
22. **Return**.

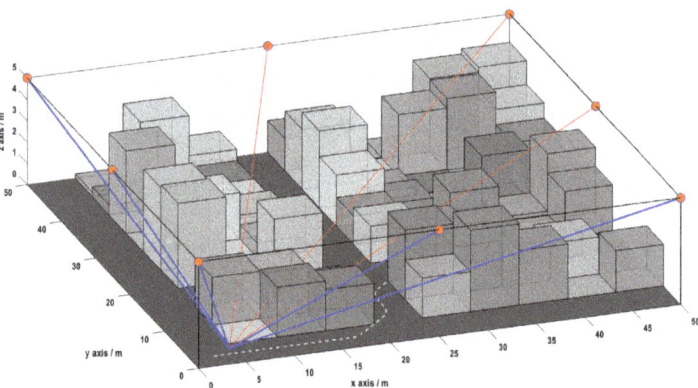

Figure 5. Schematics for the infrared receiver validation check. Five blue infrared beams together with corresponding receivers are valid because they are not occluded by the cargoes.

The first step is to identify the left and right bounds that surround the coarse path derived in layer three. As depicted in Figure 6, the left and right bounds are determined using an incremental check. The identified left and right bounds are denoted as functions of s, namely, $ub(s)$ and $lb(s)$. The optimization-based path planning task involves identifying a function $l(s)$ between the left and right bounds subject to kinematic constraints and collision-avoidance constraints. In our concerned task, the decision variables include $l(s)$, $dl(s)$, $ddl(s)$, and $dddl(s)$, which obey the following kinematic constraints:

$$\begin{aligned}
\frac{dl(s)}{ds} &= dl(s), \\
\frac{ddl(s)}{ds} &= ddl(s), \\
\frac{dddl(s)}{ds} &= dddl(s), \\
-dl_{max} &\leq dl(s) \leq dl_{max}, \\
-ddl_{max} &\leq ddl(s) \leq ddl_{max}, \\
-dddl_{max} &\leq dddl(s) \leq dddl_{max}, \\
s &\in [s_{start}, s_{end}].
\end{aligned} \quad (6)$$

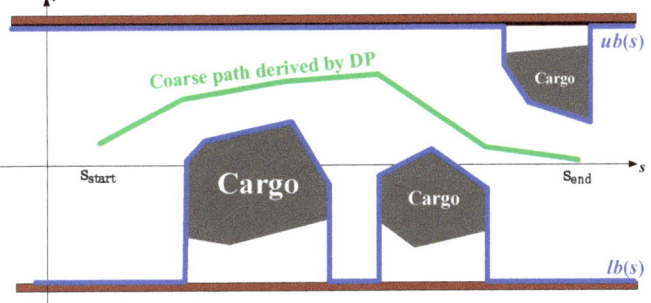

Figure 6. Schematics for the construction of left/right bounds in EM planner.

Herein, dl_{max}, ddl_{max}, and $dddl_{max}$ are parameters that determine how fast l changes with s, and thus are related to path smoothness. Empirically, the bounding parameters in (6) should not be set strictly, otherwise the vehicle kinematics easily become in conflict with the collision-avoidance constraints that will be introduced later. At this point, we believe that the path smoothness can be enhanced via the cost function without suffering from the infeasibility risk [23].

The two-point boundary constraints are written as

$$\begin{aligned}
l(s_{start}) &= L_{start}, \; dl(s_{start}) = DL_{start}, \; ddl(s_{start}) = 0, \; dddl(s_{start}) = 0, \\
l(s_{end}) &= L_{end}, \; dl(s_{end}) = DL_{end}, \; ddl(s_{end}) = 0, \; dddl(s_{end}) = 0.
\end{aligned} \quad (7)$$

Herein, the parameters L_{start}, DL_{start}, L_{end}, and DL_{end} reflect the assigned initial and goal configurations. Particularly, DL_{start} and DL_{end} are related to the vehicle orientation angles at $s = s_{start}$ and s_{end}, respectively.

Regarding the collision-avoidance constraints, the vehicle body should not collide with $ub(s)$ or $lb(s)$. Setting the vehicle body as rectangular is too complex, and thus we use a series of same-radius circles centered along the longitudinal axle of the ego vehicle to cover the rectangular vehicle body (Figure 7), and then require that each circle lies between $lb(s)$ and $ub(s)$. For a circle biased from the vehicle's rear axle by η, the collision-avoidance constraints are defined as

$$\begin{aligned}
\eta \cdot \tan \theta_s + l(s) + 0.5 \cdot L_B &\leq ub(s+\eta), \\
\eta \cdot \tan \theta_s + l(s) - 0.5 \cdot L_B &\geq lb(s+\eta).
\end{aligned} \quad (8)$$

The complete collision-avoidance constraints are formed by imposing (8) for any $\eta \in [-L_R \cos \theta_s, (L_W + L_F) \cos \theta_s]$. Herein, $\theta_s(s)$ denotes the vehicle's orientation angle within the Frenet frame. Inherently, $\theta_s(s)$ stands for the difference between the ego vehicle's heading direction and the tangent direction along the reference line at s. $|\theta_s(s)|$ is small because the ego vehicle's heading direction, in most cases, is not much biased from the reference line. Thus, we have

$$\begin{aligned}
\tan \theta_s &\equiv \frac{dl(s)}{ds} \equiv dl(s), \\
\cos \theta_s &\approx 1.
\end{aligned} \quad (9)$$

This yields the following constraints:

$$\eta \cdot dl(s) + l(s) + 0.5 \cdot L_B \leq ub(s+\eta),$$
$$\eta \cdot dl(s) + l(s) - 0.5 \cdot L_B \geq lb(s+\eta), \quad (10)$$
$$\forall \eta \in [-L_R, L_W + L_F].$$

The cost function is defined as

$$J = w_1 \cdot \int_{s=s_{start}}^{s_{end}} (l(s) - l_{DP}(s))^2 ds + w_2 \cdot \int_{s=s_{start}}^{s_{end}} dl^2(s) ds + w_3 \cdot \int_{s=s_{start}}^{s_{end}} ddl^2(s) ds + w_4 \cdot \int_{s=s_{start}}^{s_{end}} dddl^2(s) ds, \quad (11)$$

wherein w_1, w_2, w_3, and w_4 are weighting parameters, and $l_{DP}(s)$ denotes the coarse path derived by DP in layer three. An OCP is formed by combining (6), (7), (10), and (11). The discretized version of this OCP is a QP, which is easily solved using a QP solver, such as osqp [51] and qpOASES [52]. The resultant path, after being converted back to the Cartesian frame, may be infeasible if its curvature exceeds the allowable bounds. The infeasibility is inevitable because the vehicle kinematics cannot be accurately modeled within the Frenet frame [44]. As a remedy for this, we check the resultant path for violations of curvature limits in the Cartesian frame; if an infeasible solution is derived, w_1 is set smaller before the QP problem is solved again. This process continues until a curvature-feasible path is derived.

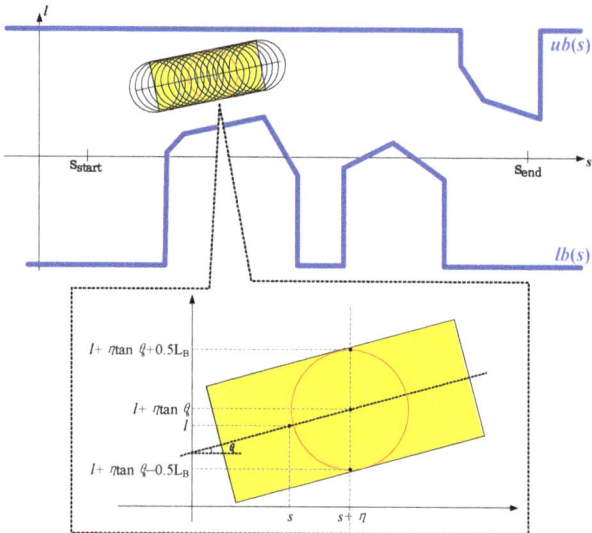

Figure 7. Schematics for the formulation of collision-avoidance constraints in EM planner (zoom in to see more clearly).

4. Simulation Results and Discussion

This section discusses the efficacy, occlusion awareness, and closed-loop tractability of our proposed path planner.

4.1. Simulation Setup

Simulations were implemented in a MATLAB + CarSim platform and executed on an Intel Core i9-9900 CPU with 32 GB RAM that runs at 3.10 × 2 GHz. We define a 50 m × 50 m warehouse with eight infrared receivers that are evenly distributed along the four edges of the rectangular ceiling, the height of which is 5 m. The geometric size of each cargo is 5 m × 5 m × h, wherein h is a random value ranging from 0 to 5 m. Other parametric settings are presented in our source codes, which are available at https://github.com/libai1943/OcclusionAwarePathPlanningForAGV (accessed on 8 December 2021).

4.2. On the Efficacy of the Proposed Planner

The planning results of two typical simulation cases are depicted in Figure 8, which shows the efficiency of each layer. In each of the two cases, the finally derived path is collision free and kinematically feasible. Both properties can be reflected by the footprints and curvature profiles plotted in Figure 9.

4.3. On the Occlusion Awareness of the Proposed Planner

This subsection investigates the occlusion awareness of the paths planned by the proposed method. Figure 10 shows the results with/without the positioning-related cost $J_{positioning}$ in layer three. When the cost term $J_{positioning}$ is discarded, the rate of good positioning distance along the entire path is 86.5% and 92.5% in the aforementioned two typical simulation cases, respectively. By contrast, with the cost term included, the rate grows to 97.0% and 96.5%. This comparative result clearly shows that our proposed planner can efficiently reduce the positioning inaccuracy caused by occlusions.

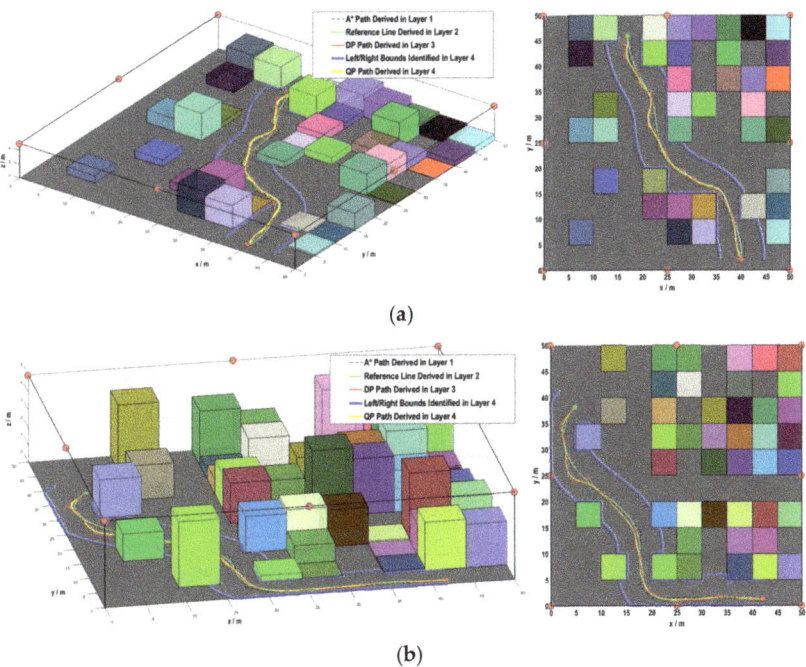

Figure 8. Path planning results of typical simulation cases in side view and bird-eye view: (**a**) Case 1; (**b**) Case 2 (zoom in to see more clearly).

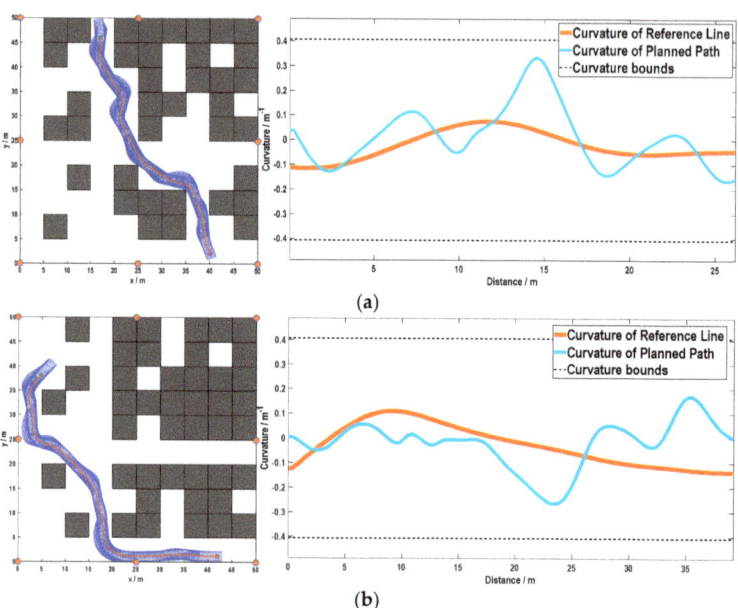

Figure 9. Path planning performance, w.r.t. collision avoidance, and kinematic feasibility: (**a**) Case 1; (**b**) Case 2.

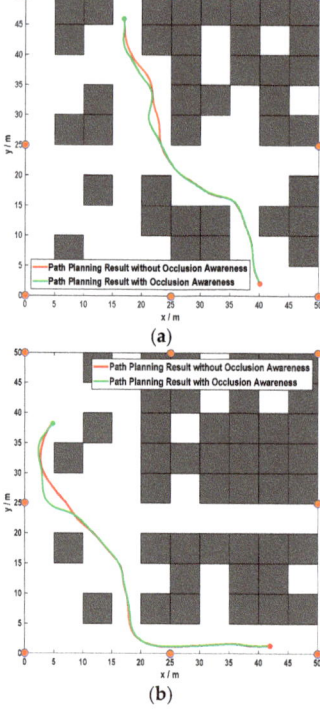

Figure 10. Comparative path planning results, w.r.t. occlusion awareness: (**a**) Case 1; (**b**) Case 2.

4.4. On the Closed-Loop Tractability of the Proposed Planner

This subsection reports the closed-loop tracking performance in CarSim when following the open-loop paths planned by the proposed planner (Figure 11). A linear quadratic regulator (LQR) is adopted as the controller. As illustrated in Figure 12, the open-loop and closed-loop paths do not differ much, which indicates that the planned paths are sufficiently smooth and thus easy to track. The concrete closed-loop tracking simulation results are presented in the following video link: https://www.bilibili.com/video/av677126688 (accessed on 8 December 2021).

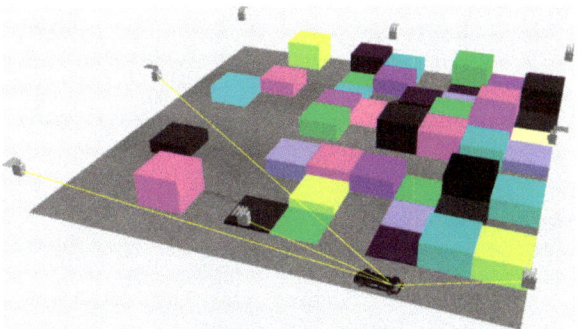

Figure 11. CarSim simulation scenario layout and screenshot of simulation process.

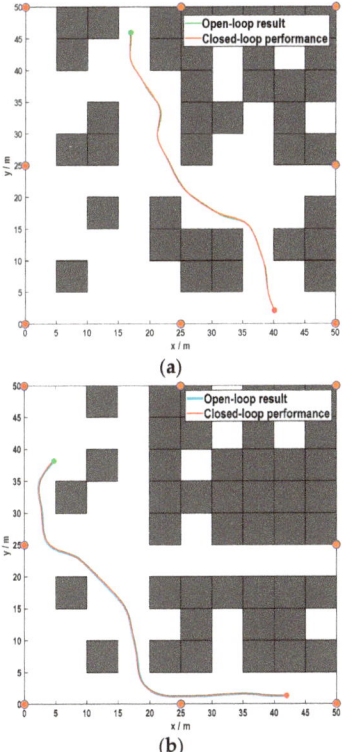

Figure 12. Path planning results, w.r.t. tractability: (**a**) Case 1; (**b**) Case 2.

5. Conclusions

This paper has introduced a path planning method for an autonomous vehicle in a warehouse, wherein the cargoes may occlude the inflated signals emitted by the ego vehicle for inflated positioning. According to our conducted simulations, the proposed planner is efficient according to its w.r.t. collision avoidance, kinematic feasibility, occlusion awareness, and tractability.

Author Contributions: Conceptualization, B.L. and Y.Z.; methodology, B.L.; validation, S.T. and X.Z. All authors have read and agreed to the published version of the manuscript.

Funding: This research was funded by Fundamental Research Funds for the Central Universities of China, grant number 531118010509, National Natural Science Foundation of China, grant number 62103139, and Natural Science Foundation of Hunan Province, China, grant number 2021JJ40114.

Conflicts of Interest: The authors declare no conflict of interest.

References

1. Jia, S.; Ma, L.; Yang, S.; Qin, D. Semantic and context-based image retrieval method using a single image sensor for visual indoor positioning. *IEEE Sens. J.* **2021**, *21*, 18020–18032. [CrossRef]
2. Cengiz, K. Comprehensive analysis on least-squares lateration for indoor positioning systems. *IEEE Internet Things J.* **2021**, *8*, 2842–2856. [CrossRef]
3. Lin, P.T.; Liao, C.; Liang, S. Probabilistic indoor positioning and navigation (PIPN) of autonomous ground vehicle (AGV) based on wireless measurements. *IEEE Access* **2021**, *9*, 25200–25207. [CrossRef]
4. Seo, H.; Kim, H.; Kim, H.; Hong, D. Accurate positioning using beamforming. In Proceedings of the 2021 IEEE 18th Annual Consumer Communications & Networking Conference (CCNC), Las Vegas, NV, USA, 9–12 January 2021; pp. 1–4.
5. Chavez-Burbano, P.; Guerra, V.; Rabadan, J.; Perez-Jimenez, R. Optical camera communication system for three-dimensional indoor localization. *Optik* **2019**, *192*, 162870. [CrossRef]
6. Wang, Z.; Liu, B.; Huang, F.; Chen, Y.; Zhang, S.; Cheng, Y. Corners positioning for binocular ultra-wide angle long-wave infrared camera calibration. *Optik* **2020**, *206*, 163441. [CrossRef]
7. Aparicio-Esteve, E.; Hernández, Á.; Ureña, J. Design, calibration and evaluation of a long-range 3D infrared positioning system based on encoding techniques. *IEEE Trans. Instrum. Meas.* **2021**, *70*, 1–13. [CrossRef]
8. Yao, W.; Ma, L. Research and application of indoor positioning method based on fixed infrared beacon. In Proceedings of the 2018 37th Chinese Control Conference (CCC), Wuhan, China, 25–27 July 2018; pp. 5375–5379.
9. Martínez-Barberá, H.; Herrero-Pérez, D. Autonomous navigation of an automated guided vehicle in industrial environments. *Robot. Comput.-Integr. Manuf.* **2010**, *26*, 296–311. [CrossRef]
10. Khedkar, A.; Kajani, K.; Ipkal, P.; Banthia, S.; Jagdale, B.N.; Kulkarni, M. Automated Guided Vehicle System with Collision Avoidance and Navigation in Warehouse Environments. *Sensors* **2020**, *7*, 5442–5448.
11. Li, B.; Li, L.; Acarman, T.; Shao, Z.; Yue, M. Optimization-based maneuver planning for a tractor-trailer vehicle in a curvy tunnel: A weak reliance on sampling and search. *IEEE Robot. Autom. Lett.* **2021**. accepted. [CrossRef]
12. Cao, H.; Zhao, S.; Song, X.; Li, M. Toward high automatic driving by a dynamic optimal trajectory planning method based on high-order polynomials. *SAE Tech. Pap.* **2020**, *1*, 1–10. [CrossRef]
13. Li, X.; Sun, Z.; Cao, D.; He, Z.; Zhu, Q. Real-time trajectory planning for autonomous urban driving: Framework, algorithms, and verifications. *IEEE/ASME Trans. Mechatron.* **2016**, *21*, 740–753. [CrossRef]
14. Cao, H.; Zhao, S.; Song, X.; Bao, S.; Li, M.; Huang, Z.; Hu, C. An optimal hierarchical framework of the trajectory following by convex optimisation for highly automated driving vehicles. *Veh. Syst. Dyn.* **2019**, *57*, 1287–1317. [CrossRef]
15. Jian, Z.; Chen, S.; Zhang, S.; Chen, Y.; Zheng, N. Multi-model-based local path planning methodology for autonomous driving: An integrated framework. *IEEE Trans. Intell. Transp. Syst.* **2020**. accepted. [CrossRef]
16. Dolgov, D.; Thrun, S.; Montemerlo, M.; Diebel, J. Path planning for autonomous vehicles in unknown semi-structured environments. *Int. J. Robot. Res.* **2010**, *29*, 485–501. [CrossRef]
17. Li, X.; Sun, Z.; Cao, D.; Liu, D.; He, H. Development of a new integrated local trajectory planning and tracking control framework for autonomous ground vehicles. *Mech. Syst. Signal Process.* **2017**, *87*, 118–137. [CrossRef]
18. Kuwata, Y.; Teo, J.; Fiore, G.; Karaman, S.; Frazzoli, E.; How, J.P. Real-time motion planning with applications to autonomous urban driving. *IEEE Trans. Control Syst. Technol.* **2009**, *17*, 1105–1118. [CrossRef]
19. Wang, Y.; Li, S.; Cheng, W.; Cui, X.; Su, B. Toward efficient trajectory planning based on deterministic sampling and optimization. In Proceedings of the 2020 Chinese Automation Congress (CAC), Shanghai, China, 6–8 November 2020; pp. 1318–1323.
20. Fan, H.; Zhu, F.; Liu, C.; Zhang, L.; Zhuang, L.; Li, D.; Zhu, W.; Hu, J.; Li, H.; Kong, Q. Baidu Apollo EM motion planner. *arXiv* **2018**, arXiv:1807.08048.

21. Ajanovic, Z.; Lacevic, B.; Shyrokau, B.; Stolz, M.; Horn, M. Search-based optimal motion planning for automated driving. In Proceedings of the 2018 IEEE/RSJ International Conference on Intelligent Robots and Systems (IROS), Madrid, Spain, 1–5 October 2018; pp. 4523–4530.
22. Ma, L.; Xue, J.; Kawabata, K.; Zhu, J.; Ma, C.; Zheng, N. Efficient sampling-based motion planning for on-road autonomous driving. *IEEE Trans. Intell. Transp. Syst.* **2015**, *16*, 1961–1976. [CrossRef]
23. Li, B.; Kong, Q.; Zhang, Y.; Shao, Z.; Wang, Y.; Peng, X.; Yan, D. On-road trajectory planning with spatio-temporal RRT* and always-feasible quadratic program. In Proceedings of the 2020 16th IEEE International Conference on Automation Science and Engineering (CASE), Hongkong, China, 20–21 August 2020; pp. 943–948.
24. Tian, F.; Zhou, R.; Li, Z.; Li, L.; Gao, Y.; Cao, D.; Chen, L. Trajectory planning for autonomous mining trucks considering terrain constraints. *IEEE Trans. Intell. Veh.* **2021**, *6*, 772–786. [CrossRef]
25. Lim, W.; Lee, S.; Sunwoo, M.; Jo, K. Hierarchical trajectory planning of an autonomous car based on the integration of a sampling and an optimization method. *IEEE Trans. Intell. Transp. Syst.* **2018**, *19*, 613–626. [CrossRef]
26. Chen, J.; Liu, C.; Tomizuka, M. FOAD: Fast optimization-based autonomous driving motion planner. In Proceedings of the 2018 Annual American Control Conference (ACC), Milwaukee, WI, USA, 27–29 June 2018; pp. 4725–4732.
27. Li, B.; Zhang, Y. Fast trajectory planning in Cartesian rather than Frenet frame: A precise solution for autonomous driving in complex urban scenarios. *IFAC-PapersOnLine* **2020**, *53*, 17065–17070. [CrossRef]
28. Chen, J.; Zhan, W.; Tomizuka, M. Autonomous driving motion planning with constrained iterative LQR. *IEEE Trans. Intell. Veh.* **2019**, *4*, 244–254. [CrossRef]
29. Luo, S.; Li, X.; Sun, Z. An optimization-based motion planning method for autonomous driving vehicle. In Proceedings of the 2020 3rd International Conference on Unmanned Systems (ICUS), Harbin, China, 27–28 November 2020; pp. 739–744.
30. Ziegler, J.; Bender, P.; Dang, T.; Stiller, C. Trajectory planning for Bertha—A local, continuous method. In Proceedings of the 2014 IEEE Intelligent Vehicles Symposium (IV), Dearborn, MI, USA, 8–11 June 2014; pp. 450–457.
31. Eiras, F.; Hawasly, M.; Albrecht, S.; Ramamoorthy, S. A two-stage optimization-based motion planner for safe urban driving. *IEEE Trans. Robot.* **2021**, accepted. [CrossRef]
32. Zhang, Y.; Chen, H.; Waslander, S.; Gong, J.; Xiong, G.; Yang, T.; Liu, K. Hybrid trajectory planning for autonomous driving in highly constrained environments. *IEEE Access* **2018**, *6*, 32800–32819. [CrossRef]
33. Lim, W.; Lee, S.; Sunwoo, M.; Jo, K. Hybrid trajectory planning for autonomous driving in on-road dynamic scenarios. *IEEE Trans. Intell. Transp. Syst.* **2021**, *22*, 341–355. [CrossRef]
34. Hegedüs, F.; Bécsi, T.; Aradi, S.; Gáldi, G. Hybrid trajectory planning for autonomous vehicles using neural networks. In Proceedings of the 2018 IEEE 18th International Symposium on Computational Intelligence and Informatics (CINTI), Budapest, Hungary, 21–22 November 2018; pp. 25–30.
35. Acerbo, F.; Auweraer, H.; Son, T. Safe and computational efficient imitation learning for autonomous vehicle driving. In Proceedings of the 2020 American Control Conference (ACC), Denver, CO, USA, 1–3 July 2020; pp. 647–652.
36. Marchesini, E.; Farinelli, A. Discrete deep reinforcement learning for mapless navigation. In Proceedings of the 2020 IEEE International Conference on Robotics and Automation (ICRA), Paris, France, 31 May–31 August 2020; pp. 10688–10694.
37. Chen, L.; Hu, X.; Tang, B.; Cheng, Y. Conditional DQN-based motion planning with fuzzy logic for autonomous driving. *IEEE Trans. Intell. Transp. Syst.* **2020**, accepted. [CrossRef]
38. Jaritz, M.; Charette, R.; Toromanoff, M.; Perot, E.; Nashashibi, R. End-to-end race driving with deep reinforcement learning. In Proceedings of the 2018 IEEE International Conference on Robotics and Automation (ICRA), Brisbane, QLD, Australia, 21–25 May 2018; pp. 2070–2075.
39. Li, B. Occlusion-aware on-road autonomous driving: A trajectory planning method considering occlusions of Lidars. *Optik* **2021**, *243*, 167347. [CrossRef]
40. Li, B.; Acarman, T.; Zhang, Y.; Kong, Q. Occlusion-aware on-road autonomous driving: A path planning method in combination with honking decision making. In Proceedings of the 33rd Chinese Control and Decision Conference (CCDC), Kunming, China, 22–24 May 2021; pp. 7403–7408.
41. Manzinger, S.; Pek, C.; Althoff, M. Using reachable sets for trajectory planning of automated vehicles. *IEEE Trans. Intell. Veh.* **2021**, *6*, 232–248. [CrossRef]
42. Chen, J.; Li, S.; Tomizuka, M. Interpretable end-to-end urban autonomous driving with latent deep reinforcement learning. *IEEE Trans. Intell. Transp. Syst.* **2020**, accepted. [CrossRef]
43. Werling, M.; Ziegler, J.; Kammel, S.; Thrun, S. Optimal trajectory generation for dynamic street scenarios in a Frenet frame. In Proceedings of the 2010 IEEE International Conference on Robotics and Automation (ICRA), Anchorage, AK, USA, 3–7 May 2010; pp. 987–993.
44. Li, B.; Ouyang, Y.; Li, L.; Zhang, Y. Autonomous driving on curvy roads without reliance on Frenet frame: A cartesian-based trajectory planning method. *IEEE Trans. Intell. Transp. Syst.* **2021**, under review.
45. Li, B.; Acarman, T.; Zhang, Y.; Ouyang, Y.; Yaman, C.; Kong, Q.; Zhong, X.; Peng, X. Optimization-based trajectory planning for autonomous parking with irregularly placed obstacles: A lightweight iterative framework. *IEEE Trans. Intell. Transp. Syst.* **2021**. accepted. [CrossRef]
46. Hart, P.E.; Nilsson, N.J.; Raphael, B. A formal basis for the heuristic determination of minimum cost paths. *IEEE Trans. Syst. Sci. Cybern.* **1968**, *4*, 100–107. [CrossRef]

47. Li, B.; Shao, Z. A unified motion planning method for parking an autonomous vehicle in the presence of irregularly placed obstacles. *Knowl.-Based Syst.* **2015**, *86*, 11–20. [CrossRef]
48. Li, B.; Shao, Z. Simultaneous dynamic optimization: A trajectory planning method for nonholonomic car-like robots. *Adv. Eng. Softw.* **2015**, *87*, 30–42. [CrossRef]
49. Wächter, A.; Biegler, L.T. On the implementation of an interior-point filter line-search algorithm for large-scale nonlinear programming. *Math. Program.* **2006**, *106*, 25–57. [CrossRef]
50. Xu, W.; Pan, J.; Wei, J.; Dolan, J.M. Motion planning under uncertainty for on-road autonomous driving. In Proceedings of the 2014 IEEE International Conference on Robotics and Automation (ICRA), Hong Kong, China, 31 May–7 June 2014; pp. 2507–2512.
51. Stellato, B.; Banjac, G.; Goulart, P.; Bemporad, A.; Boyd, S. OSQP: An operator splitting solver for quadratic programs. *Math. Program. Comput.* **2020**, *12*, 637–672. [CrossRef]
52. Ferreau, H.J.; Kirches, C.; Potschka, A.; Bock, H.G.; Diehl, M. qpOASES: A parametric active-set algorithm for quadratic programming. *Math. Program. Comput.* **2014**, *6*, 327–363. [CrossRef]

Article

A Path Planning Method for Underground Intelligent Vehicles Based on an Improved RRT* Algorithm

Hao Wang [1,2], Guoqing Li [1,2,*], Jie Hou [1,2], Lianyun Chen [1,3] and Nailian Hu [1,2]

1. College of Civil and Resource Engineering, University of Science and Technology Beijing, Beijing 100083, China; haowang@xs.ustb.edu.cn (H.W.); houjie@ustb.edu.cn (J.H.); chenlianyun@sd-gold.com (L.C.); hnl@ustb.edu.cn (N.H.)
2. Key Laboratory of Efficiency Mining and Safety of Metal Mines, Ministry of Education, Beijing 100083, China
3. Shandong Gold Group Co., Ltd., Jinan 250014, China
* Correspondence: qqlee@ustb.edu.cn

Abstract: Path planning is one of the key technologies for unmanned driving of underground intelligent vehicles. Due to the complexity of the drift environment and the vehicle structure, some improvements should be made to adapt to underground mining conditions. This paper proposes a path planning method based on an improved RRT* (Rapidly-Exploring Random Tree Star) algorithm for solving the problem of path planning for underground intelligent vehicles based on articulated structure and drift environment conditions. The kinematics of underground intelligent vehicles are realized by vectorized map and dynamic constraints. The RRT* algorithm is selected for improvement, including dynamic step size, steering angle constraints, and optimal tree reconnection. The simulation case study proves the effectiveness of the algorithm, with a lower path length, lower node count, and 100% steering angle efficiency.

Keywords: underground intelligent vehicles; path planning; RRT* algorithm; articulated vehicles; unmanned driving

1. Introduction

In recent years, major mining groups have increased their investment in intelligent mining, and the mining industry is gradually entering the era of being remote, smart, and unmanned [1–5]. Intelligent vehicles are the most important pieces of equipment for intelligent mining with unmanned driving. Path planning is one of the key technologies for autonomous driving of intelligent unmanned vehicles. A reasonable path planning algorithm helps vehicles optimize the running trajectory, avoid obstacles according to the environment, and realize safe and efficient driving. The intelligent vehicles include drilling rigs, charging jumbo, load–haul–dump (LHD), trucks, scaling jumbo, and bolting jumbo, etc., the goal of which is to achieve intelligent mining processes by autonomous positioning, autonomous navigation, autonomous driving, and autonomous operation. These underground intelligent vehicles are shown in Figure 1.

The path planning of underground intelligent vehicles is one of the branches of research of unmanned ground vehicles (UGV) and unmanned aerial vehicles (UAV). With the advancement of technology, they have been widely used in the fields of logistics, transportation, disaster relief, etc., [6,7]. The research into UGV automatic driving in underground mining can be traced back to the 1960s [8,9]. The USA, Canada, Sweden, etc., have researched the remote control of vehicles, but due to the limitations of communications and sensors, the application progress was slow. With the technological revolution, such as the Internet of Things (IoT) and machine learning, unmanned mining operation has become a research hotspot in the mining field again. The European Union (EU) initiated the "Robominers" project to develop bionic robots for underground mining operations in harsh environments [10]. Rio Tinto launched the "Mine of the Future" program, which aims to remote control more than 10 mines in Pilbara from Perth to realize unmanned mining operations [11]. The Swedish Mining Automation Group (SMAG) also proposed a plan to lead the automation upgrading of the mining industry [12]. The main research interest in this paper is the path planning of underground intelligent vehicles. Based on the known environmental map, starting point, and target, we use the path planning algorithm to obtain an appropriate path that accords with mining operation and vehicle kinematics. More generally, we research global path planning algorithms.

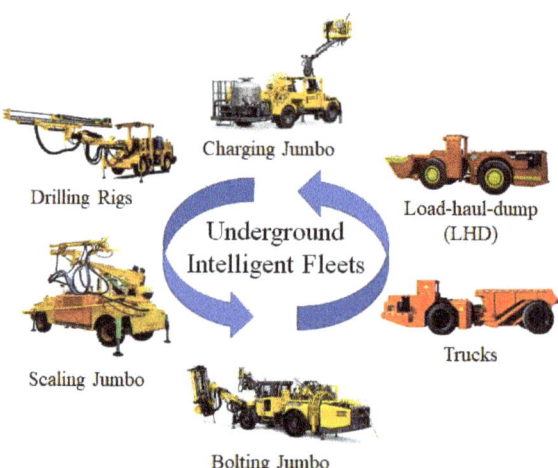

Figure 1. Underground intelligent vehicles.

In addition to the characteristics of common UGVs, the control of underground vehicles has strong industry specificity, which leads to more complicated path planning. First, the mechanical structure of the underground vehicles is more complicated, which is different from the common four-wheeled UGVs on the ground and UAVs in the air [13]. Thus, the underground vehicles are more difficult to control from kinematics and need a defined path. Second, compared to roads on the ground, the underground space is narrow and curved, with many irregular surfaces. Path planning for underground vehicles needs to focus more on passing narrow points and turns. Finally, there is no GPS underground, and the communication is worse than that on the ground. The path is required to be relatively simple, which reduces the control commands. Above all, the path planning method for UAVs or UGVs will not totally accord with that of underground vehicles. Therefore, it is necessary to upgrade the existing path planning method to adapt to underground intelligent vehicles.

RRT* is a sampling-based algorithm with probabilistic and complete resolution, high speed, and smooth results. For the 2D finite space of underground vehicles, it has a higher probability to create a path through narrow points and turns, which is closer to the underground requirements. Therefore, the RRT* algorithm was selected as the basic algorithm in this paper. With the aim of intelligent mining operation, by considering the kinematics of the intelligent vehicles and the drift environment, three improvements are proposed, including dynamic step size, steering angle constraints, and optimal tree reconnection. The algorithm improves the effectiveness of obstacle avoidance and shortens the distance while ensuring efficiency, which provides a feasible path planning method for intelligent vehicles.

Overall, this paper proposes a path planning method based on an improved RRT* algorithm to solve the problem of path planning for underground intelligent vehicles under articulated structures and drift environment conditions. Fully considering the environmental and equipment characteristics of underground mines is also an important feature.

The remainder of this paper is organized as follows. In Section 2, the related works are reviewed and the necessary preliminaries of intelligent mining are presented. In Section 3, the constraints of intelligent mining are formulated, including the drift environment formulation and the kinematics of vehicles. In Section 4, the process of the classic RRT* algorithm is analyzed and three improvement measures are proposed to adapt to underground intelligent vehicles. In Section 5, the case study by simulation method is presented, and the results are discussed. In Section 6, the paper is concluded.

2. Related Works

2.1. Underground Intelligent Vehicles

Autonomous vehicle driving is one of the key technologies of intelligent mining, and its main sensors and operating modes are shown in Figure 2. The intelligent vehicles collect their states and environmental information by laser lidar, inertial measurement unit (IMU), camera, RFID, and other sensors and calculate their current position and state by using edge computing. Then, they interact with cloud computing through wireless communication to obtain driving paths and complete the current driving process.

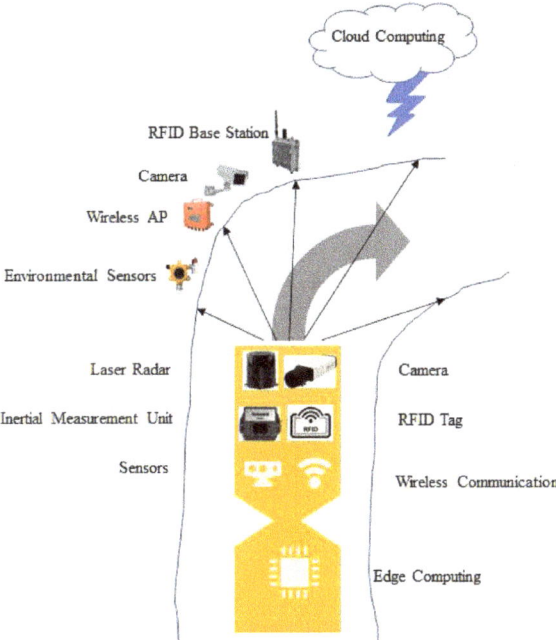

Figure 2. The main sensors and operating modes of autonomous driving vehicles.

Intelligent vehicles for underground mining can be divided into either an integral type or an articulated type according to their structure. The integral type has the advantage of a simple structure but has the disadvantage of often having insufficient power. It is mainly used in pick-up trucks, small LHDs, and other small vehicles. The articulated type has the advantages of a small steering radius and sufficient power and is more suitable for the narrow environment of underground mining [14]. Therefore, it is widely used in heavy equipment such as underground large LHDs, trucks, and jumbos. Articulated vehicles are more suitable in underground mines [13]. However, articulated vehicles have more complex structures than four-wheeled cars. For these reasons, articulated vehicles were selected as the main research object of this paper.

Large vehicles have a higher production capacity, but increasing the size of the drift increases the development cost. The size of the drift is often selected to meet the minimum specifications for vehicles, which put forward strict requirements for the running trajectory of vehicles. Therefore, the core of path planning for intelligent vehicles is to coordinate the environment of the drift and the kinematics of vehicles to obtain an optimal path trajectory that guides the vehicles to drive autonomously.

2.2. Path Planning Methods

Based on high-precision positioning and unmanned driving technology, many unmanned equipment path planning algorithms have been studied, which are mainly divided into artificial potential field methods, graph search algorithms, and sampling-based algorithms. The artificial potential field, proposed by Khatib, is a virtual force method, which makes the equipment subject to the repulsive forces from obstacles and gravity and from the target at the same time [15–18]. This method is simple to calculate, and the obtained path is safe and smooth, but it easily falls into a local optimal solution. The graph search algorithm converts the map for path planning into a graph and obtains the optimal path through graph theory, including the Dijkstra algorithm, A* algorithm, etc. [19–24]. This method takes into account both efficiency and completeness, but the map needs to be rasterized to complete the graph conversion, resulting in poor path smoothness. The sampling-based algorithm narrows the search space by discrete sampling in a continuous space. It is a Monte Carlo method with uniform space, including the Probabilistic Road Map Method (PRM), Rapid Random Extended Tree Method (RRT), etc. [25–29]. It has the advantages of fast search speed and simple environment modeling, but it cannot obtain a global optimal solution, and its efficiency is greatly affected by its step size and sampling mode.

The RRT algorithm [30] was proposed by Lavalle et al. in 1998. It is a random sampling algorithm that uses incremental growth to achieve rapid search in non-convex high-dimensional spaces. The RRT algorithm does not need to rasterize the search space and has the advantage of high search space coverage. It is suitable for dealing with scenes containing obstacles and motion constraints. Therefore, it is widely used in path planning for intelligent devices. The RRT algorithm is a Monte Carlo method. It usually takes the starting point as the root node and generates a random extended tree through random sampling. When the child node reaches the target area, the sampling is completed, and a feasible path is obtained.

The sampling of the RRT algorithm is random, and the generated path is a feasible path rather than an optimal path. Therefore, a variety of improved methods are derived. The RRT* algorithm [31] was improved based on the RRT algorithm, and the goal is to improve the performance of the RRT algorithm in order to ascertain the optimal path. The RRT* algorithm continuously optimizes the path during the search process by reselecting the parent node and rerouting. With the increase in iterations, the obtained path gradually approaches the optimal path.

There is relatively little research on path planning in underground mining, and currently it is mainly focused on underground disaster relief, surveying, and mapping. Ma et al. [32] proposed a path planning method considering gas concentration distributions. The global working path for a coal mine robot was planned based on the Dijkstra algorithm and the ant colony algorithm, then local path adjustments were carried out. The research object was coal mine robots, and the scene was disaster relief. Mauricio [33] proposed a strategy of exploration and mapping for multi-robot systems in underground mines where toxic gas concentrations are unknown. The principal algorithm was behavior control. Papachristos et al. [34] considered the challenge of autonomous navigation, exploration, and mapping in underground mines using aerial robots, and proposed an optimized multimodal sensor fusion approach combined with a local environment morphology-aware exploration path planning strategy. The research objects were four-rotor drones, and the scene was underground surveying and mapping. Gamache et al. [35] set up a shortest-path algorithm for solving the fleet management problem in underground mines with consideration for dispatching, routing, and scheduling vehicles. The solution approach was based on a shortest-path algorithm. They considered all single-lane bi-directional road segments of the haulage network. The research focused more on mining scheduling than vehicle path planning. The solution provided the direction of the vehicle at an intersection, rather than the trajectory of a single device. It can be considered as a form of cooperative scheduling, which relates to the upper-level control of intelligent vehicles. Larsson [36] developed a new flexible infrastructure-less guidance system for autonomous tramming of center-articulated underground mining vehicles. The results showed that it was capable of autonomous navigation in tunnel-like environments. However, the process of path planning was not described. Tian [37] presented a novel strategy for autonomous graph-based exploration path planning in subterranean environments. Yuan [38] focused more on path planning and an obstacle avoidance mechanism under the complex geological conditions of a coal mine. Dang [39] presented a novel strategy for autonomous graph-based exploration path planning in subterranean environments. Song [40] considered both the distance of the path and some hybrid costs to obtain a global path. Bai [41] proposed a multisensor data fusion algorithm based on genetic algorithm optimization of the variably structured fuzzy neural network. Ma [42] improved both the distance function and the selection of child nodes. The feature of this paper is the full consideration of the environment with a vectorized map and the articulated kinematics of underground mines. A comparison between some typical underground mine path planning studies is shown in Table 1.

Table 1. Comparison of typical underground mine path planning research.

Research	Algorithms	Scenarios	Path Type	Map Type	Equipment
[32]	Dijkstra, Ant colony	Rescue	Global	Rasterized	Mine robots
[33]	Scanning algorithms	Dangerous environment in coal mines	Local	Real-time sensing	Multi-robot systems
[34]	Optimized multimodal sensor fusion approach	Navigation, mapping	Navigation	Real-time sensing	Aerial robots
[35]	Enumeration algorithm	Production	Global	Topological	Underground vehicles
[36]	Feature detection algorithm	Production	Navigation	Real-time sensing	Underground articulated vehicles
[37]	Artificial potential field	Rescue	Global	Rasterized	Mine robots and UAVs
[38]	A* algorithm	Production	Global and local	Rasterized	Underground four-wheeled vehicles
[39]	Graph-based exploration path planning	Exploration, mapping	Global and local	Real-time sensing	UAVs
[40]	Ant colony algorithm	Not mentioned	Global	Rasterized	Mine robots
[41]	Genetic algorithm	Rescue	Navigation	Real-time sensing	Rescue snake robot
[42]	D* algorithm	Not mentioned	Global	Rasterized	Mine robots
This paper	Improved RRT* algorithm	Production	Global	Vectorized	Underground articulated vehicles

3. Constraints Formulation

Autonomously driven underground intelligent vehicles initiate a process of interaction between the underground environment and the vehicle. Before path planning, it is necessary to establish the environment, vehicle features, and interaction constraints.

3.1. Drift Environment Formulation

Drifts are the main environments for underground intelligent vehicles. These intelligent vehicles start at the stope filled with ore, drive through the drifts, then reach the orepass, and offload the ore. The point cloud is a common method for intelligent mine environmental modeling, which is generated by laser scanners [43]. Figure 3 shows the point cloud data obtained through SLAM, which is a typical drift environment. A typical design profile of a drift is shown in Figure 4 [44]. Where vehicles are required to travel through drifts, the vehicle cross-section will fix the dimensions of the opening. Underground intelligent vehicles do not make vertical movements, so it is possible to process 3D point cloud data into a 2D map by extracting the waistline and then converting the map into a graph for the path planning algorithm.

Figure 3. Typical drift environment point cloud data by SLAM.

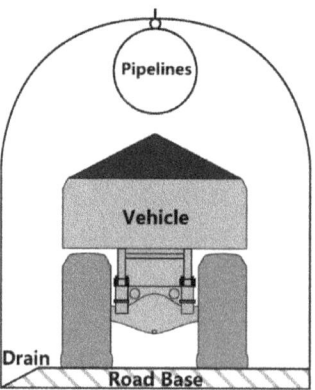

Figure 4. A typical design profile of a drift.

This paper uses the vectorized method instead of the common rasterization method as the map preprocessing method. The drifts are narrow and long with complicated surfaces. In the process of rasterization, the grid size has a great influence. Large-size grids cannot express the small edges and corners of the drifts well, resulting in a lack of detailed map information, and collisions during driving of the vehicles. Small grids lead to large total grids, which result in calculations being carried out in increments and a reduction in efficiency. Therefore, the rasterization method has certain limitations in processing the drift environment. A vectorized map can effectively improve these shortcomings. It expresses map information such as points, lines, and areas by recording coordinates. The points are represented by the north coordinate and east coordinate. The lines are represented by a series of ordered coordinates. The surfaces are represented by a series of ordered and closed coordinates. We recorded the coordinates of the scattered points on the map boundary through dense interpolation and connected them to form lines. The dataset included coordinate points, lines, and polygons, named as $Polygon_{map}$ in the following. The effect comparison between the rasterized map and vectorized map is shown in Figure 5. The rasterized map used a 22 × 41 matrix, and the dataset was 26.4 kb, as shown in Figure 5a. The vectorized map included 17 points, 17 lines, and 1 polygon. The dataset was 0.9 kb, as shown in Figure 5b. The vectorized map has great advantages in map refinement and data size.

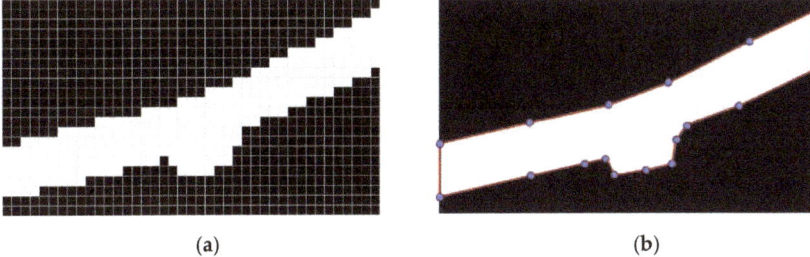

Figure 5. Comparison of the rasterized map and vectorized map. (**a**) The rasterized map; (**b**) the vectorized map.

3.2. Kinematics of Vehicles

Intelligent vehicle path planning needs to consider the kinematics to realize steering and obstacle avoidance. Articulated vehicles are considered in this paper, which are composed of a front body and rear body, and the vehicle bodies are connected through the articulated points. Articulated vehicles in underground mines are usually rear-wheel drives, and the steering is completed by controlling the relative position between the front and rear bodies through the expansion and contraction of the steering cylinders. Non-articulated vehicles can be abstracted as articulated vehicles with a rear body length of 0 to achieve the universality of this article. Assuming that the tire and the ground have pure rolling friction, the movement process of the vehicles can be simplified to rigid body plane movement, as shown in Figure 6.

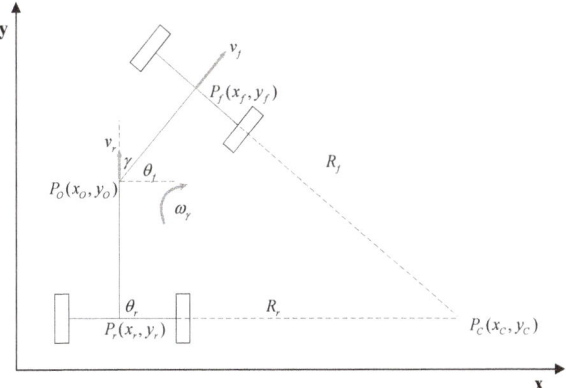

Figure 6. The simplified movement process of the vehicles.

In order to describe the position of the vehicle, the kinematics formula needs to be established. We established a Cartesian coordinate system for articulated vehicles. We set the instantaneous steering center as $P_C(x_C, y_C)$, the axle center of the front body as $P_f(x_f, y_f)$, the axle center of the rear body as $P_r(x_r, y_r)$, the articulated point as $P_O(x_O, y_O)$, the front and rear linear velocities as v_f and v_r, respectively, the headings as θ_f and θ_r, the radius of the front and rear body as R_f and R_r, respectively, the steering angle as γ, and the angular velocity as ω_γ. Then the kinematics model of the articulated vehicles can be described as follows [45].

$$\begin{bmatrix} \dot{x}_r \\ \dot{y}_r \\ \dot{\theta}_f \\ \dot{\gamma}_f \end{bmatrix} = \begin{bmatrix} \cos\theta_f \\ \sin\theta_f \\ \frac{\sin\gamma}{R_f\cos\gamma + R_r} \\ 0 \end{bmatrix} v_f + \begin{bmatrix} 0 \\ 0 \\ \frac{R_r}{R_f\cos\gamma + R_r} \\ 1 \end{bmatrix} \omega_r \quad (1)$$

Then, the position equation of the vehicles can be derived to avoid collision with drifts or obstacles; that is, the collision detection should be performed on the geometric shape of the vehicles, drifts, and obstacles. The Oriented Bounding Box (OBB) method was used to transform each entity into multiple bounding boxes in different directions for intersection testing. On the premise of authenticity, it is assumed that the front and rear bodies of the articulated vehicles are two rectangles, the width is w, and the length of the front and rear bodies are l_f and l_r, respectively, as shown in Figure 7.

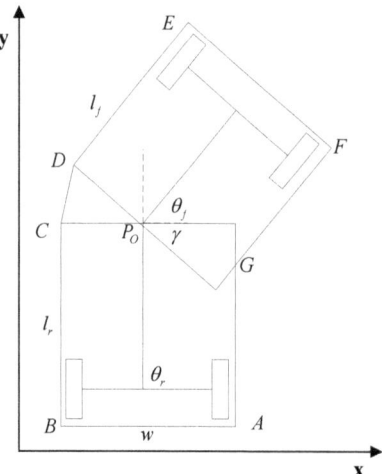

Figure 7. The geometric movement state of vehicles.

The OBB of the articulated vehicles ($Polygon_{car}$) can be represented by polygon ABCDEFG. According to the geometric relationship, the coordinates of each point of $Polygon_{car}$ can be expressed as the formulas

$$x_A = x_o - l_r\cos\theta_r + \frac{w\sin\theta_r}{2} \quad (2)$$

$$y_A = y_o - l_r\sin\theta_r - \frac{w\cos\theta_r}{2} \quad (3)$$

$$x_B = x_o - l_r\cos\theta_r - \frac{w\sin\theta_r}{2} \quad (4)$$

$$y_B = y_o - l_r\sin\theta_r + \frac{w\cos\theta_r}{2} \quad (5)$$

$$x_C = x_o - \frac{w\sin\theta_r}{2} \qquad (6)$$

$$y_C = y_o + \frac{w\cos\theta_r}{2} \qquad (7)$$

$$x_D = x_o - \frac{w\sin\theta_f}{2} \qquad (8)$$

$$y_D = y_o + \frac{w\cos\theta_f}{2} \qquad (9)$$

$$x_E = x_o + l_f\cos\theta_f - \frac{w\sin\theta_f}{2} \qquad (10)$$

$$y_E = y_o + l_f\sin\theta_f + \frac{w\cos\theta_f}{2} \qquad (11)$$

$$x_F = x_o + l_f\cos\theta_f + \frac{w\sin\theta_f}{2} \qquad (12)$$

$$y_F = y_o + l_f\sin\theta_f - \frac{w\cos\theta_f}{2} \qquad (13)$$

$$x_G = \frac{w\cos\theta_f{}^2\cos\theta_r - w\cos\theta_f\cos\theta_r{}^2 - w\cos\theta_f\sin\theta_r{}^2 + w\cos\theta_r\sin\theta_f{}^2}{2(\cos\theta_f\sin\theta_r - \cos\theta_r\sin\theta_f)} + \frac{x_o\cos\theta_f\sin\theta_r - x_o\cos\theta_r\sin\theta_f}{(\cos\theta_f\sin\theta_r - \cos\theta_r\sin\theta_f)} \qquad (14)$$

The collision can be detected by the intersection area between the map and the OBB of vehicles. If the vehicle is just within the feasible area of the map, it can be defined as no collision. We constructed the collision detection formula of the vehicle according to the $Polygon_{map}$ and $Polygon_{car}$, as shown in Formula (15).

$$Collision = \begin{cases} 1, & \varnothing < Polygon_{map} \cap Polygon_{car} < Polygon_{car} \\ 1, & Polygon_{map} \cap Polygon_{car} = \varnothing \\ 0, & Polygon_{map} \cap Polygon_{car} = Polygon_{car} \end{cases} \qquad (15)$$

4. Improved RRT* Algorithm for Intelligent Vehicles

The RRT* algorithm has great advantages in search efficiency and search quality and has been successfully applied in unmanned vehicle driving, UAV navigation, etc.

For underground mines, the application of the RRT* algorithm must consider the following aspects:

(1) The underground drift is long and narrow, and the available area of the entire map is small. The RRT* algorithm uses fixed-step full-map sampling, which results in low sampling efficiency in the scene of the drift map;
(2) Drifts are usually constructed by a drilling and blasting method, and their surface will inevitably be irregular. As a result, the map of drifts cannot be as smooth as a regular road map, which will affect the smoothness of the solution path;
(3) Underground vehicles are usually large in size, and the steering radius should be strictly controlled during their driving. Due to the randomness of the expansion, the RRT* algorithm cannot guarantee a path that meets the steering radius of the vehicles.

Aiming to adopt the use of intelligent vehicles in underground mines, this paper makes the following improvements to the RRT* algorithm:

(1) Dynamic step size

The classic RTT* algorithm adopts a fixed step size expansion strategy. When the step size is small, the convergence speed is slow. When the step size is large, the vehicle easily collides with the drift wall, causing sampling failure and indirectly affecting the solution speed. The strategy of a fixed step size is: first, we randomly sampled x_{rand} in the map; secondly, we obtained its neighbor x_{near}; then, we connected x_{rand} and x_{near}, and took the length of the StepSize from x_{near} to obtain the point x_{new}; if the collision detection was valid, an expansion was completed. Collision detection with a fixed step has a higher probability of failure. To solve this problem, a dynamic step size strategy was designed, and the step size was taken as a dynamic random function of CollisionSize (the distance from x_{near} to the collision point). When far from the obstacle, a larger step size was taken to ensure the speed of convergence; when the obstacle was closer, a smaller step size was taken to ensure the effectiveness of collision detection, as shown in Formula (16) and Figure 8.

$$DynamicSize = \begin{cases} StepSize & Collision = false \\ CollisionSize \times U[0,1] & Collision = true \end{cases} \quad (16)$$

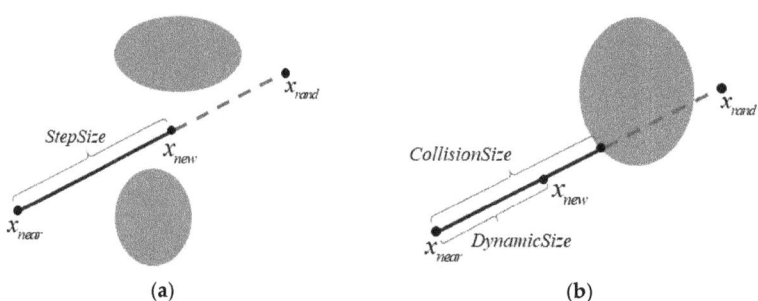

Figure 8. Comparison of fixed step size and dynamic step size. (a) Fixed step size; (b) dynamic step size.

(2) Steering angle constraints

The steering process of vehicles is strictly constrained by the max steering angle β. Therefore, during the sampling process of the RRT* algorithm, the angle θ between the new path and the parent path should be less than β, as shown in Formula (17) and Figure 9.

$$\theta = |\dot{\gamma}_f| = \arccos\left(\frac{\overrightarrow{x_{parent}x_{near}} \cdot \overrightarrow{x_{near}x_{new}}}{\left|\overrightarrow{x_{parent}x_{near}}\right|\left|\overrightarrow{x_{near}x_{new}}\right|}\right) \leq \beta \quad (17)$$

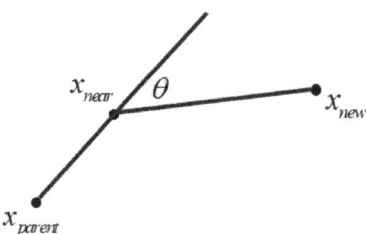

Figure 9. Geometric relation of steering angle constraints.

(3) Optimal tree reconnection

The classic RRT* algorithm uses random sampling, so the obtained path usually has certain turns, which lead to the deceleration of vehicles. Therefore, unnecessary turns should be avoided to lead the vehicles to drive straight. This will reduce the control difficulty of unmanned driving while reducing the path distance. The optimal tree reconnection process is as follows: we straightened and optimized the feasible path when the RRT* algorithm found a solution; we continuously traversed from the root node to the child node; we searched for the child nodes that were directly connected without obstacles; we connected the two nodes and deleted the intermediate nodes. This process turned the path into a curve by reducing the number of nodes, as shown in Figures 10 and 11.

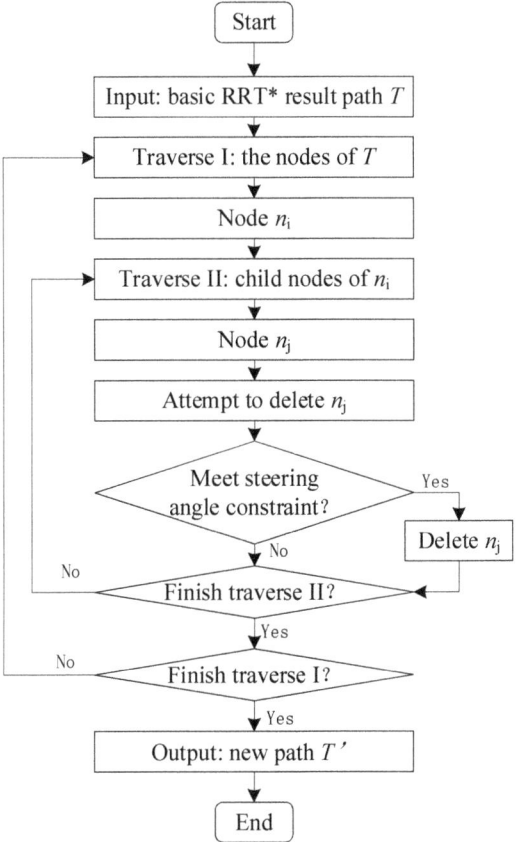

Figure 10. Process of optimal tree reconnection.

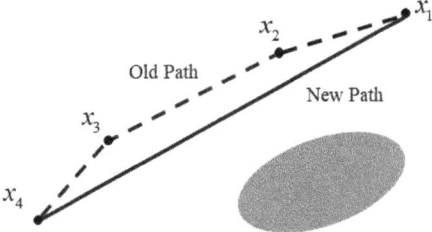

Figure 11. Geometric relation of optimal tree reconnection.

The pseudo code of the improved Algorithm 1 is as follows:

Algorithm 1 Improved RRT* Algorithm

Input: x_{start}, x_{goal}, Map
Output: A path T from x_{start} to x_{goal}
1 T.initalize();
2 **for** $i = 1$ to n **do**
3   **while** true **do**
4          $x_{rand} \leftarrow$ Sample(Map);
5          $x_{near} \leftarrow$ Near(x_{rand}, T);
6   DynamicSize $\leftarrow$ CollisionCheck(x_{near}, Map);
7          $x_{new} \leftarrow$ Steer(x_{rand}, x_{near}, DynamicSize);
8   **if** CollisionFree(x_{new}, Map) and Turnable(x_{new}, x_{near}, x_{parent}) **then**
9   break;
10   **end**
11   **end**
12     $X_{near\_neighbours} \leftarrow$ NearNeighbour(x_{new}, T)
13   **foreach** $x_{near\_neighbour} \in X_{near\_neighbours}$ **do**
14   Test_dis $\leftarrow$ Cost(x_{new}) + Distance(x_{new}, $x_{near\_neighbour}$)
15   **if** CollisionFree(x_{new}, $x_{near\_neighbour}$, Map) and Test_dis < Cost($x_{near\_neighbour}$) **then**
16             $x_{parent} \leftarrow$ Parent($x_{near\_neighbour}$);
17   Update(T);
18   **end**
19   **end**
20   **if** $x_{new} = x_{goal}$ **then**
21         $T \leftarrow$ OptimalTreeReconnection(T);
22   success();
23   **end**
24 **end**

5. Simulation Analysis

5.1. Simulation Environment

In order to verify the adaptability of the improved RRT* algorithm, the classic RRT, the classic RRT*, and the improved RRT* algorithms are simulated and verified in the underground ore transportation scenario. The parameters of the vehicles come from the Scooptram ST3.5 diesel LHD, as shown in Figure 12 and Table 2. The verification map comes from a large underground mine in China, as shown in Figure 13a. The design size of the drifts was 4.4 m × 3.9 m. The ore is transported by an LHD from Stope #1 to Orepass #1. The map was preprocessed, and only the route of the LHD was retained. The simplified map is shown in Figure 13b.

Figure 12. Scooptram ST3.5 diesel LHD.

Table 2. Parameters of the Scooptram ST3.5 diesel LHD.

Parameter	Value
Max steering angle	42.5°
Width	2120 mm
Front body length	4130 mm
Rear body length	4330 mm

Data Source: Epiroc official website.

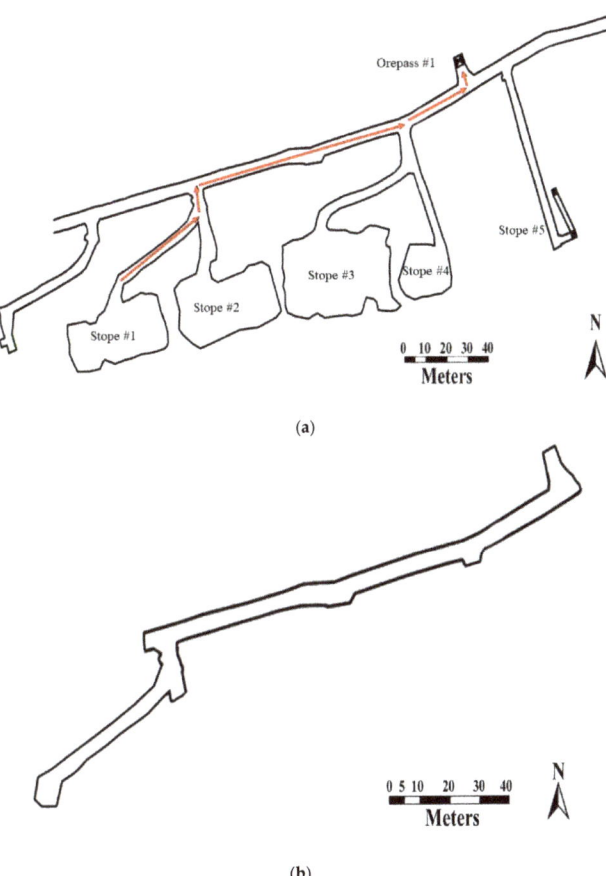

Figure 13. The map of the case study. (**a**) The original map; (**b**) the simplified map.

The case study simulated the operation process of the LHD from the stope to the orepass and verified the algorithm's ability to plan a feasible path in a long and narrow space. The LHD is required to complete ore transportation with the minimum distance under safe conditions and kinematic constraints. The simulation process was developed with Python 3.7, the operating system was Windows 10 × 64 bit, the CPU was Intel Core i7-8550U, and the memory was 16 GB. The simulation environment included Scipy 1.6.2, Shapely 1.8.0, and Matplotlib 3.3.4. Scipy was used to create the formulas. Shapely was used to calculate the OBB of vehicles and map polygons. Matplotlib was used to show the path.

5.2. Simulation Results

Comparative simulation experiments of the classic RRT algorithm, classic RRT* algorithm, and improved RRT* algorithm were carried out, and the results are shown in Figure 14. The red "X" represents the starting point and end point of the path planning, the blue line represents the wall of the drifts, and the horizontal and vertical axes represent the east and north coordinates. The yellow line represents the result of the classic RRT algorithm, the green line represents the result of the classic RRT* algorithm, and the red line represents the result of the improved RRT* algorithm.

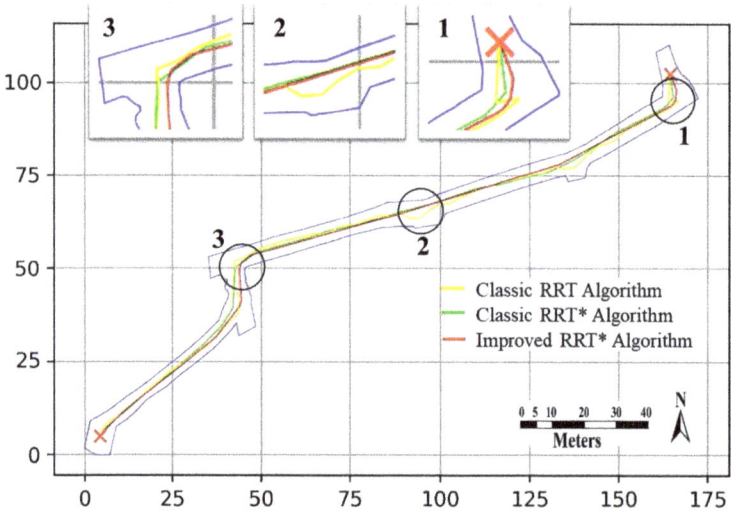

Figure 14. The simulation results.

It can be seen from Figure 13 that the path generated by the classic RTT algorithm had robust randomness, and there were a lot of irregular corners, such as Circle 1 and Circle 2. In contrast, the smoothness of the path generated by the classic RRT* algorithm was greatly improved, but the steering angle at the bend of the drift was too sharp, which was not suitable for the steering angle of the vehicles, such as Circle 2 and Circle 3.

Ten independent random simulations were performed on each algorithm to offset the random deviation of a single experiment. The results are shown in Table 3. The average path length obtained by the improved RRT* algorithm was much lower than that of the classic RRT algorithm but had only a small reduction compared with the classic RRT* algorithm. The main reason is that the reconnection in the classic RRT* algorithm can quickly approach the theoretically shortest time. The improved RRT* algorithm inherited this feature, and there was no more room for improvement. For the average search time, the performance of the improved RRT* algorithm was between the classic RRT algorithm and the classic RTT* algorithm. The same reason also led to the increment in average search nodes. Due to the optimal tree reconnection, the improved RRT* algorithm had a significant advantage over the classic algorithm in terms of average path nodes. This parameter reduced the control points during vehicle driving and reduced the difficulty of automatic driving. The steering angle constraints made the improved RRT* algorithm result fully meet the steering requirements, and the optimal tree reconnection increased the smoothness of the path, so the device can directly follow the path without further adjustment, avoiding multiple calculations. In general, the improved RRT* algorithm greatly improved the quality of the path while appropriately sacrificing the solution speed.

Table 3. Statistics of 10 independent random simulations.

Parameters	Classic RRT	Classic RRT*	Improved RRT*
Average path length (m)	211.11	189.86	189.54
Average search time (s)	168.94	44.16	86.12
Average of search node count	561.60	267.30	360.00
Average of path node count	32.00	28.80	16.20
Effective ratio of steering angle	81.87%	92.71%	100.00%

Obstacles in underground drifts are common, such as faulty vehicles and stacked materials. Further verification was conducted with known obstacles, as shown in Figure 15. Two scenarios were considered with both avoidable obstacles and unavoidable obstacles in the drift. The red line represents the final result, the yellow line represents the invalid leaf of a random tree, and the blue point represents the obstacle. For avoidable obstacles, the algorithm could pass them using a smooth curve without more additional sampling being necessary. For unavoidable obstacles, the algorithm stopped sampling after a certain number of samples.

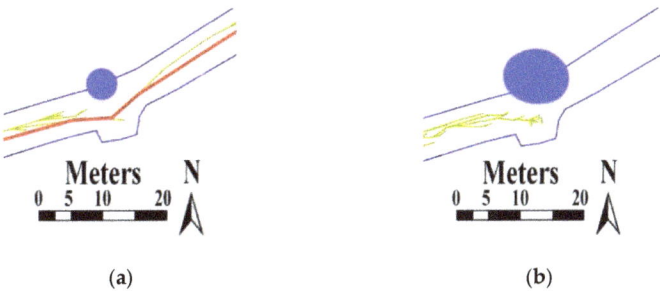

Figure 15. The simulation result with known obstacles. (a) With avoidable obstacles; (b) with unavoidable obstacles.

The kidnapping problem of intelligent vehicles might occur due to navigation failure or other reasons. For the verification of the kidnapping problem, we assumed that the vehicle planned to reach point B from point A but reached point B' for kidnapping reasons. Two scenarios were considered with both turnable kidnapping and unturnable kidnapping for the vehicle, as shown in Figure 16. For turnable kidnapping, it will reach the front point of the original path by the maximum steering angle. For unturnable kidnapping, it will drive astern to the back point of the original path by the maximum steering angle.

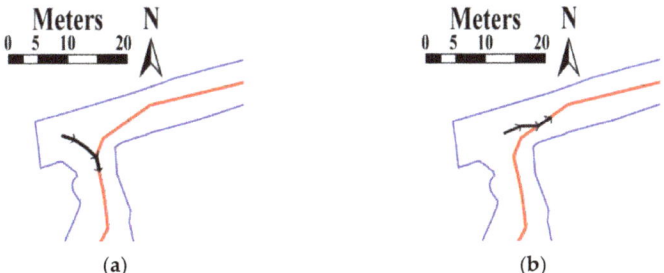

Figure 16. The simulation result for the kidnapping problem. (a) With turnable kidnapping; (b) with unturnable kidnapping.

5.3. Discussion

With the aim of the unmanned driving of intelligent vehicles in underground mines, we improved the path planning algorithm to adapt to the complex drift environment based on the RRT* algorithm. Many existing algorithms have to rasterize the map, but rasterized maps are not suitable for the drift environment. We constructed a vectorized drift environment map and then selected the RRT* algorithm to improve it. The vectorized map can effectively restore the details of the roadway environment and can also reduce the dataset. Combined with the articulated structure of underground intelligent vehicles, the dynamic characteristics were analyzed, and then the constraints were constructed. It strengthened the consideration of complex vehicle structures in this field. The process of the classical RRT* algorithm was analyzed, and then its shortcomings in adaptability to underground mining were extracted. On this basis, three improvements were proposed: a dynamic step size solved the algorithm efficiency problem; steering angle constraints solved the vehicle dynamics problem; optimal tree reconnection solved the control difficulty problem. By way of a simulation case study, the improved RRT* algorithm obtained a path suitable for underground intelligent vehicles within a reasonable time. Its results increased the effective ratio of the steering angle to 100%, fully met the vehicle's requirements, eliminated the secondary optimization of the path, greatly reduced the average number of path nodes, and simplified the vehicle's automatic driving control. Many existing algorithms have to rasterize the map.

However, we must admit that in order to achieve the path planning effect, a large number of invalid samples were discarded, which led to an increase in calculation time. This algorithm can improve the sampling efficiency and shorten the calculation time through parallel calculation. This will be improved in future research to further reduce the calculation time. In addition, the simulation case study was completed in this paper, but no on-site industrial experiment was carried out. The unmanned driving design of underground intelligent vehicles coordinates with multiple modules, including communication, sensors, SLAM, mechanical control, etc. It is also necessary to shut down some mining operations to ensure the safety of the experiment area. Due to these difficulties, this research only completed the path planning algorithm module, and in the future, an on-site industrial experiment will be completed after the preparation of each module.

6. Conclusions

This paper proposed a path planning method based on an improved RRT* algorithm for solving the problem of path planning for underground intelligent vehicles on an articulated structure and in drift environment conditions. Through a vectorized drift map and using the kinematics of vehicles, the constraints of articulated underground intelligent vehicles can be ascertained. The RRT* algorithm is an efficient sampling-based path planning algorithm, but it cannot meet the constraints of articulated underground intelligent vehicles. To solve this problem, this paper proposed an improved RRT* algorithm, including dynamic step size, steering angle constraints, and optimal tree reconnection. A simulation case study proved that the algorithm was effective and could solve the problem of underground intelligent vehicle path planning.

However, the method in this paper still has limitations, and future research will focus on the following aspects. (1) The solution time is still unsatisfactory because 86.12s cannot meet the application requirement for underground unmanned driving. Vehicles need to obtain a path within several seconds. A parallel calculation will be used to increase the solution speed and further reduce the calculation time. (2) There is still no joint debugging with intelligent vehicles. After the preparation of the industrial site, it will be combined with other modules to complete on-site industrial experiments and test the gap between the simulated and actual performance.

Author Contributions: Conceptualization, N.H. and H.W.; methodology, H.W.; software, J.H. and L.C.; validation, H.W., G.L. and L.C.; formal analysis, H.W.; data curation, L.C.; writing—original draft preparation, H.W.; writing—review and editing, G.L. and N.H; visualization, J.H.; supervision, N.H.; funding acquisition, G.L. All authors have read and agreed to the published version of the manuscript.

Funding: This research was funded by the Natural Science Foundation of China, grant number 52074022 and National Key R&D Program of China, grant number 2018YFC0604400.

Data Availability Statement: The parameters of the Scooptram ST3.5 diesel LHD were obtained from https://www.epiroc.com/en-us/products/loaders-and-trucks/diesel-loaders/scooptram-st3-5 (accessed on 12 December 2021). The source code associated with the algorithms introduced in the paper is available from the corresponding author upon request. The environment data are not available according to the privacy policy.

Acknowledgments: The authors would like to thank SD Gold for drift environment data support and Epiroc for LHD parameters support.

Conflicts of Interest: The authors declare no conflict of interest.

References

1. Guo, H.; Zhu, K.; Ding, C.; Li, L. Intelligent optimization for project scheduling of the first mining face in coal mining. *Expert Syst. Appl.* **2010**, *37*, 1294–1301. [CrossRef]
2. Li, J.G.; Zhan, K. Intelligent Mining Technology for an Underground Metal Mine Based on Unmanned Equipment. *Engineering* **2018**, *4*, 11. [CrossRef]
3. Lw, J.; Abrahamsson, L.; Johansson, J. Mining 4.0—The Impact of New Technology from a Work Place Perspective. *Min. Metall. Explor.* **2019**, *36*, 701–707.
4. Sánchez, F.; Hartlieb, P. Innovation in the Mining Industry: Technological Trends and a Case Study of the Challenges of Disruptive Innovation. *Min. Metall. Explor.* **2020**, *37*, 1385–1399. [CrossRef]
5. Chen, W.; Wang, X. Coal Mine Safety Intelligent Monitoring Based on Wireless Sensor Network. *IEEE Sens. J.* **2020**, *21*, 25465–25471. [CrossRef]
6. Roberge, V.; Tarbouchi, M.; LaBonte, G. Comparison of Parallel Genetic Algorithm and Particle Swarm Optimization for Real-Time UAV Path Planning. *Ind. Inform. IEEE Trans.* **2012**, *9*, 132–141. [CrossRef]
7. Tokekar, P.; Hook, J.V.; Mulla, D.; Isler, V. Sensor Planning for a Symbiotic UAV and UGV System for Precision Agriculture. *IEEE Trans. Robot.* **2016**, *32*, 1498–1511. [CrossRef]
8. Chugh, Y.; Parkinson, H. Automation trends in room-and-pillar continous-mining systems in the USA. *J. S. Afr. Inst. Min. Metall.* **1981**, *81*, 57–65.
9. Scoble, M. Canadian mining automation evolution: The digital mine en route to minewide automation. *Int. J. Rock Mech. Min. Sci. Geomech. Abstr.* **1995**, *32*, 351A.
10. Kot-Niewiadomska, A. ROBOMINERS Project—The future of mining. In Proceedings of the 30th International Conference on the Present and Future of the Mining and Geology, Demanovska Dolina, Slovakia, 3–4 October 2019; Robominers: Demanovska Dolina, Slovakia, 2019; pp. 1–12.
11. Bearne, G. Innovation in mining: Rio Tinto's Mine of the Future (TM) programme. *Alum. Int. Today* **2014**, *26*, 15.
12. Mullan, L. *Spearheading a Digital Transformation in the Canadian Mining Industry*; Swedish Mining Automation Group: Toronto, ON, Canada, 2017; pp. 1–3.
13. Altafini, C. Why to use an articulated vehicle in underground mining operations? In Proceedings of the 1999 IEEE International Conference on Robotics and Automation, Detroit, MI, USA, 10–15 May 1999; pp. 3020–3025.
14. Rodrigo, P. *Mining Haul Roads: Theory and Practice*; CRC Press: Boca Raton, FL, USA, 2019.
15. Khatib, O. Real-Time Obstacle Avoidance System for Manipulators and Mobile Robots. *Int. J. Robot. Res.* **1986**, *5*, 90–98. [CrossRef]
16. Luo, G.; Yu, J.I.; Mei, Y.O.; Zhang, S.Y. UAV Path Planning in Mixed-Obstacle Environment via Artificial Potential Field Method Improved by Additional Control Force. *Asian J. Control* **2015**, *17*, 1600–1610. [CrossRef]
17. Malone, N.; Chiang, H.T.; Lesser, K.; Oishi, M.; Tapia, L. Hybrid Dynamic Moving Obstacle Avoidance Using a Stochastic Reachable Set-Based Potential Field. *IEEE Trans. Robot.* **2017**, *33*, 1124–1138. [CrossRef]
18. Yang, W.; Wu, P.; Zhou, X.; Lv, H.; Liu, X.; Zhang, G.; Hou, Z.; Wang, W. Improved Artificial Potential Field and Dynamic Window Method for Amphibious Robot Fish Path Planning. *Appl. Sci.* **2021**, *11*, 2114. [CrossRef]
19. Seder, M.; Baotic, M.; Petrovic, I. Receding Horizon Control for Convergent Navigation of a Differential Drive Mobile Robot. *IEEE Trans. Control Syst. Technol.* **2017**, *25*, 653–660. [CrossRef]
20. Saad, M.; Salameh, A.I.; Abdallah, S.; El-Moursy, A.; Cheng, C.T. A Composite Metric Routing Approach for Energy-Efficient Shortest Path Planning on Natural Terrains. *Appl. Sci.* **2021**, *11*, 6939. [CrossRef]
21. Biyela, P.; Rawatlal, R. Development of an optimal state transition graph for trajectory optimisation of dynamic systems by application of Dijkstra's algorithm. *Comput. Chem. Eng.* **2019**, *125*, 569–586. [CrossRef]
22. Goerzen, C.; Kong, Z.; Mettler, B. A Survey of Motion Planning Algorithms from the Perspective of Autonomous UAV Guidance. *J. Intell. Robot. Syst.* **2010**, *57*, 65. [CrossRef]

23. Christie, G.A.; Shoemaker, A.; Kochersberger, K.B.; Tokekar, P.; Mclean, L.; Leonessa, A. Radiation search operations using scene understanding with autonomous UAV and UGV. *J. Field Robot.* **2017**, *34*, 1450–1468. [CrossRef]
24. Zhuang, H.; Dong, K.; Qi, Y.; Wang, N.; Dong, L. Multi-Destination Path Planning Method Research of Mobile Robots Based on Goal of Passing through the Fewest Obstacles. *Appl. Sci.* **2021**, *11*, 7378. [CrossRef]
25. Kavraki, L.E.; Svestka, P.; Latombe, J.C.; Overmars, M.H. Probabilistic Roadmaps for Path Planning in High-Dimensional Configuration Spaces. *IEEE Trans. Robot. Autom.* **1996**, *12*, 566–580. [CrossRef]
26. Park, B.; Chung, W.K. Efficient environment representation for mobile robot path planning using CVT-PRM with Halton sampling. *Electron. Lett.* **2012**, *48*, 1397–1399. [CrossRef]
27. LaValle, S.M.; Kuffner, J.J., Jr. Randomized kinodynamic planning. *Int. J. Robot. Res.* **2001**, *20*, 378–400. [CrossRef]
28. Cui, R.; Yang, L.; Yan, W. Mutual Information-Based Multi-AUV Path Planning for Scalar Field Sampling Using Multidimensional RRT*. *IEEE Trans. Syst. Man Cybern. Syst.* **2017**, *46*, 993–1004. [CrossRef]
29. Shome, R.; Solovey, K.; Dobson, A.; Halperin, D.; Bekris, K.E. dRRT*: Scalable and informed asymptotically-optimal multi-robot motion planning. *Auton. Robot.* **2020**, *44*, 443–467. [CrossRef]
30. Lavalle, S.M. Rapidly-Exploring Random Trees: A New Tool for Path Planning. *Res. Rep.* **1998**, *RT98-11*, 1–4.
31. Karaman, S.; Frazzoli, E. Optimal Kinodynamic Motion Planning using Incremental Sampling-based Methods. In Proceedings of the IEEE Conference on Decision & Control, Atlanta, GA, USA, 15–17 December 2010; pp. 7681–7687.
32. Ma, X.; Mao, R. Path planning for coal mine robot to avoid obstacle in gas distribution area. *Int. J. Adv. Robot. Syst.* **2018**, *15*, 172988141775150. [CrossRef]
33. Mauricio, A.; Nieves, A.; Castillo, Y.; Hilasaca, K.; Fonseca, C.; Gallardo, J.; Rodríguez, R.; Rodríguez, G. Multi-robot exploration and mapping strategy in underground mines by behavior control. In *Multibody Mechatronic Systems*; Springer: Berlin/Heidelberg, Germany, 2015; pp. 101–110.
34. Papachristos, C.; Khattak, S.; Mascarich, F.; Dang, T.; Alexis, K. Autonomous Aerial Robotic Exploration of Subterranean Environments relying on Morphology–aware Path Planning. In Proceedings of the 2019 International Conference on Unmanned Aircraft Systems (ICUAS), Atlanta, GA, USA, 11–14 June 2019; pp. 299–305.
35. Gamache, M.; Grimard, R.; Cohen, P. A shortest-path algorithm for solving the fleet management problem in underground mines. *Eur. J. Oper. Res.* **2005**, *166*, 497–506. [CrossRef]
36. Larsson, J.; Broxvall, M.; Saffiotti, A. Flexible infrastructure free navigation for vehicles in underground mines. In Proceedings of the 2008 IEEE 4th International IEEE Conference Intelligent Systems, Varna, Bulgaria, 6–8 September 2008; pp. 2-45-2-50.
37. Tian, Z.; Gao, X.; Zhang, M. Path planning based on the improved artificial potential field of coal mine dynamic target navigation. *J. China Coal Soc.* **2016**, *41* (Suppl. S2), 589–597.
38. Yuan, X.; Hao, M. Research on key technology of coal mine auxiliary transportation robot. *Ind. Mine Autom.* **2020**, *46*, 8–14.
39. Dang, T.; Mascarich, F.; Khattak, S.; Papachristos, C.; Alexis, K. Graph-based path planning for autonomous robotic exploration in subterranean environments. In Proceedings of the 2019 IEEE/RSJ International Conference on Intelligent Robots and Systems (IROS), Macau, China, 3–8 November 2019; pp. 3105–3112.
40. Song, B.; Miao, H.; Xu, L. Path planning for coal mine robot via improved ant colony optimization algorithm. *Syst. Sci. Control Eng.* **2021**, *9*, 283–289. [CrossRef]
41. Bai, Y.; Hou, Y.B. Research of environmental modeling method of coal mine rescue snake robot based on information fusion. In Proceedings of the 2017 IEEE 20th International Conference on Information Fusion (Fusion), Xi'an, China, 10–13 July 2017; pp. 1–8.
42. Ma, T.; Lv, J.; Guo, M. Downhole robot path planning based on improved D* algorithmn. In Proceedings of the 2020 IEEE International Conference on Signal Processing, Communications and Computing (ICSPCC), Macau, China, 21–24 August 2020; pp. 1–5.
43. Liu, Z. Application of handheld 3D laser scanner based on slam technology in tongkuangshan mine. *China Mine Eng.* **2021**, *50*, 13–16. [CrossRef]
44. Peter, D. *SME Mining Engineering Handbook*; Socieity for Mining, Metallurgy, and Exploration: Englewood, CO, USA, 2011; pp. 1148–1149.
45. Bai, G.; Liu, L.; Meng, Y.; Luo, W.; Gu, Q.; Ma, B. Path Tracking of Mining Vehicles Based on Nonlinear Model Predictive Control. *Appl. Sci.* **2019**, *9*, 1372. [CrossRef]

Article

A Hybrid and Hierarchical Approach for Spatial Exploration in Dynamic Environments

Qi Zhang [1,†], Yukai Song [1,2,†], Peng Jiao [1] and Yue Hu [1,*]

1. College of Systems Engineering, National University of Defense Technology, Changsha 410003, China; zhangqiy123@nudt.edu.cn (Q.Z.); songyukai14@163.com (Y.S.); crocus201@163.com (P.J.)
2. Unit.63880 of PLA, Luoyang 471000, China
* Correspondence: huyue11@nudt.edu.cn
† These authors are both first authors.

Abstract: Exploration in unknown dynamic environments is a challenging problem in an AI system, and current techniques tend to produce irrational exploratory behaviours and fail in obstacle avoidance. To this end, we present a three-tiered hierarchical and modular spatial exploration model that combines the intrinsic motivation integrated deep reinforcement learning (DRL) and rule-based real-time obstacle avoidance approach. We address the spatial exploration problem in two levels on the whole. On the higher level, a DRL based global module learns to determine a distant but easily reachable target that maximizes the current exploration progress. On the lower level, another two-level hierarchical movement controller is used to produce locally smooth and safe movements between targets based on the information of known areas and free space assumption. Experimental results on diverse and challenging 2D dynamic maps show that the proposed model achieves almost 90% coverage and generates smoother trajectories compared with a state-of-the-art IM based DRL and some other heuristic methods on the basis of avoiding obstacles in real time.

Keywords: spatial exploration; hierarchical framework; deep reinforcement learning; intrinsic motivation; path planning; obstacle avoidance

1. Introduction

Spatial cognitive behaviour modelling is the basic content of human cognitive behaviour modelling, and is one of the hottest topics in the field of neuroscience and computer science. At its core, the agent in an AI system needs to explore the environment to gain enough information about the spatial structure. The possible applications include, for example, search and rescue (SAR) missions, intelligence, surveillance and reconnaissance (ISR), and planetary exploration. Therefore, it is important to design an efficient and effective exploration strategy in unknown spaces.

At present, autonomous spatial exploration falls into two main categories: traditional rule-based exploration and intelligent machine-learning-based exploration. The rule-based exploration methods such as frontier-based method [1] is simple, convenient and efficient. This kind of approach rely on an expert feature of maps, expanded the exploration scope by searching for the next optimal frontier point which is between free points and unknown points according to the explored map. However, the locomotion of the agent driven by this method is mechanical and rigid, and it is also difficult to balance between exploration efficiency and computational burden. As an effective tool for autonomous learning strategies, deep reinforcement learning (DRL) has been more and more widely used in spatial exploration. However, DRL suffers much from the inherent "exploration-exploitation" dilemma, resulting in sampling inefficiency if the extrinsic rewards are sparse or even nonexistent. To solve the problem of sparse rewards, many recent DRL approaches incorporate the concept of intrinsic motivation (IM) from cognitive psychology to produce intrinsic rewards to make the rewards denser. However, intrinsic motivation based enhancement

is insufficient for efficient exploration in unknown spaces. The main reason is that IM treats all unseen states indiscriminately and ignores the structural regularities of physical spaces. In addition, it is difficult for end-to-end DRL agent to simultaneously learn obstacle avoidance, path planning and spatial exploration from raw sensor data.

To this end, we extend our previous work [2] and propose a three-tiered hierarchical autonomous spatial exploration model, named Intrinsic Rewards based Hierarchical Exploration with Soft-adaptive Finite-time Velocity Obstacle (IRHE-SFVO), to explore unknown static and dynamic 2D spaces. This model consists of two parts: a Global Exploration Module (GEM) and a Local Movement Module (LMM). GEM is used to learn an exploration policy to produce a sequence of target points that will maximize the information gain about the spatial structure based on the location of the agent, the trace of the agent, and the explored portions as its spatial memory. Specifically, to make the motion pattern of the agent more like human beings, GEM is not concerned with the immediate neighbourhood of the agent, but determines a distant yet reasonably reachable target to be explored next. Selected based on intrinsic rewards, such targets are usually those with a lot of unexplored areas around them.

In the local movement phase, this paper designs a hierarchical framework to control the movement to the target point. We separate this phase into two parts: planning and controlling. In the planning stage, an optimistic A* path planning algorithm, which can conduct self-adaptive path planning in a partially known environment, is used to compute a shortest path between the current location of the agent and the target point. It assumes that unknown areas are freely reachable and decides whether to replan the global path according to the ongoing perception. In the controlling stage, we use the improved Finite-time Velocity Obstacle (FVO), called Self-adaptive Finite-time Velocity Obstacle (SFVO), and design an optimal velocity function to drive the agent to avoid moving obstacles in real-time. This allows the agent to reach the target point quickly while avoiding collision with moving obstacles at the same time.

Working in synergy, the modules in the three levels apply a long-horizon decision-making paradigm instead of the step-by-step or state-by-state way used by some other exploration methods [3]. This segmentation not only reduces the training difficulty, but also tends to generate smooth movements between targets instead of unnatural trajectories. In summary, the main novelties and technical contributions of this paper include: (a) a hierarchical framework for spatial exploration that well exploits the structural regularities of unknown environments, (b) an information-maximal intrinsic reward function for determining the next best target to be explored, (c) a hierarchical framework for local movement that combines the global path planning with the local path planning for reaching the target point rapidly and safely and (d) an optimal velocity function for choosing the best velocity in collision-avoidance velocity set.

This paper is organized as follows. Section 2 describes related works in automatic exploration, the DRL based on IM and real-time obstacle avoidance. Section 3 formulates automatic exploration. Then, we present the details of our proposed algorithm and hyperparameter setting in Section 4. In Section 5, we compare our approach against several popular competitors in a series of simulation experiments, showing that IRHE-SFVO is promising for spatial exploration. Finally, in Section 6, we summarize our work this paper and discuss future work.

2. Related Work

In this section, we will describe and analyse the research status and development trends of autonomous spatial exploration, reinforcement learning based on IM and various velocity obstacle methods in this section.

2.1. Autonomous Spatial Exploration

At present, the research on autonomous spatial exploration mainly includes two categories: traditional rule-based autonomous spatial exploration and intelligent machine-learning-based autonomous spatial exploration. The mainstream of rule-based method is frontier-based method proposed by Yamauchi in 1997 [1]. This method detects the "frontier", that is, the edges between the free area and the unknown area, then selects the best "frontier point" by some principles, and the agent moves from the current position to the selected "frontier point" by path planning and locomotion, so as to finally achieve the purpose of exploring the whole map. The frontier-based exploration strategy is similar to the NBV (Next Best View) problem in computer vision and graphics. Similarly, there is a lot of literature on the second step of frontier-based exploration strategy, i.e., evaluating and choosing the best frontier. There are generally three types of metrics: (a) cost-based which select the next target based on the path length or time cost [4–7], (b) utility-based which select the next target based on the information gain [8,9] and (c) the mixture [10]. Another typical traditional rule-based method is associated with information theory. These methods leverage some metrics such as entropy [11] or mutual information (MI) [12] to evaluate the uncertainty of the agent's position and the evidence grid map to control the agent to move in the direction of maximizing the information gain. In general, although the rule-based approach is simple and efficient, the movement mode of the agent driven by them is mechanical and rigid, and it is also difficult to balance exploration efficiency with computational burden.

Due to the recent significant advance in DRL, a number of researchers have tried to solve the exploration problem as an optimal control problem. Tai Lei and Liu Ming [13] proposed an improved DQN framework to train robots to master obstacle avoidance strategies in unknown environments through supervised learning based on convolution neural networks (CNN). However, they only solved the collision avoidance problem and failed to finish the spatial exploration task. Zhang et al. [14] trained an Asynchronous Advantage Actor-Critic (A3C) agent that can learn from perceptual information and construct a global map by combining it with a memory module. Similarly, an A3C network in [15] receives the current map, the agent's location and orientation as input, and returns the next visiting direction, given that the space around the agent is equally divided into six sectors. Chen et al. [16] designed a module of spatial memory and used the coverage area gain as an intrinsic reward, and accelerated the convergence of policy through imitation learning. Razin et al. [17] used Faster R-CNN to avoid collision and used double deep Q-learning (DDQN) model to explore unknown space. However, although DRL can solve the problem of limited dimensions, it has difficulty training in end-to-end control.

To solve these problems, Niroui et al. [18] and Shrestha et al. [19] combined DRL with a frontier-based method to enable robots to learn exploration strategies from their own experience. Li et al. [20] proposed a modular framework for robot exploration based on decision, planning and mapping modules. This framework used DQN to learn a policy for selecting the next exploration target in the decision module and used an auxiliary edge segmentation task to speed up training. Chaplot et al. [21] used the Active Neural SLAM module to address the exploration in 3D environments under the condition of perception noises. We draw some inspiration from these two works but are more interested in exploration in 2D environments.

2.2. RL Based on Intrinsic Motivation

To solve the notorious reward-sparse problem, many recent DRL approaches incorporate the intrinsic motivation from cognitive psychology. Intrinsic motivation is produced from human's natural interest in all kinds of activities that can provide novelty, surprise, curiosity, or challenge [22], without any external rewards such as food, money or punishment.

Applying IM to the RL means that the agent generates an "intrinsic reward" by itself during the interaction with the environment. The formulation of intrinsic rewards can be roughly divided into three categories, (a) visit count and uncertainty evaluation-based methods, (b) knowledge and information gain-based methods, and (c) competence-based methods. The first class of methods, based on upper confidence bound (UCB), estimate the counts of state visitation in high-dimensional feature space and large-scale state space, to encourage the agent to visit poorly known states. This genre includes the density-based methods [23,24], state generalization-based methods [25–28] and inference calculation-based methods [29]. Second, the knowledge and information gain-based methods generally establish a dynamics model of the unknown environment and measures the intrinsic rewards using the model's increased accuracy as the exploration progresses. The specific formal models of this type include predict inconsistencies based model [30–32], prediction error based model [3,33–36], learning process based model [37] and information theory based model [36,38,39]. The third class formulates the intrinsic rewards by measuring the agent's competence to control the environment or the difficulty and cost of completing a task [40]. At present, the DRL based on IM has made great progress relative to the classic RL in applications with complex state spaces and difficult exploration (such as Atari-57 games) [41].

2.3. Velocity Obstacle

A crucial problem in exploration is how to avoid static and dynamic obstacles in real time. The known static obstacles are usually considered in global path planning, while unknown or dynamic obstacles are the focus of local path planning. The common collision avoidance methods include artificial aperture method (APF) [42], dynamic window method (DWA) [43] and behaviour method [44]. These methods have strong adaptability and high efficiency, so many researchers often combine the intelligent control algorithms with these methods for obstacle avoidance [45,46]. Besides, lazy rapidly-exploring random tree method (RRT) [47] method is also used for local path planning. However, these methods above cannot avoid collisions completely with moving obstacles or have certain randomness which leads to low efficiency of obstacle avoidance such as [47]. Alternatively, Velocity Obstacle (VO), first proposed by Fiorini et al. [48], is a simple and efficient algorithm that can avoid static and moving obstacles completely. It generates a conical velocity obstacle space in the agent velocity space. As long as the current velocity vector is outside the VO space, the agent will not collide with obstacles at any time in the future. However, the basic VO has many disadvantages. First, if the agent and moving obstacles or other agents use VO for local path planning at the same time, it will lead to oscillatory motion on both sides [49]. Secondly, the VO space excludes every velocity that may lead to collision, that is, a velocity that can cause a collision after a long time will also be excluded. This leads to the reduction of the range of optional collision-avoidance velocities in some scenarios, or even no optional velocity. To overcome these problems, Abe and Matsuo [50] proposed a common velocity obstacle (CVO) method, which provides collision detection between moving agents and enables agents to share collision information without explicit communication. This information allows agents to use the general VO method for implicit cooperation, so as to achieve the effect of avoiding collision. Guy et al. [51] proposed the finite time velocity obstacle algorithm (FVO), which expands the optional velocity vector of the traditional VO algorithm by calculating the collision velocity cone within a certain time. In order to solve the local oscillation problem Fulgenzi et al. [49] proposed reciprocal velocity obstacles (RVO) by considering the velocity change of both sides of the agents.

The proposed model in this paper combines the DRL, intrinsic motivation and velocity obstacle. Due to the features of exploration in 2D dynamic spaces, we reshape the generating paradigm of intrinsic reward. In order to ensure the safe and fast movement of the agent, we propose another hierarchical approach that combines a variation of the A* path planning method (called optimistic A*) and improved FVO (called self-adaptive FVO, SFVO).

3. Problem Formulation

Before giving the details of the proposed model, this section first formulates the exploration problem in a 2D environment.

Definition 1. *A Working Space, denoted as WS_M, represents a 2D grid map of the size $M \times M$. Any element in WS_M can be represented as (x, y), $1 \leq x, y \leq M$. Each cell in the grid is represented by $T(x, y)$: $T(x, y) = 0$ means a free cell while 1 is for a location occupied by an obstacle. Besides, we assume that the area of each cell is 1.*

Definition 2. *Definition 2 Observation Range (ObsR) of an agent is the set of any point whose vertical and horizontal distance to the current position of the agent is not more than the observation radius (n):*

$$ObsR(x_i, y_i) = \{(x, y) | (|x - x_i| \leq n, |y - y_i| \leq n)\} \tag{1}$$

Definition 3. *Exploration Range (ExpR) of an agent is the set of any point whose vertical and horizontal distance to the current position of the agent is not more than the exploration radius(m), and it can be covered more than half of the area by the 'radar wave' emitted by agent:*

$$ExpR(x_i, y_i) = \{(x, y) | (|x - x_i| \leq m, |y - y_i| \leq m), S((x_i, y_i) \to (x, y)) > \frac{1}{2}\} \tag{2}$$

$S((x_i, y_i) \to (x, y))$ means the covered area by the "radar wave" emitted from (x_i, y_i) to (x, y). A specific example is shown in Figure 1.

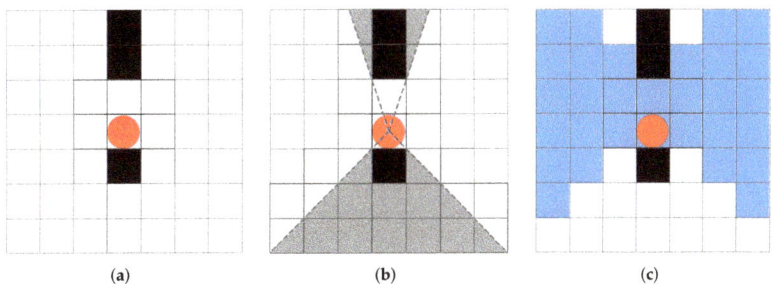

Figure 1. An example of Exploration Range. (**a**) shows the obstacles around the agent, where the red solid circle represents the agent, and the black squares represent two obstacles. (**b**) shows the range that can be covered by the "radar wave" emitted from the agent. The gray shaded areas indicate that these areas are not covered by the "radar wave". (**c**) shows whether each cell in this scenario can be regarded as an explored area under Definition 3 when $m = 3$. The blue cells are the areas that the agent has been explored, while the agent has not explored the white areas.

Note that the region observed by the agent does not represent where it has been explored. As a simple example, imagine that we are searching for gold that cannot be seen from the earth's surface, so that we should use a gold detector to explore the region as far as it can extend into. We cannot find gold using our eyes, but the detector can. In general, the "detection range" (m) should not be greater than the "length of field of view" (n), i.e., $m \leq n$ and $ExpR(x_i, y_i) \subseteq ObsR(x_i, y_i)$.

4. The Proposed Model

This paper combines the advantages of DRL algorithms, traditional non-learning planning algorithms and real-time collision avoidance algorithms, and propose a novel approach to solve the exploration problem in the 2D dynamic grid. The proposed model is modular and hierarchical so that it cannot only exploit the structural regularities of the environment but also improve the training efficiency of DRL methods. The overall structure of our model is shown in Figure 2. GEM determines the next long-term target point to be explored based on a spatial map m_t maintained by the agent. LMM takes the next target point as input and computes the specific action to reach the target point. We use t_g to index the step of selecting the next target in only GEM. For example, we select a target point at initial time, $t = t_g = 1$, and we assume the agent takes 10 steps to reach this target and select a next target point, then $t = 11$ and $t_g = 2$ at this moment.

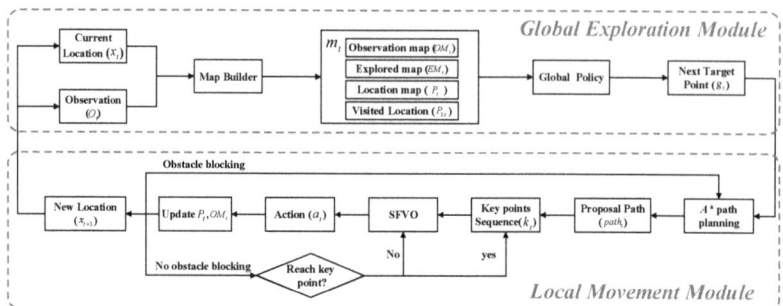

Figure 2. The overview of our IRHE-SFVO. In Global Exploration Module (GEM), the agent uses the current location and observation to build a spatial map m_t, then input m_t into Global Policy and output the next target point that will be explored. The Local Movement Module (LMM) determines the specific action to reach the target point quickly and safely based on the agent's current location, the next target point, and the obstacle map maintained by the agent.

4.1. Global Exploration Module

We want to learn an exploration policy π_g that enables the agent to select a location to explore so that the information gain about the environment can be maximized. For this purpose, we design an intrinsic reward function, favouring states where the agent can increase its exploration range at a fastest speed. Proximal Policy Optimization (PPO) [52] is used for training π_g. Importantly, π_g is learned on a set of training maps and tested on another set of unseen maps. This setting is to demonstrate the desirable generalization of our method across different environments.

4.1.1. Spatial Map Representation

First, as shown in the top block in Figure 2, GEM maintains a four-channel spatial map, m_t, as the input of the global policy. Then, the policy network outputs a next target point (g_{t_g}) that will be explored. To be specific, the spatial map contains four matrices of the same size, i.e., $m_t = \{0,1\}^{4 \times M \times M}$, where M is the height and width of the explored maps. Each element in the first channel represents whether the location is an obstacle (OM_t): 0 is for a free cell and 1 is for a blocked one. In the beginning, $OM_0 = \{0\}^{M \times M}$ based on the assumption of free space. Each element in the second channel represents whether the location has been explored (EM_t). The third channel encodes the current location (P_t) in a one-hot manner, i.e., the element corresponding to the agent location is set to be 1, and the others are 0. The fourth channel labels the visited locations ($P_{1:t}$) from the initial time to the current time. The rationality of establishing these four channels is that the agent can fully exploit all spatiotemporal information useful for target decision-making. In particular, this elegant design is: (a) to enable the agent to use the structural regularities of the spatial environment to make correct decisions, (b) to prevent the agent from selecting the points that have already been explored when choosing the next target point, and (c) to make the agent select the best next target point based on the current location, considering the time cost and exploration utility comprehensively.

4.1.2. Network Architecture

The policy network takes m_t as input and outputs a $g_{t_g} \mapsto \pi_g(m_t; \theta_g)$, where θ_g are the parameters of the global policy. As shown in Figure 3, the spatial map m_t is first passed through an embedding layer and the layer outputs a four-dimensional tensor of size $4 \times N \times M \times M$, where N represents the length of each embedding vector. Then, add the four constituent 3D tensors along their first dimension and we get a tensor with rich information whose size is $N \times M \times M \times M$. Then, this 3D tensor is passed through three convolution layers and three fully connected layers successively, and finally outputs a next target point: g_{t_g}. Note that the embedding layer is essential for preserving information embedded in the input because its input is 0–1 matrices of size $4 \times M \times M \times M$, which are all very sparse. Although the convolutional and pooling operations can extract spatial structure information, they will result in loss of many valuable information, and ignore the association between the overall and part as well if we send a matrix to the CNN and pooling layer directly. Therefore, to ensure the integrity of the information, it is necessary to map the m_t to a higher-dimensional vector first.

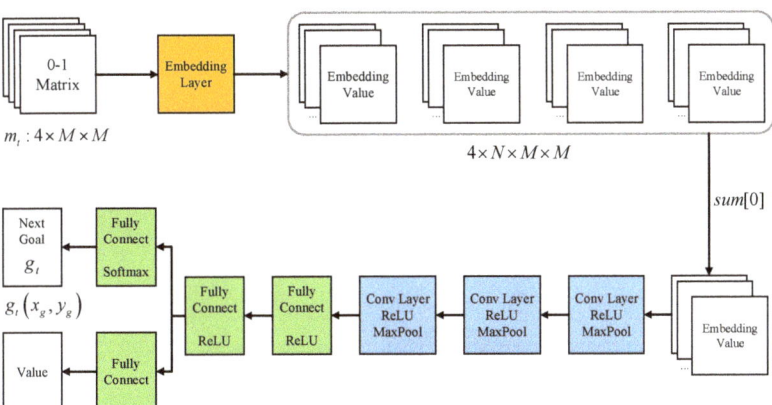

Figure 3. The structure of the actor-critic network in GEM. N represents the size of each embedding vector.

4.1.3. Intrinsic Reward

The effectiveness of DRL relies on rewards heavily. However, the exploration task is a reward-sparse RL problem. To alleviate the problem, we design an intrinsic reward (denoted by $r^i_{t_g}$) and combine it with the external rewards (denoted by $r^e_{t_g}$) given by the environment, i.e., $r_{t_g} = r^i_{t_g} + r^e_{t_g}$, so that the rewards along the exploration trajectory becomes denser. This is critically helpful to speed up the convergence of the policy and for the emergence of directed exploration. In literature, possible IM formulations include "curiosity" [34], "novelty" [53] or "empowerment" [40] to generate intrinsic rewards as described in Section 2. However, these approaches use blackbox models that cannot be initialized at each episode because the weights of neural networks cannot be reset in different episodes, resulting in the intrinsic reward getting smaller and smaller after each episode under the same scenario. To solve this problem, we design a simple yet effective intrinsic reward function that resets r^i at each episode. We use the increase of the explored area deduced from EM_t when the agent arrives at a new target point as the intrinsic rewards $r^i_{t_g}$.

4.2. Local Movement Module

To be able to explore in dynamic spaces, the agent needs both to reach the target point quickly and avoid colliding with moving obstacles. To achieve this goal, we design another hierarchical framework in local movement module including two levels: planning and controlling. In the planning stage, we use optimistic A* algorithm to plan an optimal path under partial observability, and then divides the path into several segments according to some rules. The end point of each segment is called a key point. In the controlling stage, we design an SFVO (Self-adaptive FVO) for the agent to reach these key points sequentially, and finally completes the movement along the path.

4.2.1. Planning Stage

There are many global path planning algorithms such as breadth first search , depth first search and Dijkstra. Instead of using the less efficient Generalized Dijkstra's algorithm to solve the Shortest Path Problem (SPP) in [54], we use A* algorithm which has better search efficiency to plan the optimal global path. The basic A* algorithm performs well in fully observable environments, but it does not work directly in our task since the OM_t dose not reflect the whole map. So we use a variation of A*, called optimistic A* algorithm. We assume that all unknown cells of the obstacle map are traversable and then plan a path between the current position of the agent and the target point. If the agent observes some static obstacles while moving, then it will replan the path using A* algorithm.

Once an optimal path is computed, we select several key points on this path to guide the agent reach the target point. For the motion controller, presented below, to drive the agent to move between them. As shown in Figure 4, this paper categorizes three types of key points: (a) turning points on the path, (b) boundary points on the path that crosses the known and unknown region and (c) the destination point of the path, i.e., the target point.

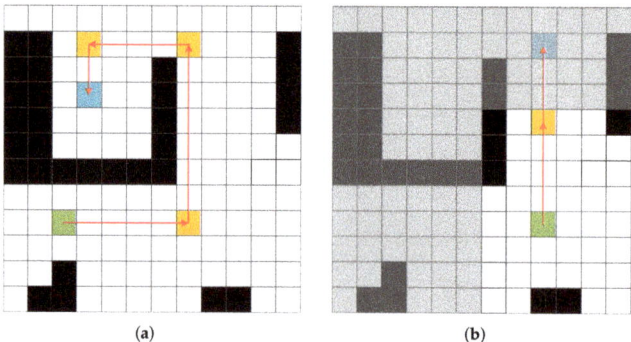

(a)   (b)

Figure 4. Examples of key points. The left figure shows the first and third types of key points, while the right figure shows the second and third types of key points. The green squares are the current locations of the agent. The blue squares represent the target points, which is also the third type of key points. The red lines represent the optimal path that was generated by A* algorithm and the orange squares are the first or second type of key points. The shaded area represents the unknown range of the agent while the other area represents the known range that has been observed by the agent.

Note that, the second type of key points are selected in the known area. Otherwise, if we select the boundary point in unknown area (the neighbour square above the orange square in Figure 4b, an obstacle might be selected as the key point.

In particular, the rationality of the selection strategy is that: (a) each segment of the path between key points is straight without considering dynamic obstacles, so that it is convenient for the controller to control the movement; (b) it is applicable to unknown spatial exploration problems under the partial observation conditions. We always choose the locations known for the agent as the key point, making its performance more similar to human exploration behaviour.

4.2.2. Controlling Stage

To avoid colliding with moving obstacles, we propose Self-adaptive Finite-time Velocity Obstacle algorithm (SFVO) built on FVO. Let A be the agent and B be an obstacle. For ease of calculation, we assume the agent and obstacles are dish-shaped. We use $D(\mathbf{p}, r)$ to represent a circular region with center $\mathbf{p}$ and radius r:

$$D(\mathbf{p}, r) = \{\mathbf{q} | \; \| \mathbf{q} - \mathbf{p} \| < r\} \tag{3}$$

The finite-time velocity obstacle $FVO^\tau_{A|B}$ represents the set of relative velocity values between A and B that will cause the collision in time τ in the future:

$$FVO^\tau_{A|B} = \{\mathbf{v} | \exists t \in [0, \tau], t\mathbf{v} \in D(\mathbf{p}_B - \mathbf{p}_A, r_A + r_B)\} \qquad (4)$$

The collision-avoidance velocity (CA) of the agent is:

$$CA^\tau_{A|B}(V_B) = \{\mathbf{v} | \mathbf{v} \notin FVO^\tau_{A|B} \oplus V_B\} \qquad (5)$$

According to the features of autonomous spatial exploration in dynamic environments, we change the fixed time τ into into adaptive dynamic time τ^d, i.e., $\tau^d_0 = \tau_{max}$ and $\tau^d(n) = \tau_{max} - \Delta\tau \cdot n$, n represents the number of rounds of a cycle, $\Delta\tau$ represents the reduction of finite time. This method is called Self-adaptive Finite-time Velocity Obstacle (SFVO). Figure 5 tells that the larger τ^d we set, the larger range of $FVO^{\tau^d}_{A|B} \oplus V_B$, and the smaller range of $CA^{\tau^d}_{A|B}(V_B)$. Therefore, we will adaptively adjust the velocity obstacle range of the agent by decreasing τ^d gradually. Specifically, at the beginning, the agent calculates the $CA^{\tau^d}_{A|B}(V_B)$ under the condition of $\tau^d = \tau_{max}$. If $CA^{\tau^d}_{A|B}(V_B) = \emptyset$, decrease the τ^d by a fix time interval $\Delta\tau$, and then calculate the collision-avoidance velocity again. If there is still no alternative velocity when $\tau^d = 0$, the agent stays idle until the next time step to continue the process above. The pseudo-code of SFVO is shown in Algorithm 1.

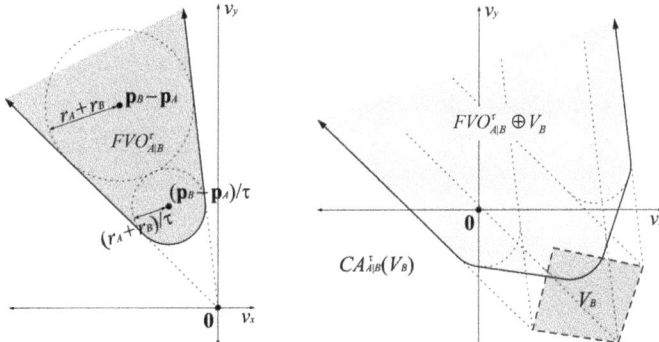

Figure 5. FVO algorithm diagram. In the left figure, the shaded area shows the relative velocity that will cause collision in time in the future. The right figure shows the collision velocity (shaded area) and collision-avoidance velocity (white area) of the agent given the velocity of the obstacle.

Algorithm 1 SFVO

1: $\tau^d \leftarrow \tau_{max}$
2: **for** $\tau^d > 0$ **do**
3:     $FVO^\tau_{A|B} = \{\mathbf{v} | \exists t \in [0, \tau], t\mathbf{v} \in D(\mathbf{p}_B - \mathbf{p}_A, r_A + r_B)\}$
4:     $CA^\tau_{A|B}(V_B) = \{\mathbf{v} | \mathbf{v} \notin FVO^\tau_{A|B} \oplus V_B\}$
5:     **if** $CA^{\tau^d}_{A|B}(V_B) = \emptyset$ **then**
6:         $\tau^d \leftarrow \tau^d - \Delta\tau$
7:         **continue**
8:     **else**
9:         **return** $CA^\tau_{A|B}(V_B)$
10:     **end if**
11: **end for**
12: **return** $\emptyset$

Based on the SFVO, the agent can avoid static and dynamic obstacles in real time, and its collision-avoidance velocity is $\mathbf{v}_A \in CA_{A|B}^{\tau d}(V_B)$. However, if there is more than one element in set $\mathbf{v}_A$, how can we choose an optimal velocity that not only drives the agent to reach the target point quickly, but also minimizes the risk of collision.

Inspired by Kim et al. [55], we design an optimal velocity evaluation function which consists of two parts: Expected Velocity Direction Evaluation Function (f_v) and Relative Vertical Distance Evaluation Function (f_d). As shown in Figure 6. The target point of the agent is known, so the direction of its expected velocity ($\mathbf{v}_v$) is the direction from the agent's current position points to the target point. So, f_v can be expressed as Equation (6).

$$f_v = k_v |\mathbf{v}_v - \mathbf{v}_A| \Rightarrow k_v \cos\langle \mathbf{v}_v, \mathbf{v}_A \rangle = k_v \frac{\mathbf{v}_v \cdot \mathbf{v}_A}{|\mathbf{v}_v||\mathbf{v}_A|} \tag{6}$$

Note that, the action space of the agent in our task is discrete, and the agent moves one unit at each step, i.e., the length of its velocity is fixed. So $\cos\langle \mathbf{v}_v, \mathbf{v}_A \rangle$ is equivalent to $|\mathbf{v}_v - \mathbf{v}_A|$.

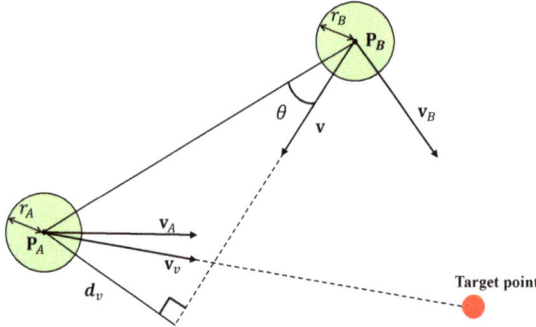

Figure 6. Schematic diagram of the expected velocity and the relative vertical distance. The green disks represent the agent and the obstacle, respectively, and the red point represents the target point.

The relative vertical distance (d_v) is defined as Equation (7).

$$d_v = |\mathbf{P}_A - \mathbf{P}_B| \sin\theta = \frac{\mathbf{v} \times (\mathbf{P}_A - \mathbf{P}_B)}{|\mathbf{v}|} \tag{7}$$

$\mathbf{v}$ is the relative velocity of the agent and the obstacle, i.e., $\mathbf{v} = \mathbf{v}_B - \mathbf{v}_A$. A smaller d_v means that the obstacle is prone to collide with the agent. Note that d_v can be negative according to Equation (7), indicating that the obstacle is moving away from the agent and there is no danger for the agent. And the large $|d_v|$ is, the safer the agent is. According to the analysis, f_d is designed as Equation (8):

$$f_d = \begin{cases} -d_v & d_v \leq 0 \\ -\frac{1}{d_v} & d_v > 0 \end{cases} \tag{8}$$

Finally, an overall evaluation function can be defined to be a weighted sum of f_v and f_d, as shown in Equation (9). The importance of each part can be regulated by the weights k_1 and k_2.

$$f = k_1 f_v + k_2 f_d \tag{9}$$

When increasing k_1 and decreasing k_2, f_v dominates and the agent is more inclined to approach the target. In contrast, the agent takes priority in obstacle avoidance.

4.2.3. The Hybrid Algorithm

The optimistic A* algorithm can calculate the shortest global path in the partially known environment, but it cannot timely avoid the moving obstacles in dynamic environments. The SFVO with optimal velocity evaluation function can avoid collision in real time, but it lacks global guidance and has only one expected direction. More specifically, in a space with many moving obstacles, it is easy to fall into local minima, resulting in planning longer paths or even failing to plan. With this problem in mind, we combine the planning and controlling methods described above. The workflow of the hybrid algorithm is shown in Figure 7.

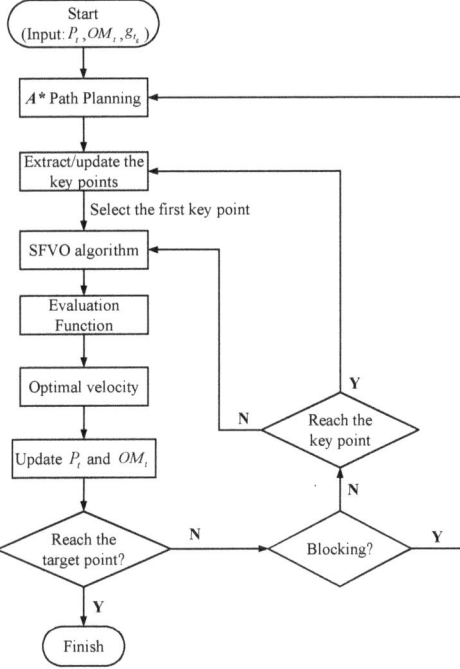

Figure 7. Flowchart of the hybrid algorithm.

At the time t, the algorithm puts the target point (g_{t_g}), which is produced by the Global Exploration Module P_t and OM_t into A* algorithm, plans a global path: $path_t = f_{A*}(P_t, g_{t_g}|OM_t)$. And extract the key points on the $path_t$: $\{K_1, K_2, ..., K_n\}$. Then, put the set of collision-avoidance velocities V_A which is calculated by the SFVO algorithm (Algorithm 1) $V_A = CA_{A|B}^\tau(V_B)$, the observation (O_t) of the agent and the sequence of key points into evaluation function (f), and calculate an optimal velocity of the agent $v_{opt} \leftarrow f(V_A, K_1, K_2, ..., K_n|O_t, OM_t, P_t)$. Then, the agent moves one step at this velocity, and updates $OM_t \rightarrow OM_{t+1}$ and $P_t \rightarrow P_{t+1}$ at the same time. At time $t+1$, if there is an obstacle on $path_t$, i.e., $\exists(i,j) \in path_t, OM[i,j] = 1$, conduct the A* path planning again: $path_{t+1} = f_{A*}(P_{t+1}, g_{t_g}|OM_{t+1})$, and continue with the above process. Otherwise, $path_{t+1} = path_t$ and determine whether the agent has reached the key point K_1. If so, the sequence of key points is updated. Otherwise, the agent continues to use SFVO algorithms for local movement control. When the agent reaches the target point (g_{t_g}), the LMM stops running. Then the GEM chooses a new next target point (g_{t_g+1}).

In addition, to make the motion trajectory smoother and reduce unnecessary local oscillation, the agent can be regarded as reaching the key points as long as it reaches the eight adjacent cells around the key point.

These are all the functional modules of the IRHE-SFVO above. As we can see in Figure 2, we use the Global Policy in GEM for generating next target points which the agent will go to, then plan a best path between the current position and the target point and extract the key points in the planning stage. Then, in the controlling stage, we use the SFVO algorithm (Algorithm 1) and evaluation function (Equation (9)) to decide a specific action of the agent, and then update the current knowledge of the spatial structure which decides whether to replan the A* algorithm or update the key points sequence. We run the above functional modules sequentially until the exploration is completed.

5. Empirical Evaluation

The goal of this paper is to build agents that can autonomously explore novel 2D dynamic environments with moving obstacles. To verify the effectiveness of the proposed method, we implement mentioned components for performance evaluation.

5.1. Experimental Setup

In order to evaluate the effect of the proposed model, we construct 2D grid maps by referring to reference [56] to represent the layout of indoor scenes such as offices. The maps are sized of $M = 40$, as shown in Figure 8. The first six maps make up the training set, and the rest are test maps. These maps have different spatial layouts and there is no intersection between the training set and the test set.

We use the ratio of the explored region as the metrics, which is calculated by dividing the coverage area by the total area that can be explored. It is defined as:

$$ExpRatio = \frac{C(EM_t)}{\sum_{x,y=0}^{x,y=M}(1 - T(x,y))} \qquad (10)$$

where $C(EM_t)$ represents the total area that is explored.

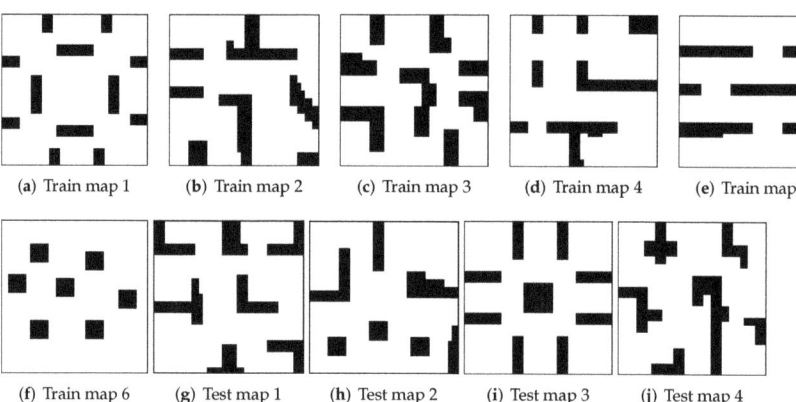

Figure 8. The different 2D grid maps without moving obstacles.

In these 2D grid maps, we set several obstacles which move independently of each other. The initial positions and moving directions of the moving obstacles are randomly selected, their movement mode is similar to that of the intelligent sweeping robots. As shown in Figure 9, the obstacles move straight until they collide with the obstacle or touch the boundary of the map, and then change the moving direction randomly.

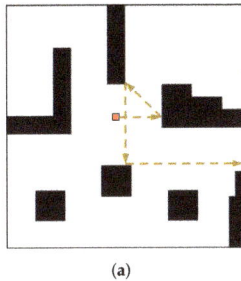
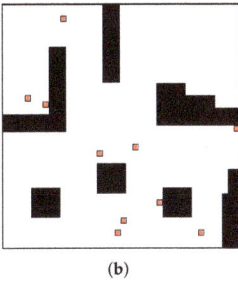

(a) (b)

Figure 9. The diagram of the initial distribution of moving obstacles and their movement trajectory. The red cells represent the dynamic obstacles, and the yellow dotted line represents the possible movement trajectory of this obstacle in left figure. When the environment is initialized, moving obstacles are randomly generated in the blank area of the map, and each obstacle is independent of each other. The right figure shows the initialization of the moving obstacles.

To simplify the calculation, both the agent and the moving obstacle are regarded as circles with a radius of 0.5. Table 1 shows the parameters used in this experiment.

Table 1. Dynamic environment parameter details.

Parameter	Value
The weight/height of grid maps (M)	40
Number of moving obstacles (i)	10
Observation range of the agent (n)	5
Exploration range of the agent (m)	2
Physical radius of the agent (r_A)	0.5
Physical radius of the moving obstacles (r_B)	0.5
The maximum of finite time in SFVO (τ_{max})	2
The reduction of finite time in SFVO ($\Delta\tau$)	1
Total steps the agent moves (T)	800

Training Details. We use multi-process paralleled PPO to train the global policy in GEM, with a different process for each map. The hyperparameters of PPO and global policy network are shown in Tables 2 and 3, respectively.

Table 2. PPO hyperparameter details.

Hyperparameter	Value
Number of parallel environment	6
Number of minibatches	12
Number of episodes	100,000
Number of optimization epochs	4
Learning rate	0.0001
Optimization algorithm	Adam
Entropy coefficient	0.001
Value loss oefficient	0.5
λ	0.95
γ	0.99
ϵ/Clip range	$0.1/[0.9, 1.1]$
Max norm of gradients	0.5

Table 3. Global policy network details.

Layer	Parameters
Embedding	Size of embedding vector $= 16$
Conv1	Output $= 32$, Kernel $= 3$, Stride $= 1$, Padding $= 1$
Conv2	Output $= 64$, Kernel $= 3$, Stride $= 1$, Padding $= 1$
Conv3	Output $= 16$, Kernel $= 3$, Stride $= 1$, Padding $= 1$
MaxPool	Kernel size $= 2$
Linear1	Output size $= 64$
Linear2	Output size $= 32$

Baselines. We use some classical methods and end-to-end DRL methods as baselines, and all methods were tested 15 times on four test maps with random initial positions:

- RND-PPO: A popular IM based DRL approach. We adapt the source code from [3] to the problem settings in this paper. RND is a SOTA (state-of-the-art) DRL method based on prediction error, which has outstanding performance in Atari games. The network of PPO is similar to the proposed model, and an LSTM module [57] is added. The intrinsic discount factor $\gamma_i = 0.999$ and the other hyperparameters as the same as the proposed model. The target and prediction network consist of 3 fully connected layers and the learning rate of optimizing the prediction network $lr_{RND} = 0.0025$. In addition, we design an external reward that is given a negative reward (-10) when the agent collides with an obstacle or moves out of the map;
- Straight: This method is widely used in intelligent sweeping robots. It works by moving the agent in a straight line and performing a random turn when a collision will occur in next time step [58];
- Random: The agent takes a sequence of random actions to exploration.
- Frontier: A method which is based on geometric features to decided its next best frontier, drives the agent always goes to unknown spaces [59].

The RND-PPO is an end-to-end method, taking the observation as input and outputting a specific action of the agent. This kind of methods are hard to train for a desirable policy. Compared with RND-PPO and Random, the Straight is more stable, as it changes the velocity of the agent only when the agent will collide with an obstacle. Besides, the frontier-based method is also hierarchical as ours, whose workflow is still to select a position and then move to it, and we find that the SOTA DRL exploration methods are also difficult to achieve its performance in terms of exploration ratio [16].

5.2. Local Real-Time Obstacle Avoidance

We first verify the effectiveness of SFVO and the optimal velocity evaluation function for real-time obstacle avoidance, as shown in Figures 10–12. The green squares represent the positions of the agent at the current time, and the blue squares represent the target points. The red squares represent the moving obstacle with downward velocity, and the orange squares represent key points. The task of the agent is to reach the target point quickly while avoiding static and dynamic obstacles at the same time.

Because of the existence of key points and the four-direction (up, down, left and right) action space, each part of the path that between two key points forms in a straight line, so the agent has to move perpendicular to the line or in the opposite direction in order to avoid the moving obstacles. As a result, the expected velocity direction evaluation function f_v of the agent during obstacle avoidance is not greater than 0. Then, the agent completely depends on the relative vertical distance evaluation function f_d to select the optimal velocity. On other words, if the theory described above is correct, no matter how large k_2 is, it will always play a role in obstacle avoidance. Then, after obstacle avoidance, the agent needs to change its velocity to approach the key point, and the velocity selection at this time completely depends on the expected velocity direction evaluation function f_v. That is to say, no matter how large k_1 is, it always plays a role in the velocity selection of approaching the key point after completing obstacle avoidance. Therefore, the combination of weighting coefficients in the evaluation function set in this experiment (Equation (9)) is relatively single. We set three groups of different weighting coefficients: $k_1 = 0$, $k_2 = 1$; $k_1 = 1$, $k_2 = 0$; $k_1 = 1$, $k_2 = 1$. The trajectories of the agent under the three groups of coefficients are shown in Figures 10–12.

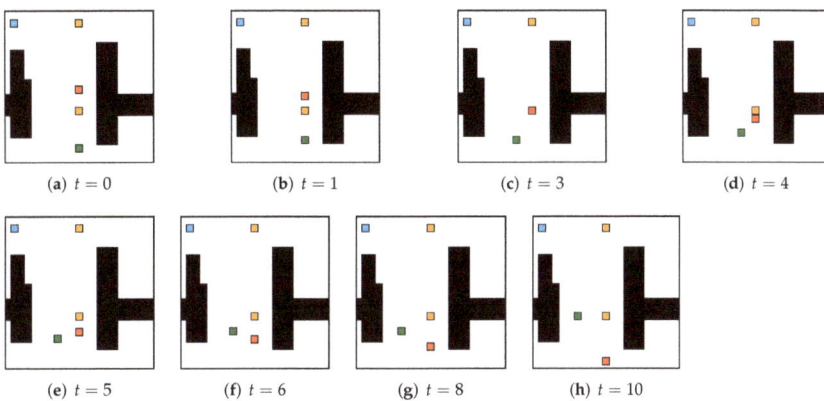

Figure 10. The trajectory of the agent under the condition of $k_1 = 0, k_2 = 1$ set in f.

As shown in Figure 10, when $k_1 = 0$, $k_2 = 1$, the moving trajectory of the agent is more and more away from the obstacle, but does not move towards key points. In essence, the agent is still taking random movements on the basis of avoiding collision with obstacles.

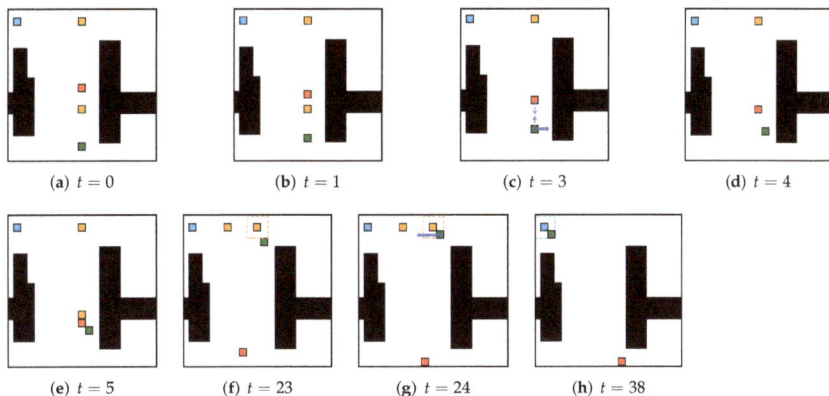

Figure 11. The trajectory of the agent under the condition of $k_1 = 1, k_2 = 0$ set in f.

When $k_1 = 1$, $k_2 = 0$, the agent will ignore the risk of collisions while moving. As shown in Figure 11, at $t = 3$, the agent judges that it will collide with the moving obstacle in the next two steps with the current motion direction through SFVO algorithm, so that it needs to make obstacle avoidance action. The agent is close to the static obstacle on the right and far from the one on the left. Therefore, the best obstacle avoidance action of the agent at this time should be to move left, which can avoid colliding with the moving obstacles and reduce the risk of collision with other obstacles, as shown in Figure 12. However, the agent does not consider the relative vertical distance d_v from the obstacle under the condition of

$k_1 = 1$, $k_2 = 0$, nd has a 50% probability of moving right, as shown in Figure 11, which increases the risk of collision with other obstacles.

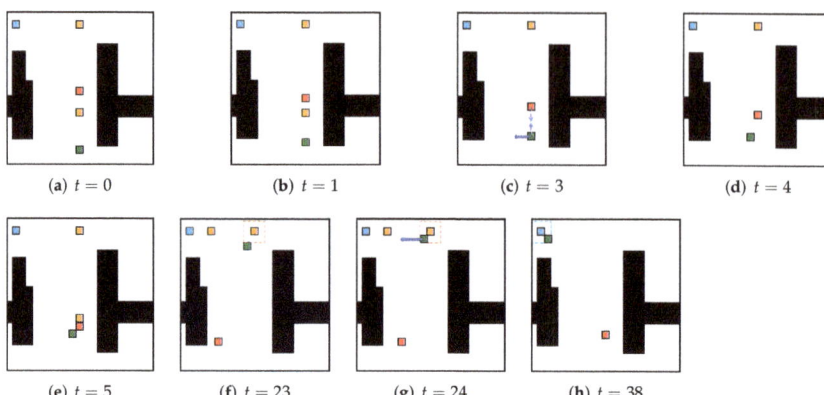

Figure 12. The trajectory of the agent under the condition of $k_1 = 1$, $k_2 = 1$ set in f.

Figures 11 and 12 not only show the effectiveness of SFVO algorithm and evaluation function for obstacle avoidance, but also demonstrate the efficaciousness of the hybrid algorithm for path planning. At $t = 0$, the orange square which close to the agent is the second type of key points, the square that is far from the agent is the first type, and the blue square is the third type. When the agent reaches a key point or one of its adjacent eight squares, it is deemed to have reached the key point, so it continues to select the subsequent key points for local path planning. Finally, it guides the agent to the target point.

5.3. Comparison with Baselines on Spatial Exploration

We test the IRHE-SFVO with weighting coefficient $k_1 = 1$, $k_2 = 1$ of the evaluation function and compare it with the baselines on the test maps. The results are shown in Figure 13 and Table 4. It is worth noting that, from the perspective of safety, when the agent will collide with an obstacle, it should stop moving or change the direction immediately. However, the vanilla frontier-based strategy has no such specific design.

Table 4. The average exploration ratios of the proposed method and baselines on the four test maps. The brackets indicate the average number of times when the agent driven by the frontier-based strategy collides with moving obstacles in 15 spatial explorations on different test maps.

	IRHE-SFVO	RND-PPO	Random	Straight	Fronteir
Test map 1	0.8656	0.2258	0.2406	0.5276	0.9943 (4.53)
Test map 2	0.8552	0.2707	0.2107	0.6078	0.9992 (3.06)
Test map 3	0.8842	0.2501	0.1861	0.4721	0.9966 (5.13)
Test map 4	0.8953	0.2177	0.2287	0.5498	0.9997 (4.13)

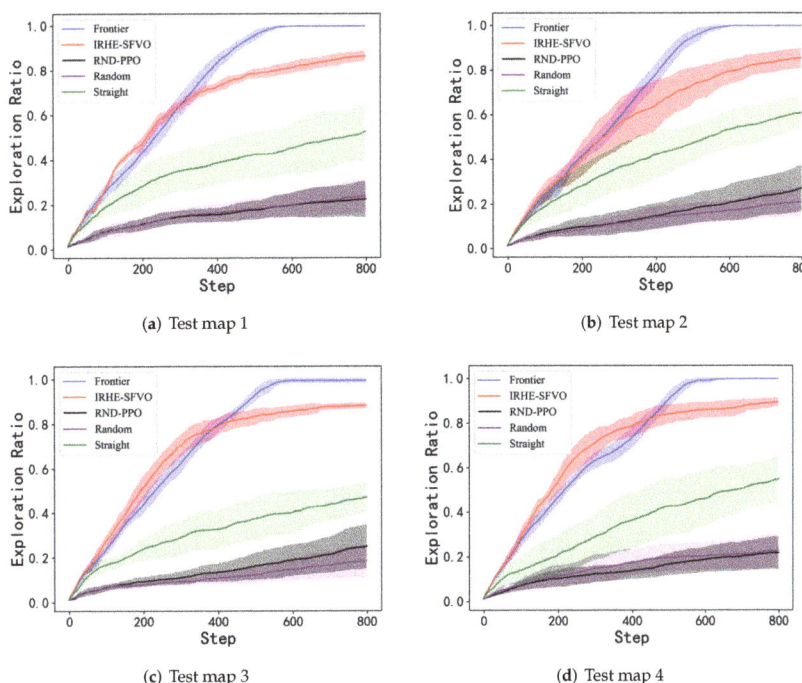

Figure 13. Coverage Performance. Policies are tested in 800 steps (15 random start locations on each of the 4 testing maps). Darker line represents mean exploration ratio and shaded area represents the standard deviation across the 15 runs.

Figure 13 shows the coverage performance of different methods on the four test maps. Combining Table 4, we can see that the order of coverage from low to high is: Random and RND-PPO, Straight, IRHE-SFVO, Frontier. Specifically, we first notice that RND-PPO and Random method have similar poor performance. The reasons are as follows: the core of the dynamics-based curiosity agent such as RND is that if a state is encountered many times, its novelty will continue declining due to network parameter updates during training. So, in these grid maps, the agent needs to come out of its familiar area which is around the initial location at each episode. However, after several episodes of training, the "novelty" of the states around the initial position drop to a low level, and the agent does not touch the high-novelty world outside so that it is difficult for the agent to walk out of its familiar area. Second, the policy learned by the RND agent lacks the ability to explore since the intrinsic reward is to encourage agents to traverse more states rather than teaching the agent to learn how to explore. To be specific, its intrinsic reward will gradually decrease as training times due to the black-box model, so that the learned policy will depend more and more on the external reward, and the policy is dependent on external rewards completely at the end of training. In our experimental setting, RND only has a collision punishment as its external reward, so the agent moves randomly and only learns to avoid collision at the end. This is why the RND-PPO algorithm is slightly better than the Random algorithm.

Then, we notice that the Straight method performs much better than Random and RND-PPO because this method takes random actions only when a collision occurs. It is more stable than the Random and RND-PPO algorithms, which move randomly at every step.

Overall, these methods perform largely worse than IRHE-SFVO because our algorithm has an instructive high-level exploration strategy and an effective local movement module.

At the beginning of training GEM, the target points are selected almost randomly. As the training proceeds, the agent, driven by the intrinsic rewards which are generated by itself, gradually choose the points that are distant but easily reachable. Intrinsic reward, generated from intrinsic motivation, can make individual feel satisfied psychologically or emotionally because of increase of obtained knowledge or control [60]. In our exploration task, the intrinsic rewards are calculated by the increase of the explored area, so the agent will fill more cheerful when it chooses the point in an unknown region that is further away from it. Furthermore, the agent will reasonably adjust the distance between itself and the selected target point as the training progresses with the fixed number of target points that will be chosen during an exploration task. For example, the distances will be larger when the number of the target points is 10, while they will be smaller when the number of the target points is 20. In addition, LMM can adapt the agent to the planned path and avoid colliding with moving obstacles according to its perception, which can reach the target points quickly and safely.

Finally, we notice that the frontier-based strategy has achieved the highest exploration ratio in all the test maps. This method selects frontier points that lie on the boundary between the known free space and unknown region according to the maps built by the agent, and the experimental environments in this paper are very realistic, without any perceived noise or action errors, which is highly favourable to the frontier-based exploration agent. In the environment with moving obstacles, although this method may miss some "frontier points" at some time, resulting in that the spaces around them are not explored for a period of time. However, at a later time, this method can always select these "frontier points" again for spatial exploration. Because the motion trajectories of moving obstacles are random, it is impossible for them to stay at the positions where the frontier method always misjudges these "frontier points", so the exploration ratio of the frontier-based method is almost unaffected by dynamic obstacles. However, safety is a crucial problem that we must consider in spatial exploration. And Table 4 shows that the frontier-based method has collided with dynamic obstacles many times during exploration, while the others do not.

In addition, we visualize the paths and coverage areas of IRHE-SFVO and baselines on test map 2, the initial position of the agent is $(1,1)$ on the bottom left. As shown in Figure 14, in each row of subfigures from left to right are the trajectories at step 0, step 40, step 200, step 400, Step 600 and step 800 respectively.

We can see that IRHE-SFVO algorithm covers almost all space, and its motion trajectory is relatively smoother and more reasonable than those of its competitors. It can be thought that it produces similar exploration strategies as human beings. Although frontier-based method has high exploration coverage, its motion trajectory is mechanical and very zigzag, such as the upper left corner and the blank area in the middle of the map. In addition, combined with Figure 13 we can also see that the exploration efficiency of IRHE-SFVO is slightly higher than that of the frontier in the initial exploration stage, because IRHE-SFVO aims to maximize the information gain of each step, but this is also the reason for its insufficient local exploration. Every time IRHE-SFVO selects a target point, it tends to select the locations where a large number of unexplored areas around it, so that the agent can quickly obtain a large amount of map information, but the exploration is insufficient for local details.

In summary, we use the DRL based on the intrinsic motivation to simulate the human high-level cognitive behavior during exploration, so that the agent always chooses those places that are particularly unknown to explore. And as for the quick and safe movement, we use a hierarchical framework including planning and controlling instead of learning methods that have difficulty in joint training to simulate the human low-level real-time response. Therefore, combining the two modules above, the IRHE-SFVO algorithm could meet the requirements of high efficiency and quasi-humanity of spatial exploration.

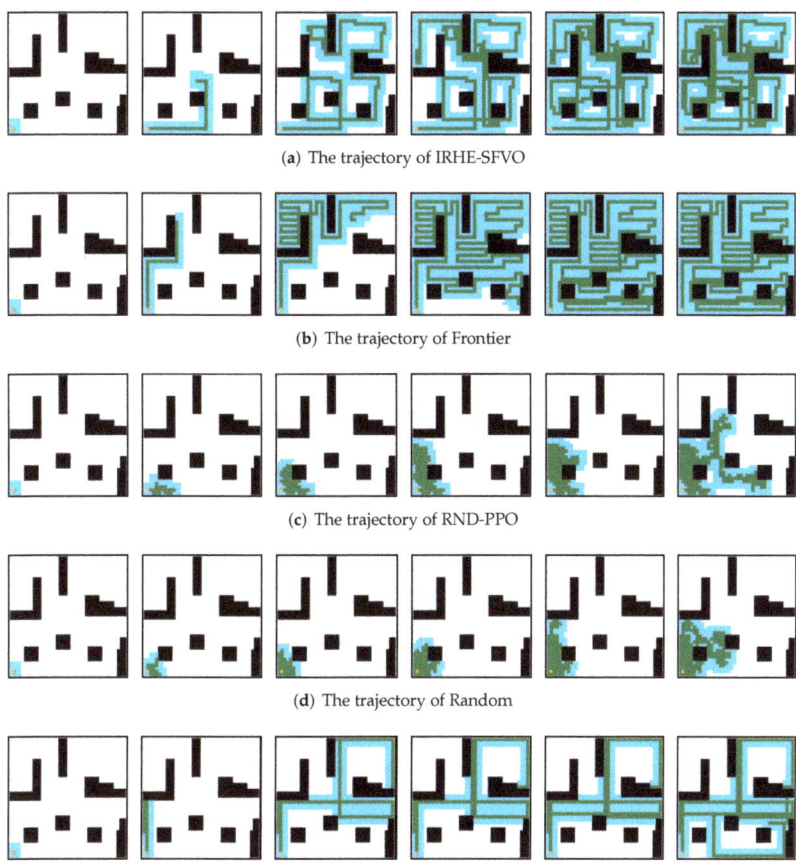

(a) The trajectory of IRHE-SFVO

(b) The trajectory of Frontier

(c) The trajectory of RND-PPO

(d) The trajectory of Random

(e) The trajectory of Straight

Figure 14. Sample trajectories of the competing approaches along with the coverage region in Test map 2. The orange point represents the initial location of the agent. The solid green lines represent the trajectories that the agent has traversed and the blue shaded region shows the explored area.

6. Conclusions and Future Work

This paper proposed a three-tiered hierarchical autonomous spatial exploration model, IRHE-SFVO, that combines a high-level exploration strategy (GEM) and a low-level module (LMM) including a planning phase and a controlling phase. This decomposition not only overcomes the disadvantage of the end-to-end training difficulty, but also generates smooth movements which makes the behaviours of agent more reasonable and safely. We showed how to design and train these modules and validated them on multiple challenging 2D maps with complex structures and moving obstacles. The results showed that the proposed model has consistently better efficiency and generality than a state-of-the-art IM based DRL and some other heuristic methods. Although the proposed approach tends to revisit explored locations in some time, resulting in the lower coverage performance compared with frontier-based method, IRHE-SFVO still meets the application requirements to a certain extent.

For future work, we would like to extend this work to the following directions. First, in order to further improve the coverage of exploration, we would like to design more complex mechanisms like incorporating spatial abstraction into the framework to improve the efficiency of exploration and the rationality of motion mode. Second, more complex constraints should be considered, such as uneven terrains, diverse surface features and the energy of the agent. Third, we would like to work on multi-agent collaborative spatial exploration, which faces the problems of non-stationary environments, incomplete observations and inefficient exploration of single agent in complex environments.

Author Contributions: Proposal and conceptualization, Q.Z. and Y.H.; methodology, Y.S.; implementation, Y.S. and P.J.; writing-original draft preparation, Y.S. and Y.H.; writing-review and editing, Q.Z. All authors have read and agreed to the published version of the manuscript.

Funding: This research was supported partially by the National Natural Science Fund of China (Grant NO. 62102432, 62103420, 62103428 and 62103425) and the Natural Science Fund of Hunan Province (Grant NO. 2021JJ40697 and 2021JJ40702).

Institutional Review Board Statement: This study does not involve humans or animals.

Informed Consent Statement: This study does not involve humans or animals.

Conflicts of Interest: The authors declare no conflict of interest.

References

1. Yamauchi, B. A frontier-based approach for autonomous exploration. In Proceedings of the 1997 IEEE International Symposium on Computational Intelligence in Robotics and Automation CIRA'97. 'Towards New Computational Principles for Robotics and Automation', Monterey, CA, USA, 11–12 July 1997; pp. 146–151.
2. Song, Y.; Hu, Y.; Zeng, J.; Hu, C.; Qin, L.; Yin, Q. Towards Efficient Exploration in Unknown Spaces: A Novel Hierarchical Approach Based on Intrinsic Rewards. In Proceedings of the 2021 6th International Conference on Automation, Control and Robotics Engineering (CACRE), Dalian, China, 15–17 July 2021; pp. 414–422.
3. Burda, Y.; Edwards, H.; Storkey, A.; Klimov, O. Exploration by random network distillation. *arXiv* **2018**, arXiv:1810.12894.
4. Wirth, S.; Pellenz, J. Exploration transform: A stable exploring algorithm for robots in rescue environments. In Proceedings of the 2007 IEEE International Workshop on Safety, Security and Rescue Robotics, Rome, Italy, 27–29 September 2007; pp. 1–5.
5. Mei, Y.; Lu, Y.H.; Lee, C.G.; Hu, Y.C. Energy-efficient mobile robot exploration. In Proceedings of the 2006 IEEE International Conference on Robotics and Automation, ICRA 2006, Orlando, FL, USA, 15–19 May 2006; pp. 505–511.
6. Juliá, M.; Gil, A.; Reinoso, O. A comparison of path planning strategies for autonomous exploration and mapping of unknown environments. *Auton. Robot.* **2012**, *33*, 427–444. [CrossRef]
7. Oßwald, S.; Bennewitz, M.; Burgard, W.; Stachniss, C. Speeding-up robot exploration by exploiting background information. *IEEE Robot. Autom. Lett.* **2016**, *1*, 716–723.
8. Basilico, N.; Amigoni, F. Exploration strategies based on multi-criteria decision making for searching environments in rescue operations. *Auton. Robot.* **2011**, *31*, 401–417. [CrossRef]
9. Niroui, F.; Sprenger, B.; Nejat, G. Robot exploration in unknown cluttered environments when dealing with uncertainty. In Proceedings of the 2017 IEEE International Symposium on Robotics and Intelligent Sensors (IRIS), Ottawa, ON, Canada, 5–7 October 2017; pp. 224–229.
10. González-Banos, H.H.; Latombe, J.C. Navigation strategies for exploring indoor environments. *Int. J. Robot. Res.* **2002**, *21*, 829–848. [CrossRef]
11. Whaite, P.; Ferrie, F.P. Autonomous exploration: Driven by uncertainty. *IEEE Trans. Pattern Anal. Mach. Intell.* **1997**, *19*, 193–205. [CrossRef]
12. Julian, B.J.; Karaman, S.; Rus, D. On mutual information-based control of range sensing robots for mapping applications. *Int. J. Robot. Res.* **2014**, *33*, 1375–1392. [CrossRef]
13. Tai, L.; Liu, M. A robot exploration strategy based on q-learning network. In Proceedings of the 2016 IEEE International Conference on Real-Time Computing and Robotics (RCAR), Angkor Wat, Cambodia, 6–10 June 2016; pp. 57–62.
14. Zhang, J.; Tai, L.; Liu, M.; Boedecker, J.; Burgard, W. Neural slam: Learning to explore with external memory. *arXiv* **2017**, arXiv:1706.09520.
15. Zhu, D.; Li, T.; Ho, D.; Wang, C.; Meng, M.Q.H. Deep reinforcement learning supervised autonomous exploration in office environments. In Proceedings of the 2018 IEEE International Conference on Robotics and Automation (ICRA), Brisbane, Australia, 21–25 May 2018; pp. 7548–7555.
16. Chen, T.; Gupta, S.; Gupta, A. Learning exploration policies for navigation. *arXiv* **2019**, arXiv:1903.01959.
17. Issa, R.B.; Rahman, M.S.; Das, M.; Barua, M.; Alam, M.G.R. Reinforcement Learning based Autonomous Vehicle for Exploration and Exploitation of Undiscovered Track. In Proceedings of the 2020 International Conference on Information Networking (ICOIN), Barcelona, Spain, 7–10 Janaury 2020; pp. 276–281.
18. Niroui, F.; Zhang, K.; Kashino, Z.; Nejat, G. Deep reinforcement learning robot for search and rescue applications: Exploration in unknown cluttered environments. *IEEE Robot. Autom. Lett.* **2019**, *4*, 610–617. [CrossRef]

19. Shrestha, R.; Tian, F.P.; Feng, W.; Tan, P.; Vaughan, R. Learned map prediction for enhanced mobile robot exploration. In Proceedings of the 2019 International Conference on Robotics and Automation (ICRA), Montreal, QC, Canada, 20–24 May 2019; pp. 1197–1204.
20. Li, H.; Zhang, Q.; Zhao, D. Deep reinforcement learning-based automatic exploration for navigation in unknown environment. *IEEE Trans. Neural Net. Learn. Syst.* **2019**, *31*, 2064–2076. [CrossRef] [PubMed]
21. Chaplot, D.S.; Gandhi, D.; Gupta, S.; Gupta, A.; Salakhutdinov, R. Learning to explore using active neural slam. *arXiv* **2020**, arXiv:2004.05155.
22. Barto, A.; Mirolli, M.; Baldassarre, G. Novelty or surprise? *Front. Psychol.* **2013**, *4*, 907. [CrossRef] [PubMed]
23. Bellemare, M.; Srinivasan, S.; Ostrovski, G.; Schaul, T.; Saxton, D.; Munos, R. Unifying count-based exploration and intrinsic motivation. *Adv. Neural Inf. Process. Syst.* **2016**, *29*, 1471–1479.
24. Ostrovski, G.; Bellemare, M.G.; Oord, A.; Munos, R. Count-based exploration with neural density models. In Proceedings of the International Conference on Machine Learning, PMLR, Sydney, NSW, Australia, 6–11 August 2017; pp. 2721–2730.
25. Tang, H.; Houthooft, R.; Foote, D.; Stooke, A.; Chen, X.; Duan, Y.; Schulman, J.; De Turck, F.; Abbeel, P. A study of count-based exploration for deep reinforcement learning. In Proceedings of the 31st Conference on Neural Information Processing Systems (NIPS), Long Beach, CA, USA, 4–9 December 2017; pp. 1–18.
26. Fu, J.; Co-Reyes, J.D.; Levine, S. Ex2: Exploration with exemplar models for deep reinforcement learning. *arXiv* **2017**, arXiv:1703.01260.
27. Choi, J.; Guo, Y.; Moczulski, M.; Oh, J.; Wu, N.; Norouzi, M.; Lee, H. Contingency-aware exploration in reinforcement learning. *arXiv* **2018**, arXiv:1811.01483.
28. Machado, M.C.; Bellemare, M.G.; Bowling, M. Count-based exploration with the successor representation. In Proceedings of the AAAI Conference on Artificial Intelligence, New York, NY, USA, 7–12 Feburary 2020; pp. 5125–5133.
29. Choshen, L.; Fox, L.; Loewenstein, Y. Dora the explorer: Directed outreaching reinforcement action-selection. *arXiv* **2018**, arXiv:1804.04012.
30. Shyam, P.; Jaśkowski, W.; Gomez, F. Model-based active exploration. In Proceedings of the International Conference on Machine Learning, PMLR, Long Beach, CA, USA, 9–15 June 2019; pp. 5779–5788.
31. Pathak, D.; Gandhi, D.; Gupta, A. Self-supervised exploration via disagreement. In Proceedings of the International conference on Machine Learning, PMLR, Los Angeles, CA, USA, 9–15 June 2019; pp. 5062–5071.
32. Ratzlaff, N.; Bai, Q.; Fuxin, L.; Xu, W. Implicit generative modeling for efficient exploration. In Proceedings of the International Conference on Machine Learning, PMLR, Vienne, Austria, 12–18 July 2020; pp. 7985–7995.
33. Stadie, B.C.; Levine, S.; Abbeel, P. Incentivizing exploration in reinforcement learning with deep predictive models. *arXiv* **2015**, arXiv:1507.00814.
34. Pathak, D.; Agrawal, P.; Efros, A.A.; Darrell, T. Curiosity-driven exploration by self-supervised prediction. In Proceedings of the International Conference on Machine Learning, PMLR, Sydney, Australia, 6–12 August 2017; pp. 2778–2787.
35. Kim, H.; Kim, J.; Jeong, Y.; Levine, S.; Song, H.O. Emi: Exploration with mutual information. *arXiv* **2018**, arXiv:1810.01176.
36. Ermolov, A.; Sebe, N. Latent World Models For Intrinsically Motivated Exploration. *arXiv* **2020**, arXiv:2010.02302.
37. Lopes, M.; Lang, T.; Toussaint, M.; Oudeyer, P.Y. *Exploration in Model-Based Reinforcement Learning by Empirically Estimating Learning Progress*; Neural Information Processing Systems (NIPS): Lake Tahoe, NV, USA, 2012.
38. Gregor, K.; Rezende, D.J.; Wierstra, D. Variational intrinsic control. *arXiv* **2016**, arXiv:1611.07507.
39. Houthooft, R.; Chen, X.; Duan, Y.; Schulman, J.; De Turck, F.; Abbeel, P. Vime: Variational information maximizing exploration. *arXiv* **2016**, arXiv:1605.09674.
40. Oudeyer, P.Y.; Kaplan, F. What is intrinsic motivation? A typology of computational approaches. *Front. Neurorobotics* **2009**, *1*, 6. [CrossRef] [PubMed]
41. Badia, A.P.; Piot, B.; Kapturowski, S.; Sprechmann, P.; Vitvitskyi, A.; Guo, Z.D.; Blundell, C. Agent57: Outperforming the atari human benchmark. In Proceedings of the International Conference on Machine Learning, PMLR, Vienne, Austria, 12–18 July 2020; pp. 507–517.
42. Khatib, O. Real-time obstacle avoidance for manipulators and mobile robots. In *Autonomous Robot Vehicles*; Springer: Berlin/Heidelberg, Germany, 1986; pp. 396–404.
43. Fox, D.; Burgard, W.; Thrun, S. The dynamic window approach to collision avoidance. *IEEE Robot. Autom. Mag.* **1997**, *4*, 23–33. [CrossRef]
44. Rezaee, H.; Abdollahi, F. A decentralized cooperative control scheme with obstacle avoidance for a team of mobile robots. *IEEE Trans. Ind. Electron.* **2013**, *61*, 347–354. [CrossRef]
45. Ali, F.; Kim, E.K.; Kim, Y.G. Type-2 fuzzy ontology-based semantic knowledge for collision avoidance of autonomous underwater vehicles. *Inf. Sci.* **2015**, *295*, 441–464. [CrossRef]
46. Cheng, Y.; Zhang, W. Concise deep reinforcement learning obstacle avoidance for underactuated unmanned marine vessels. *Neurocomputing* **2018**, *272*, 63–73. [CrossRef]
47. Celsi, L.R.; Celsi, M.R. On Edge-Lazy RRT Collision Checking in Sampling-Based Motion Planning. *Int. J. Robot. Autom.* **2021**, *36*. [CrossRef]
48. Fiorini, P.; Shiller, Z. Motion planning in dynamic environments using velocity obstacles. *Int. J. Robot. Res.* **1998**, *17*, 760–772. [CrossRef]

49. Van den Berg, J.; Lin, M.; Manocha, D. Reciprocal velocity obstacles for real-time multi-agent navigation. In Proceedings of the 2008 IEEE International Conference on Robotics and Automation, Pasadena, CA, USA, 19–23 May 2008; pp. 1928–1935.
50. Abe, Y.; Yoshiki, M. Collision avoidance method for multiple autonomous mobile agents by implicit cooperation. In Proceedings of the 2001 IEEE/RSJ International Conference on Intelligent Robots and Systems, Expanding the Societal Role of Robotics in the the Next Millennium (Cat. No. 01CH37180), IEEE, Maui, HI, USA, 29 October–3 November 2001; pp. 1207–1212.
51. Guy, S.J.; Chhugani, J.; Kim, C.; Satish, N.; Lin, M.; Manocha, D.; Dubey, P. Clearpath: Highly parallel collision avoidance for multi-agent simulation. In Proceedings of the 2009 ACM SIGGRAPH/Eurographics Symposium on Computer Animation, New Orleans, Louisiana, 1–2 August 2009; pp. 177–187.
52. Schulman, J.; Wolski, F.; Dhariwal, P.; Radford, A.; Klimov, O. Proximal policy optimization algorithms. *arXiv* **2017**, arXiv:1707.06347.
53. Conti, E.; Madhavan, V.; Such, F.P.; Lehman, J.; Stanley, K.O.; Clune, J. Improving exploration in evolution strategies for deep reinforcement learning via a population of novelty-seeking agents. *arXiv* **2017**, arXiv:1712.06560.
54. Celsi, L.R.; Di Giorgio, A.; Gambuti, R.; Tortorelli, A.; Priscoli, F.D. On the many-to-many carpooling problem in the context of multi-modal trip planning. In Proceedings of the 2017 25th Mediterranean Conference on Control and Automation (MED), Valletta, Malta, 3–6 July 2017; pp. 303–309.
55. Kim, M.; Oh, J.H. Study on optimal velocity selection using velocity obstacle (OVVO) in dynamic and crowded environment. *Auton. Robot.* **2016**, *40*, 1459–1470. [CrossRef]
56. Hu, Y.; Subagdja, B.; Tan, A.H.; Yin, Q. Vision-Based Topological Mapping and Navigation with Self-Organizing Neural Networks. *IEEE Trans. Neural Net. Learn. Syst.* **2021**. Available online: https://ieeexplore.ieee.org/document/9459468 (accessed on 20 December 2021). [CrossRef] [PubMed]
57. Hochreiter, S.; Schmidhuber, J. Long short-term memory. *Neural Comput.* **1997**, *9*, 1735–1780. [CrossRef] [PubMed]
58. Savva, M.; Kadian, A.; Maksymets, O.; Zhao, Y.; Wijmans, E.; Jain, B.; Straub, J.; Liu, J.; Koltun, V.; Malik, J.; et al. Habitat: A platform for embodied ai research. In Proceedings of the IEEE/CVF International Conference on Computer Vision, Seoul, Korea, 27 October–2 November 2019; pp. 9339–9347.
59. Mobarhani, A.; Nazari, S.; Tamjidi, A.H.; Taghirad, H.D. Histogram based frontier exploration. In Proceedings of the 2011 IEEE/RSJ International Conference on Intelligent Robots and Systems, San Francisco, CA, USA, 25–30 September 2011; pp. 1128–1133.
60. Baldassarre, G. What are intrinsic motivations? A biological perspective. In Proceedings of the 2011 IEEE International Conference on Development and Learning (ICDL), Frankfurt am Main, Germany, 24–27 August 2011; Volume 2, pp. 1–8.

Article

Real-Time Drift-Driving Control for an Autonomous Vehicle: Learning from Nonlinear Model Predictive Control via a Deep Neural Network

Taekgyu Lee, Dongyoon Seo, Jinyoung Lee and Yeonsik Kang *

Graduate School of Automotive Engineering, Kookmin University, Seoul 02707, Korea
* Correspondence: ykang@kookmin.ac.kr

Abstract: A drift-driving maneuver is a control technique used by an expert driver to control a vehicle along a sharply curved path or slippery road. This study develops a nonlinear model predictive control (NMPC) method for the autonomous vehicle to perform a drift maneuver and generate the datasets necessary for training the deep neural network(DNN)-based drift controller. In general, the NMPC method is based on numerical optimization which is difficult to run in real-time. By replacing the previously designed NMPC method with the proposed DNN-based controller, we avoid the need for complex numerical optimization of the vehicle control, thereby reducing the computational load. The performance of the developed data-driven drift controller is verified through realistic simulations that included drift scenarios. Based on the results of the simulations, the DNN-based controller showed similar tracking performance to the original nonlinear model predictive controller; moreover, the DNN-based controller can demonstrate stable computation time, which is very important for the safety critical control objective such as drift maneuver.

Keywords: data-driven control; time delay neural network; drift control; autonomous driving; nonlinear model predictive control

1. Introduction

To maximize passenger safety, future autonomous vehicles will be required to operate in various road environments and cope with various emergencies. A common emergency situation is high lateral slippage of the rear wheels on a sharply curved path or an ice-covered road, which leads to oversteering (see Figure 1). In such a situation, an autonomous vehicle should be capable of guaranteeing safety.

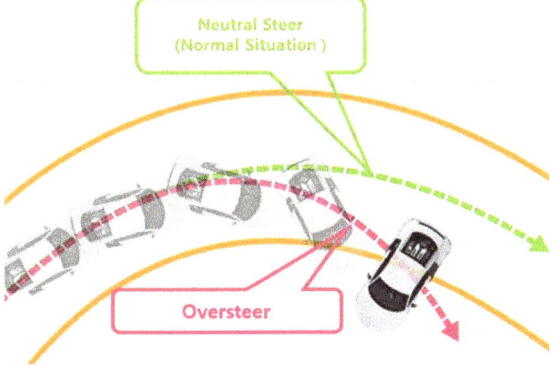

Figure 1. Schematic of the oversteering phenomenon.

Drift technology (Figure 2) is a vehicle control strategy developed for use in motorsports. This technology enables professional racecar drivers to quickly generate high yaw rates that cannot be achieved with normal steering maneuvers. Such a driving technique requires expert driving skills to handle the vehicle's behavior at its dynamic limit. Additionally, it is also used as a method for maintaining vehicle stability when an unintentional oversteering phenomenon occurs while driving.

 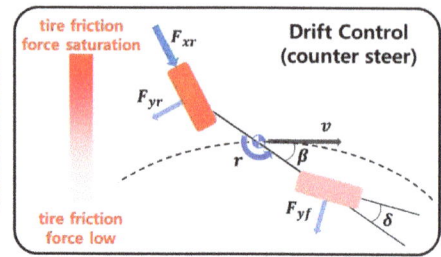

Figure 2. Drift control maneuver.

Drift control methods for autonomous vehicles have been extensively studied. For example, the University of California, Berkeley [1] and Stanford University [2,3] have been developing drift control methods for several years. However, most existing control methods are based on the drift equilibrium state derived from vehicle dynamics. In particular, methods have been proposed to control a vehicle using a counter-steering maneuver that turns the steering wheel opposite to the turning direction [4,5]. Recent studies have introduced reinforcement learning techniques for developing drift control algorithms [6].

A drift control algorithm based on a nonlinear model predictive control (NMPC) method was also developed, which is a method using real-time numerical optimization to compute the control inputs minimizing the cost function [7,8].

In general, NMPC is based on real-time optimization techniques over a finite future horizon. The NMPC approach has many advantages; for instance, it considers the input and state constraints along with the dynamics during numerical optimization. However, the unpredictable computational time of most numerical optimization algorithms has limited the performance of NMPC in real-time control applications. To overcome this limitation, this study proposes a drift controller based on a deep neural network (DNN) algorithm. The proposed controller learns from data generated using the model predictive control (MPC) technique and demonstrates similar control performance as NMPC while delivering better real-time performance.

With the continued development of algorithms and computing devices, artificial intelligence (AI) is now being applied to various industrial applications. In automated vehicle research, AI advances enhance the integrity and safety of automated vehicle software. AI is also expected to serve as a solution for critical safety scenarios that are difficult to manage with conventional approaches [9,10].

The development of AI techniques that could improve the existing control systems has been addressed in several studies in various contexts. In particular, the performance of existing control systems has been improved by learning the driving from the data, enabling shared control between a driver and an autonomous driving system [11].

Other studies have attempted to increase the online performance of proportional integral derivative controllers by learning through an artificial neural network (ANN) [12–14]. Recently, primal-dual NNs have improved the real-time performance and stability of MPC [15].

Several studies have been conducted to improve the performance of the controller by reducing the uncertainty of the model using an ANN. In particular, the performance of MPC can be significantly improved by learning from real-time data, which provide knowledge of the target model [16–27].

The present study develops an NMPC-based drift control method that accurately tracks the predefined trajectories of an automated vehicle by using an established vehicle model. The developed NMPC-based drift controller is then replaced by a DNN-based controller pretrained on the data generated from the previously designed closed-loop trajectories of the NMPC method.

By replacing the previously designed NMPC method with the proposed DNN-based controller, we avoid the need for complex numerical optimization of the vehicle control, thereby reducing the computational load. The computational time of the DNN-based controller is very small and predictable in general, once the training process is complete. However, the computational cost of the NMPC method is often high and very unpredictable because its optimization problem includes many free variables that must be explored under many constraints. By switching the iterative numerical optimization process with a fixed number of NN computational processes, real-time implementation of the final control algorithm on a cheaper controller platform can be achieved and the real-time performance of the control method can be guaranteed. The new technique is especially advantageous in safety-critical applications such as automated vehicle control [28–31].

The following sections describe the development process. Section 2 analyzes the vehicle dynamics that were used for the NMPC's design, and the simulation is introduced. The vehicle model is based on a 1:10-scaled vehicle (the test platform for future research). Section 3 presents the NMPC design process under which the automated vehicle performs the drift maneuver while following the desired curved trajectories. Section 4 illustrates the closed-loop simulation results of the designed NMPC, and Section 5 presents the design of the DNN-based controller. The research conclusions are presented in Section 6.

2. Vehicle Dynamics Analysis

2.1. Three-Degrees-of-Freedom Bicycle Model

The horizontal motion of the vehicle was computed using the bicycle model shown in Figure 3. Neglecting aerodynamic drag forces, the bicycle model is defined as follows:

$$\dot{\beta} = \frac{F_{yf}\cos(\delta) + F_{yr}}{m} - r,$$

$$\dot{r} = \frac{F_{yf}\cos(\delta) - l_r F_{yr}}{I_{zz}}, \quad (1)$$

$$\dot{v}_x = \frac{F_{xr} - F_{yf}\sin(\delta)}{m} + rv_x\beta.$$

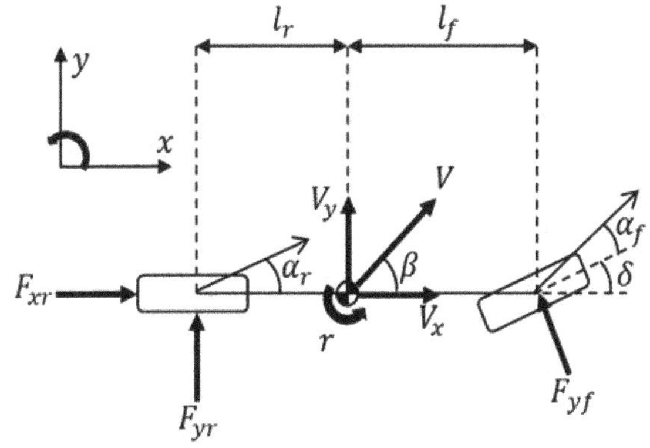

Figure 3. Bicycle model for describing the horizontal motion of the vehicle.

Table 1 defines the notations used in this study. The state variables of the vehicle model in Equation (1) are the sideslip angle (β), yaw rate (r), and forward velocity (v_x). The control inputs are the rear tire force (F_{xr}) and the steering angle (δ).

Table 1. Nomenclature of the present study.

Symbol	Meaning	Unit
F_{yf}	Front tire lateral force	N
F_{yr}	Rear tire lateral force	N
v	Vehicle velocity	m/s
v_w	Rear-wheel velocity	m/s
v_x	Longitudinal velocity	m/s
α	Tire slip angle	rad
α_f	Front tire slip angle	rad
α_r	Rear tire slip angle	rad
μ	Friction coefficient	-
μ_s	Friction coefficient of tire skids	-
r	Yaw rate	rad/s
δ	Steering angle	rad
β	Sideslip angle	rad
m	Vehicle mass	kg
l_f	Distance from the center of gravity (CG) to the front axle	M
l_r	Distance from CG to the rear axle	M
I_{zz}	Yaw moment of inertia	N·m/rad^2
C_{xr}	Rear tire longitudinal slip angle	-
κ	Tire slip ratio	-

2.2. Brush Tire Model

The longitudinal and lateral tire forces in the bicycle model are computed using a brush tire model, which constrains the maximum amount of tire force (the combined longitudinal and lateral forces) within the elliptical circle in Figure 4. A tire force curve versus the tire slip angle is illustrated in Figure 4, where the red area indicates the saturated area and the blue area denotes the unsaturated area. The brush tire model was employed using Equation (2).

Under normal driving conditions, the combined force acting on a tire remains within the elliptic region and the tire model remains in the unsaturated state. Conversely, when the magnitude of the combined force acting on the tire reaches the elliptic circle, the tire model moves to the saturated state and a large amount of slip occurs. This situation is dangerous because the vehicle can lock its wheels or skid, which increases the difficulty of controlling the vehicle.

$$F = \begin{cases} \gamma - \frac{1}{3\mu F_z}\gamma^2 + \frac{1}{27\mu^2 F_z^2}\gamma^3, & \gamma \leq 3\mu F_z \\ \mu_s F_z, & \gamma > 3\mu F_z \end{cases}$$

$$F_x = \frac{C_x}{\gamma}\left(\frac{\kappa}{1+\kappa}\right)F,$$

$$F_y = \frac{C_\alpha}{\gamma}\left(\frac{\tan\alpha}{1+\kappa}\right)F,$$

$$\gamma = \sqrt{C_x^2\left(\frac{\kappa}{1+\kappa}\right)^2 - C_\alpha^2\left(\frac{\tan\alpha}{1+\kappa}\right)^2}, \quad (2)$$

$$\alpha = \begin{cases} \alpha_f = \mathrm{atan}\left(\frac{v_y + l_f*r}{v_x}\right) - \delta \approx \mathrm{atan}\left(\beta + \frac{l_f}{v_x}*r\right) - \delta \\ \alpha_r = \mathrm{atan}\left(\frac{v_y - l_r*r}{v_x}\right) \approx \mathrm{atan}\left(\beta - \frac{l_r}{v_x}*r\right) \end{cases},$$

$$\kappa = \frac{v_w - v_y}{v_x}.$$

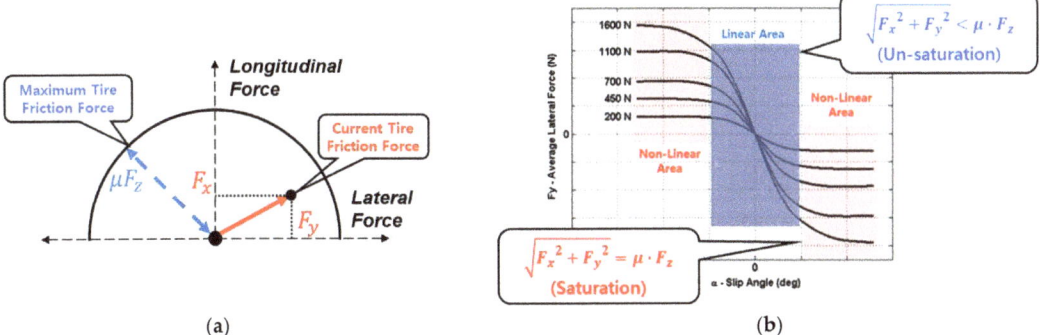

Figure 4. Saturation conditions of the brush tire model: (**a**) longitudinal and lateral forces and (**b**) slip VS tire force.

2.3. Drift Equilibrium State Analysis

The vehicle's trajectory was predicted using the bicycle model defined in Equation (1) with speed and steering angle as the control inputs. To analyze the motion and stability of the vehicle, the bicycle model was combined with the brush tire model under specific conditions (Equation (2)). When the tire slip angle remains within a specific range and the tire force is unsaturated, the vehicle's motion will remain stable. However, when the tire slip angle increases and the resulting tire force becomes saturated, the vehicle's motion will destabilize and even a slight disturbance will divert its states from equilibrium.

To maintain the drift maneuver, the vehicle must be controlled in an unstable equilibrium state. Especially on a slippery road, maintaining a drift maneuver requires a precise and agile controller.

In this study, the equilibrium states were established using Equations (1) and (2) when the time derivatives of the vehicle's states were all zero.

Figure 5 plots the β, r, and v_x equilibrium points according to the steering angle at a longitudinal speed of 1.7 m/s. Plotted are the equilibrium states during a normal driving maneuver (*) and during a drift maneuver (o, △) in the clockwise and counterclockwise directions.

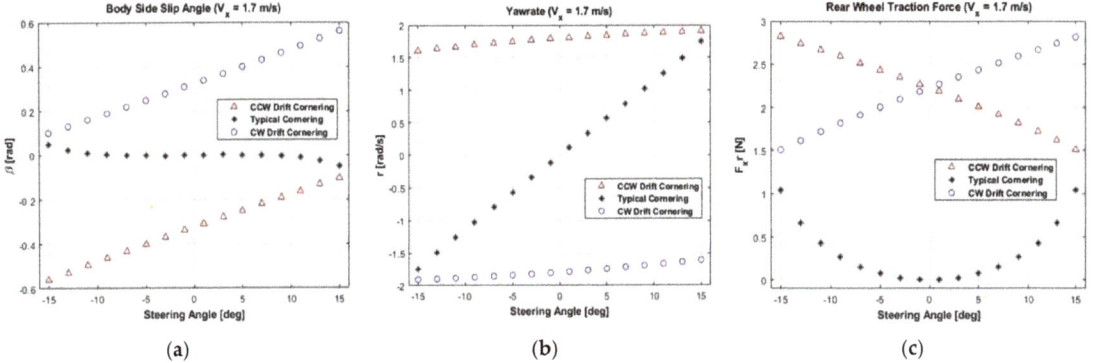

Figure 5. Equilibrium states of (**a**) sideslip angle (β), (**b**) yaw rate (r), and (**c**) rear wheel force (F_{xr}) at a vehicle speed of 1.7 m/s.

3. Design of the Nonlinear Model Predictive Controller

3.1. Vehicle State Prediction Model

Based on the dynamics of the controlled system, the NMPC method predicts the future motions of a vehicle over a fixed time horizon. In this study, the future trajectory was predicted by discretizing the model of the vehicle's dynamics (Equation (1) in Section 2). The vehicle states (X) comprise the sideslip angle (β), yaw rate (r), and speed (v_x) of the vehicle as follows:

$$X = [\beta, r, v_x]. \tag{3}$$

The control input vector (u) comprises the rear-wheel speed (v_w) and the steering angle (δ) of the vehicle.

$$u = [v_w, \delta]. \tag{4}$$

In terms of the rear-wheel speed (v_w), the rear-wheel tire force in Equation (1) is given by the following simplified tire force relation:

$$F_{xr} = \frac{C_{xr}(v_w - v_x)}{v_x}. \tag{5}$$

3.2. Nonlinear Model Predictive Controller Cost Function

The cost function for the optimization process of the NMPC method is the error vector (X_k^e) between the current vehicle state vector (X_k) and the target state vector (X_k^{ref}).

$$\begin{aligned}X_k^e &= X_k^{ref} - X_k \\ &= \left[\beta_k^{ref} - \beta_k,\ r_k^{ref} - r_k,\ v_{x_k}^{ref} - v_{x_k}\right].\end{aligned} \tag{6}$$

The cost function (J) is defined in terms of the state error vectors and the control inputs.

$$J = \frac{1}{2}(X_{k+N}^e)^T * P * X_{k+N}^e + \frac{1}{2}\sum_{j=k}^{k+N-1}\left(X_j^e\right)^T * Q * X_j^e + u_j^T R u_j. \tag{7}$$

Note that the cost function comprises a quadratic term of the final Nth step error (X_N^e), the sum of the quadratic terms of errors (X_k^e), and the quadratic terms of the control input (u_k) in future steps from k to $k + N - 1$, with weight matrices of P, Q, and R, respectively. The inputs that minimize the cost function given by Equation (7) are determined by numerical optimization based on a conjugate gradient method.

3.3. Nonlinear Model Predictive Controller System for Drift Driving

Figure 6 shows the control system of the developed NMPC-based drift control method. First, the curvature (ρ_r) and reference speed (v_r) of the driving trajectory are provided by a path-generation algorithm. The drift equilibrium state is then obtained from the three-dimensional (3D) maps shown in Figure 7. Given the vehicle speed and steering angle at each time step, the 3D maps are configured to output the equilibrium states, i.e., β_{eq}, r_{eq}, and $F_{xr_{eq}}$, based on the equilibrium analysis presented in Section 2.

The drift equilibrium points obtained from the 3D maps were assembled into the target state vector of the NMPC. The rear-wheel speed, v_w, that allows the vehicle to maintain the drift maneuver was calculated with the developed NMPC algorithm.

While maintaining the drift condition through rear-wheel control using the NMPC, an additional pure pursuit algorithm was used as the steering-angle controller to follow the desired trajectory. Similar to the NMPC, the pure pursuit algorithm inputs the current vehicle position and the target trajectory and computes the steering angle (δ) from future time steps k to $k + N$. Figure 8 illustrates the path-following implementation of the pure pursuit control algorithm.

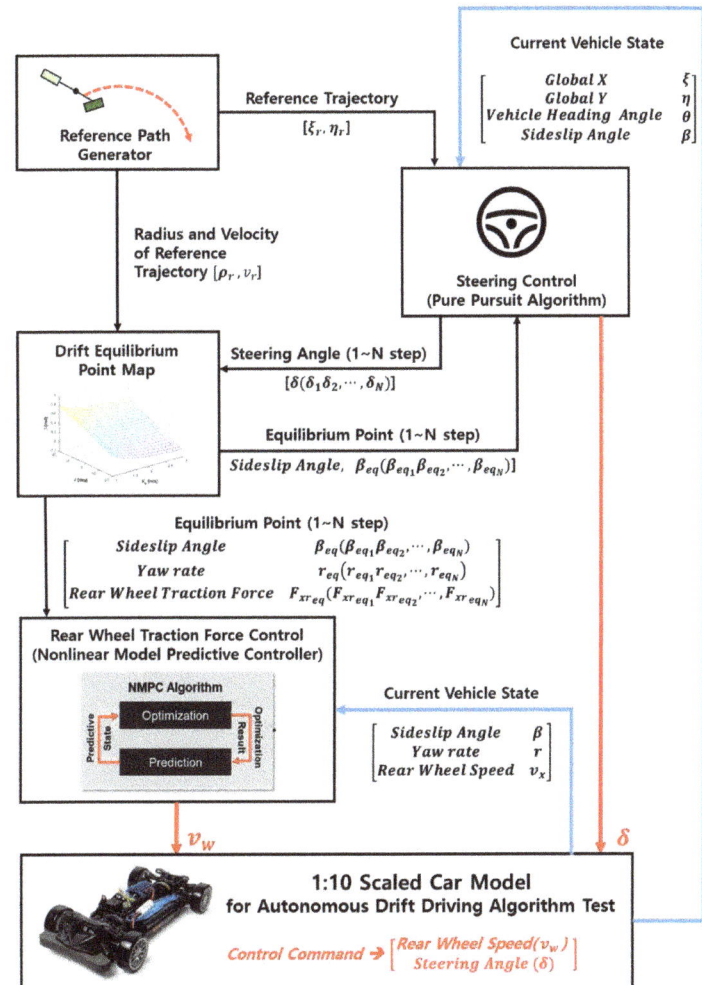

Figure 6. The 1:10-scale nonlinear model predictive control-based drift control system.

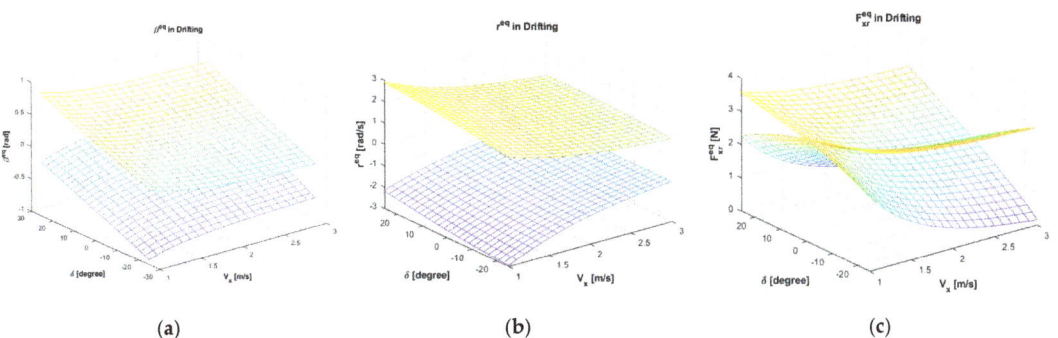

Figure 7. 3D maps of the equilibrium states of the (**a**) sideslip angle (β), (**b**) yaw rate (r), and (**c**) rear-wheel force (F_{xr}).

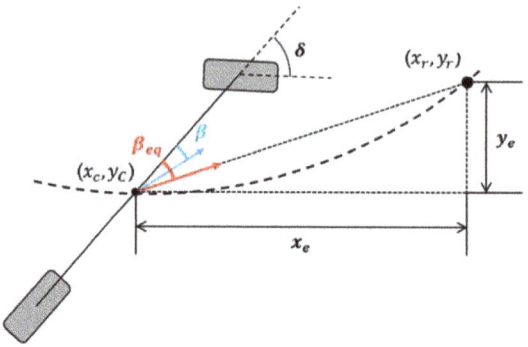

Figure 8. Vehicle state during drift driving.

In the pure pursuit control algorithm, the waypoints are determined from the center point of the rear-wheel axis, which is switched to the center point of the vehicle to simplify the control law. Each waypoint is located at distance l'_d along the straight line in the direction of the target body's sideslip angle. The vehicle's trajectory over N future steps was computed using Equation (1), and the front-wheel steering angles up to N future steps were calculated as

$$\begin{aligned}\delta_k &= \operatorname{atan}\frac{2L\sin\beta_k^{eq}}{l'_d} + k_\beta e_{\beta_k} \\ &= \operatorname{atan}\left(\frac{2L\sin\theta_k}{l'_d}\right) + k_\beta\left(\beta_k^{eq} - \beta_k\right).\end{aligned} \quad (8)$$

The first term in Equation (8) represents the control input that allows the vehicle to head toward the waypoints, and the second term represents the control input for creating the vehicle's track, i.e., β_{eq}. To obtain the future equilibrium states followed by the NMPC, the steering-angle inputs from the pure pursuit control algorithm are applied to the 3D drift equilibrium maps.

4. Drift-Driving Test of the Nonlinear Model Predictive Controller

4.1. Test Scenario

The performance of the NMPC-based drift control method was evaluated through numerical simulations. The controller was required to follow 8-shaped trajectories with diameters of 2 m (① and ②) and 2.5 m (③ and ④), as shown in Figure 9.

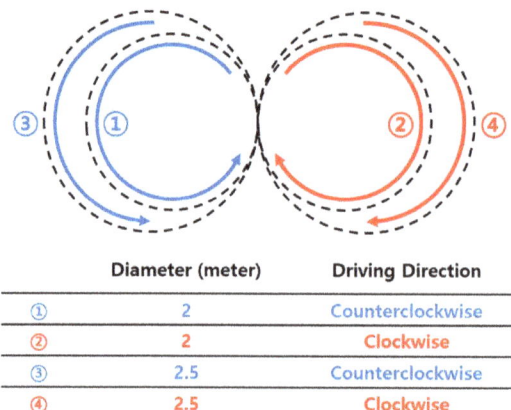

	Diameter (meter)	Driving Direction
①	2	Counterclockwise
②	2	Clockwise
③	2.5	Counterclockwise
④	2.5	Clockwise

Figure 9. Numerical drift test scenarios.

The control period and NMPC prediction period were set to 50 Hz (0.02 s) and 20 steps, respectively. Under these settings, the NMPC system can predict the maneuver for 0.4 s.

4.2. Drift Test Results

In the test scenario, the test vehicle was controlled to drive on routes ①–④ repeatedly using the drift maneuver. Figure 10 shows the sideslip angle and yaw rate (β and r, respectively) of the vehicle during the simulation. In scenarios ① and ③, the vehicle drove in the counterclockwise direction; hence, its body sideslip angle was negative and its yaw rate was positive. Conversely, in scenarios ② and ④, the vehicle drove in the clockwise direction with a positive body sideslip angle and a negative yaw rate. The designed NMPC method accurately followed the desired sideslip angle and yaw rate provided by the 3D map.

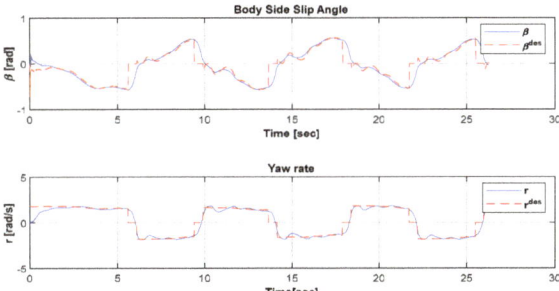

Figure 10. Sideslip angle (β) and yaw rate (r) of the vehicle during the drift maneuver. The dotted red lines and solid blue lines trace the control-target point of the drift control and the vehicle state, respectively.

As shown in Figure 11, the front tire slip never exceeded the limit but the rear tire slip did. Therefore, the front-wheel steering controller required a control-force margin to maintain the desired trajectory, whereas the rear-wheel controller successfully maintained the drift condition by following the desired angle and yaw rate.

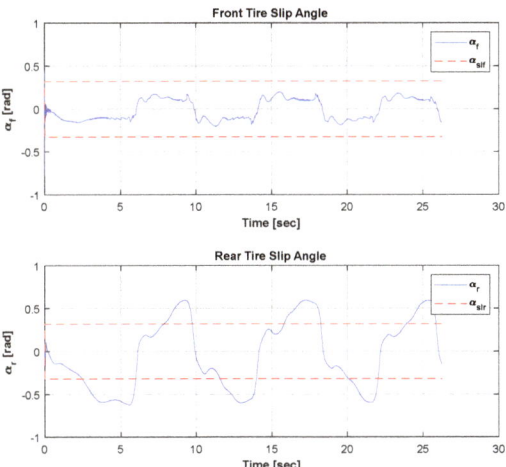

Figure 11. Front and rear tire slip angles of the vehicle during the drift maneuver. The solid blue curves in the upper and lower panels represent the front and rear tire slip angles of the vehicle, respectively, and the dotted red lines show the upper and lower saturation limits of the tires.

Figure 12 shows the driving trajectory of the NMPC-based drift-driving control method. The vehicle precisely followed the figure-eight-shaped target trajectory.

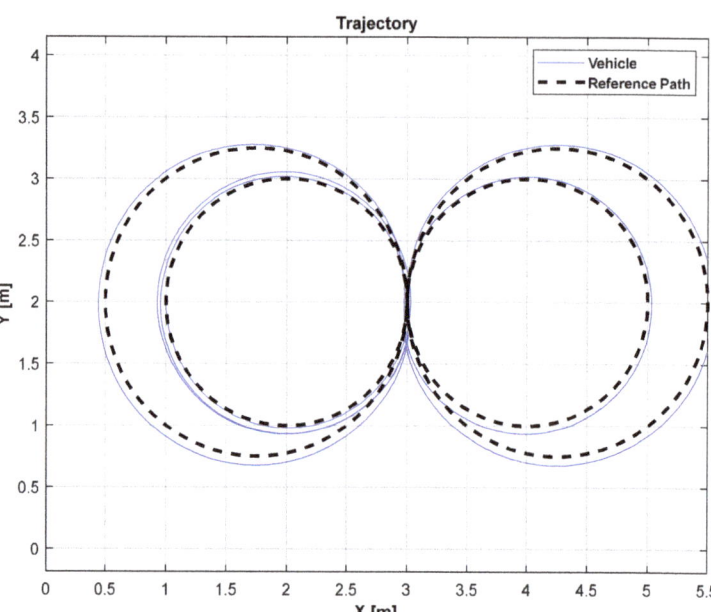

Figure 12. Trajectory of the vehicle during the drift-driving simulation.

5. Design of the Neural Network Drift Controller

The NMPC method predicts a vehicle's behavior up to a predetermined future time and derives the optimal control inputs through numerical optimization with a predesigned cost function. A notable advantage of this method is consideration of the characteristics (dynamics and constraints) during the system optimization. On the downside, accounting for these constraints significantly increases the computational time of the optimization, which is undesirable in fast real-time control applications.

To overcome these limitations while exploiting the advantages of the developed NMPC method, this study employed a DNN-based control method that uses the driving data generated by the NMPC method during drift behavior.

5.1. Training Data Preprocess

The DNN was trained on approximately 50,000 sets of simulated trajectory-driving data, collected along the 2-m-diameter path in Figure 10 (counterclockwise driving along Path ①).

Because the vehicle states, such as vehicle velocity and sideslip angle, have different units and magnitudes, the data were preprocessed by normalizing as follows:

$$x_{norm} = \frac{x - x_{min}}{x_{max} - x_{min}}, \tag{9}$$

where x represents the variable to be normalized and x_{min} and x_{max} represent the minimum and maximum values, respectively, among the sets of variables x. To increase the efficiency of the learning process, only data within the normal range were selected. The standard score z was thus defined as follows:

$$z = \frac{x - \mu}{\sigma}, \tag{10}$$

where μ and σ signify the mean and standard deviation of the data, respectively.

If the absolute value of the Z score exceeded 2, the datum was excluded from the training data because it was outside the normal range of 95% probability. In this process, the data were assumed to follow a Gaussian distribution. The data normalization results are shown in Tables 2 and 3.

Table 2. Training data for steering (lateral) control.

		Mean and Standard Deviation		Normalization Variables	
		μ	σ	Min	Max
Input Data	x_e *	−0.0307	0.1819	−0.2697	0.2674
	y †	0.0080	0.2070	−0.2688	0.2695
	β	−0.4278	0.1314	−0.5625	−0.3534
	β_{eq} °	−0.4969	0.1213	−0.6379	−0.2630
Output Data	δ	−0.1742	0.1216	−0.3876	0.0651

* Longitudinal position error with respect to the reference point; † Lateral position error with respect to the reference point; ° Sideslip angle equilibrium point.

Table 3. Training data for steering (longitudinal) control.

		Mean and Standard Deviation		Normalization Variables	
		μ	σ	Min	Max
Input Data	v_x	−0.0307	0.1819	−0.2697	0.2674
	v_y	0.0080	0.2070	−0.2688	0.2695
	β	−0.4278	0.1314	−0.5625	−0.3534
	r	−0.4969	0.1213	0.6379	0.2630
Output Data	v_w *	−0.1742	0.1216	−0.3876	0.0651

* Vehicle's rear-wheel speed.

As an example, Figure 13 presents the data before and after normalizing the sideslip angle. The data were distributed in the range of −0.48–0.43 before normalization (left panel) and the range 0 to 1 after normalization (right panel).

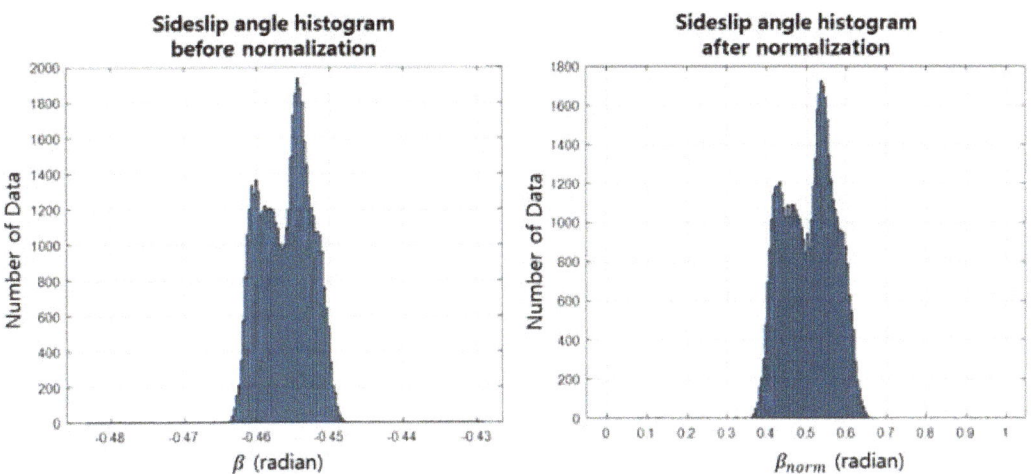

Figure 13. Results of preprocessing the training data of sideslip angle.

5.2. Neural-Network-Based Controller Architecture

The control system architecture includes two NN controllers (Figure 14). The first NN controller, based on a DNN, controls the steering wheel to drive the vehicle along the desired trajectory during a drift maneuver. The second NN controller, based on a time delay NN (TDNN), maintains the drift state of the vehicle.

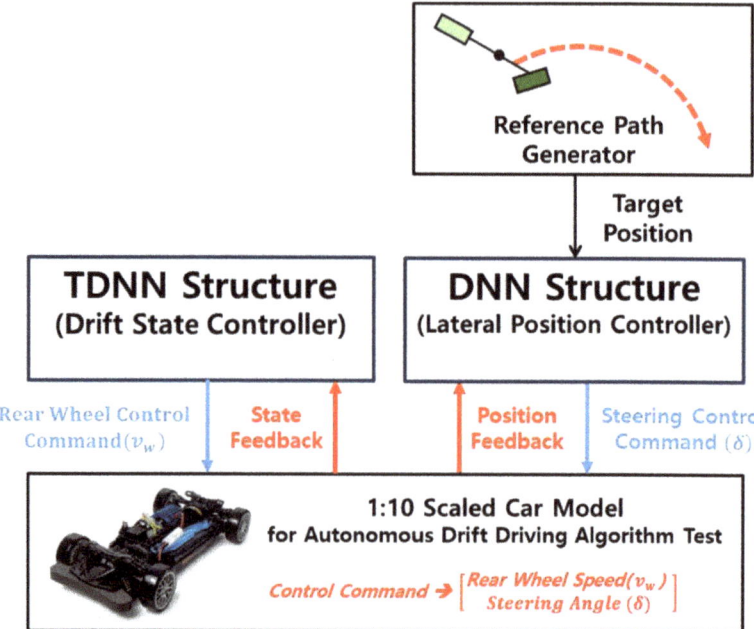

Figure 14. Neural-network-based drift control system.

5.2.1. Deep Neural-Network-Based Controller for Steering Control

A typical NN comprises an input layer, one or more hidden layers, and an output layer. To include the characteristics of the system and prevent unstable behavior due to external disturbances [26,27], the present study employed a DNN with six hidden layers. Each of the six hidden layers was configured with 20 artificial neural nodes as shown in Figure 15. The input data of the network (Table 2) include the position error (x_e, y_e) between the path point and the vehicle, the body slip angle (β), and the body slip-angle equilibrium point (β_{eq}) generated from the 3D map. The network outputs the vehicle steering angle (δ) for lateral position control.

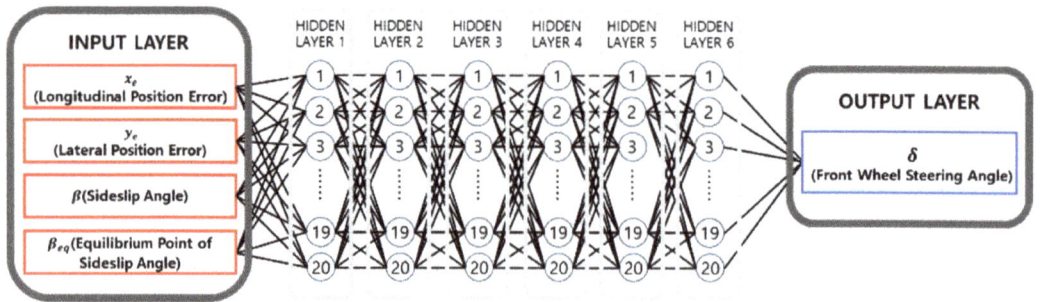

Figure 15. Deep neural network architecture for lateral positioning control.

5.2.2. Time Delay Neural-Network-Based Controller for Drift State Control

The designed NMPC method for maintaining the drift equilibrium states was replaced with a TDNN-based controller. To include the dynamic characteristics of the vehicle during the drift maneuver, the network structure must reflect the near-past vehicle states. The TDNN structure inputs the current data and the data of the past four steps (t, $t-1$, $t-2$, $t-3$, and $t-4$) as follows:

$$\text{Current States}: X^t = \left[v_x^t, v_y^t, \beta^t, r^t\right],$$

$$\text{Previous States}: X^{t-1} = \left[v_x^{t-1}, v_y^{t-1}, \beta^{t-1}, r^{t-1}\right], \qquad (11)$$

$$\vdots$$

$$X^{t-4} = \left[v_x^{t-4}, v_y^{t-4}, \beta^{t-4}, r^{t-4}\right],$$

$$\text{Input Data}: I = \left[X^t, X^{t-1}, X^{t-2}, X^{t-3}, X^{t-4}\right], \qquad (12)$$

where the number of time delay steps was set to 4. The TDNN-based drift controller (Figure 16) contains six hidden layers, each holding 20 artificial neural nodes.

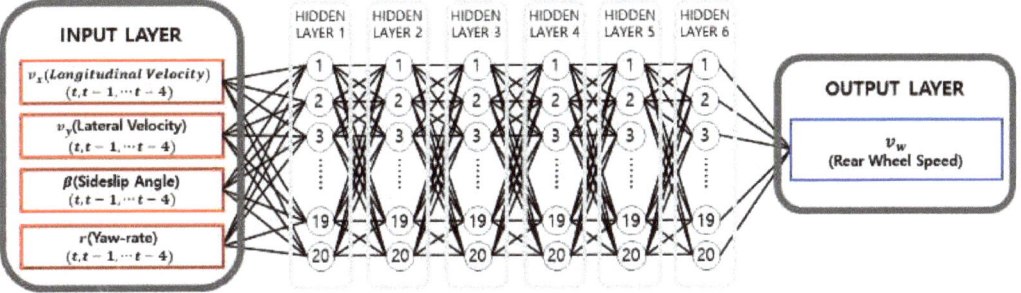

Figure 16. Time delay neural network architecture for drift state control.

The TDNN inputs were the longitudinal and lateral speeds (v_x, v_y), body slip angle (β), and rotation angular speed (r) in Table 3 and the output was the rear-wheel speed (v_w). The DNN was trained on ~50,000 sets of simulation data obtained from the trajectory-driving data (counterclockwise driving around the 2-m-diameter path; see ① in Figure 10).

6. Simulation Results of the Neural Network Drift Controller

The performance of the DNN-based drift control method was evaluated through numerical simulations of a 1:10-scale vehicle driving counterclockwise around a 1-m-radius circle. In this scenario, the vehicle speed was set to 1.7 m/s.

Figure 17 shows the sideslip angles of the front and rear wheels during the drift maneuver. The lateral slip of the front tire did not exceed the saturation limit, whereas the lateral slip of the rear tire exceeded the saturation limit while maintaining the drift condition, allowing the rapid increase of yaw rate that is necessary for following the 1-m-radius circular path. The same phenomenon was observed during the closed-loop simulation using NMPC.

Figure 18 plots the vehicle states during the drift maneuver. Although the TDNN-based rear-wheel controller did not explicitly use the 3D map information of the vehicle equilibrium states, the desired equilibrium points are also plotted as a reference.

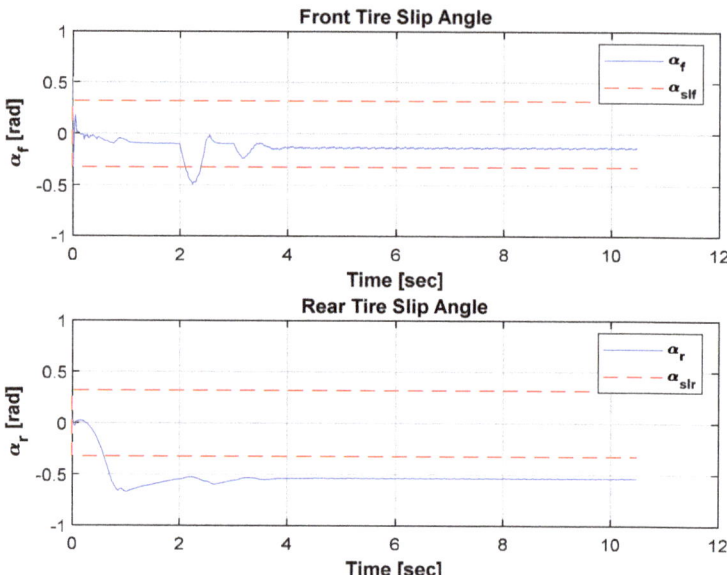

Figure 17. Front and rear tire slip angles during the drift maneuver. The solid blue lines in the upper and lower panels present the slip angles of the front and rear wheels, respectively, and the dotted red line shows the tire saturation threshold.

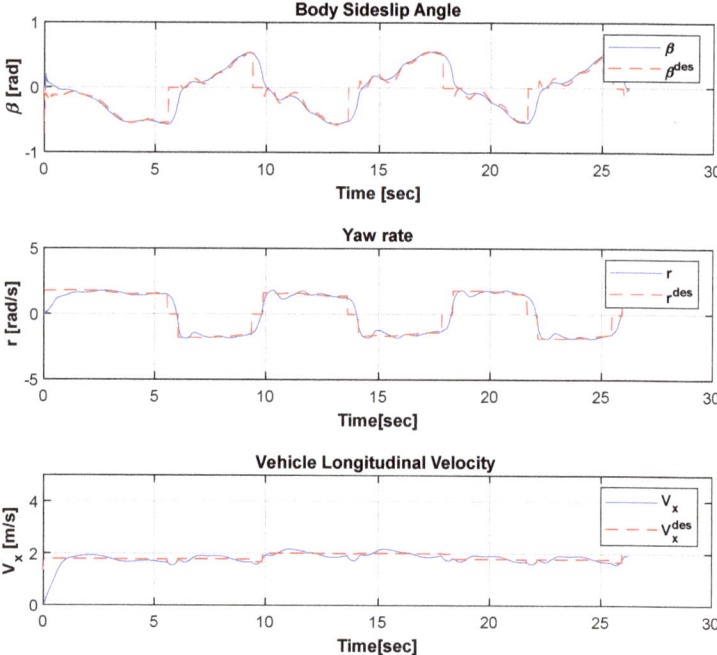

Figure 18. Vehicle states during a drift maneuver (solid blue lines). The desired equilibrium points (dotted red lines) are plotted for reference.

The vehicle's states accurately followed the desired equilibrium states of the body sideslip angle, yaw rate, and longitudinal velocity, even though the TDNN-based controller does not explicitly have information related to the 3D map. It was concluded that the closed-loop trajectory data generated by the NMPC implicitly included information on the drift equilibrium states, which was transferred to the TDNN-based controller during the learning process.

Figures 19 and 20 display the closed-loop trajectory of the vehicle controlled by the TDNN and the tracking errors, respectively. The mean lateral position error remained at ~0.06 m during the drift maneuver. The designed DNN-based steering controller accurately followed the desired trajectory.

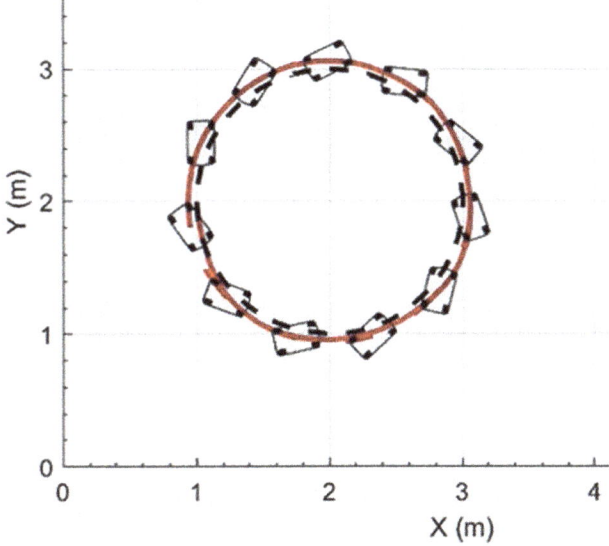

Figure 19. Vehicle trajectory during the drift maneuver.

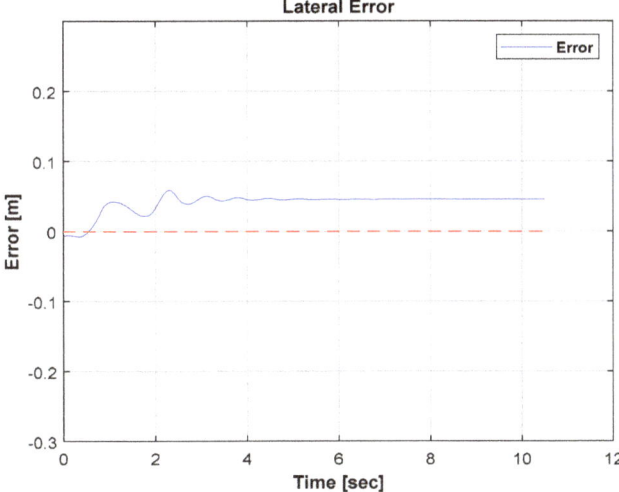

Figure 20. Lateral position error during the drift maneuver.

7. Conclusions

In this study, a drift control method for autonomous vehicles was developed as a strategy for managing a dangerous oversteer phenomenon that may occur during driving. First, a NMPC-based drift controller was designed by analyzing the tire model and vehicle dynamics during the drift maneuver. The closed-loop performance of the developed NMPC method was evaluated through numerical simulations of figure-eight-shaped vehicle trajectories with different radii.

Second, a data-driven NN-based control method was employed to overcome the limitations of the real-time performance of the existing NMPC-based drift controller. The DNN- and TDNN-based controllers incorporated the closed-loop performance of the previously designed NMPC method by learning the trajectories and input data obtained from the simulations. The performance of the developed data-driven controller was further verified through realistic numerical simulations, which confirmed the accurate tracking performance of the vehicle along a circular path.

Based on the study results, the proposed data-driven control method has the potential to be used as a controller for autonomous vehicles. The method retains the advantages of the sophisticated model-based NMPC approach for managing expert driving techniques such as drift. In addition, it can learn expert driving skills from a broad range of user data, potentially overcoming the limitations of the current rule-based autonomous driving system.

Author Contributions: Conceptualization, T.L., D.S. and J.L.; methodology, T.L.; software, T.L., D.S. and J.L.; validation D.S. and J.L.; formal analysis, T.L. and D.S.; investigation, D.S. and J.L.; resources, D.S., data curation, J.L.; writing—original draft preparation, T.L., D.S., J.L. and Y.K.; writing—review and editing, T.L., D.S. and J.L.; visualization, T.L.; supervision, Y.K.; project administration, Y.K.; funding acquisition, Y.K. All authors have read and agreed to the published version of the manuscript.

Funding: This study was supported by the Basic Science Research Program through the National Research Foundation of Korea funded by the Ministry of Education, Science and Technology (NRF2021R1A2C2003254) and the Korea Institute for Advancement of Technology(KIAT) grant funded by the Korea Government(MOTIE) (P0020536, HRD Program for Industrial Innovation).

Conflicts of Interest: The authors declare no conflict of interest.

References

1. Zhang, F.; Gonzales, J.; Li, S.E.; Borelli, F.; Li, K. Drift Control for Cornering Maneuver of Autonomous Vehicle. *Mechatronics* **2018**, *54*, 164–174. [CrossRef]
2. Hindiyeh, R.Y. Dynamics and Control of Drifting in Automobiles. Ph.D. Thesis, Department of Mechanical Engineering, Stanford University, Stanford, CA, USA, 2013.
3. Goh, J.Y.; Goel, T.; Gerdes, J.C. Towards Automated Vehicle Control Beyond the Stability Limits: Drifting Along a General Path. *J. Dyn. Syst. Meas. Control* **2020**, *142*, 02004. [CrossRef]
4. Park, M.; Kang, Y. Experimental Verification of a Drift Controller for Autonomous Vehicle Tracking: A Circular Trajectory Using LQR Method. *Int. J. Control Autom. Syst.* **2021**, *19*, 404–416. [CrossRef]
5. Kim, M.; Lee, T.; Kang, Y. Experimental Verification of the Power Slide Driving Technique for Control Strategy of Autonomous Race Cars. *Int. J. Precis. Eng. Manuf.* **2020**, *21*, 377–386. [CrossRef]
6. Cai, P.; Mei, X.; Tai, L.; Sun, Y.; Liu, M. High-Speed Autonomous Drifting with Deep Reinforcement Learning. *IEEE Robot. Autom. Lett.* **2020**, *5*, 1247–1254. [CrossRef]
7. Guo, H.; Tan, Z.; Liu, J.; Chen, H. MPC-based Steady-state Drift Control under Extreme Condition. In Proceedings of the 33rd Chinese Control and Decision Conference (CCDC), Kunming, China, 22–24 May 2021; pp. 4708–4721.
8. Xu, D.; Han, Y.; Ge, C.; Qu, L.; Zhang, R.; Wang, G. A Model Predictive Control Method for Vehicle Drifting Motions with Measurable Errors. *World Electr. Veh. J.* **2022**, *13*, 54. [CrossRef]
9. Hristozov, A. The Role of Artificial Intelligence in Autonomous Vehicles. Available online: https://www.embedded.com/the-role-of-artificial-intelligence-in-autonomous-vehicles/ (accessed on 15 July 2020).
10. Hristozov, A. Artificial Intelligence Algorithms and Challenges for Autonomous Vehicles. Available online: https://www.embedded.com/artificial-intelligence-algorithms-and-challenges-for-autonomous-vehicles/ (accessed on 3 August 2020).
11. Huang, M.; Gao, W.; Wang, Y.; Jiang, Z. Data-Driven Shared Steering Control of Semi-Autonomous Vehicles. *IEEE Trans. Hum.-Mach. Syst.* **2019**, *49*, 350–361. [CrossRef]

12. Zribi, A.; Chtourou, M.; Djemel, M. A New PID Neural Network Controller Design for Nonlinear Processes. *J. Circuits Syst. Comput.* **2018**, *27*, 1850065. [CrossRef]
13. Chertovskikh, P.; Seredkin, A.; Godyzov, O.; Styuf, A.; Pashkevich, M.; Tokarev, M. An Adaptive PID Controller with an Online Auto-tuning by a Pretrained Neural Network. *J. Phys. Conf. Ser.* **2019**, *1359*, 15–22. [CrossRef]
14. Yaadav, A.; Gaur, P. AI-based Adaptive Control and Design of Autopilot System for Nonlinear UAV. *Indian Acad. Sci.* **2014**, *39*, 765–783. [CrossRef]
15. Jhang, X.; Bujarbaruah, M.; Borrelli, F. Safe and Near-Optimal Policy Learning for Model Predictive Control Using Primal-Dual Neural Networks. In Proceedings of the IEEE American Control Conference (ACC), Philadelphia, PA, USA, 10–12 July 2019; pp. 354–359.
16. Rankovic, V.; Radulovic, J.; Grufovic, N.; Divac, D. Neural Network Model Predictive Control of Nonlinear Systems Using Genetic Algorithms. *Int. J. Comput. Commun. Control* **2012**, *7*, 540–549. [CrossRef]
17. Shi, T.; Wang, P.; Chan, C.; Zou, C. A Data Driven Method of Optimizing Feedforward Compensator for Autonomous Vehicle. *IEEE Intell. Veh. Symp.* **2019**, *1901*, 2012–2017.
18. Vasikaninová, A.; Bakošová, M. Neural Network Predictive Control of a Chemical Reactor. *Acta Chim. Slovaca* **2009**, *2*, 21–36.
19. Zhang, Z.; Wu, Z.; Rincon, D.; Christofides, P. Real-Time Optimization and Control of Nonlinear Processes Using Machine Learning. *Mathematics* **2019**, *7*, 890. [CrossRef]
20. Wong, W.; Chee, E.; Li, J.; Wang, X. Recurrent Neural Network-Based Model Predictive Control for Continuous Pharmaceutical Manufacturing. *Mathematics* **2018**, *6*, 242. [CrossRef]
21. Ramdane, H. Adaptive Neural Network Model Predictive Control. *Int. J. Innov. Comput. Inf. Control* **2013**, *9*, 1245–1257.
22. Limon, D.; Calliess, J.; Maciejowski, J.M. Learning-based Nonlinear Model Predictive Control. *IFAC-PapersOnLine* **2017**, *50*, 7769–7776. [CrossRef]
23. Afram, A.; Sharifi, F.J.; Fung, A.S.; Raahemifar, K. Artificial Neural Network (ANN) Based Model Predictive Control (MPC) and Optimization of HVAC systems: A state of the Art Review and Case Study of a Residential HVAC System. *Energy Build.* **2017**, *141*, 96–113. [CrossRef]
24. Gonzalez, L.P.; Sanchez, S.S.; Guzuman, J.G.; Boada, M.J.L.; Boada, B.L. Simultaneous Estimation of Vehicle Roll and Sideslip Angles through a Deep Learning Approach. *Sensors* **2020**, *20*, 3679. [CrossRef]
25. Mohamed, I.; Rovetta, S.; Do, T.; Dragicević, T.; Diab, A. A Neural Network Based Model Predictive Control of Three-Phase Inverter with an Output LC Filter. *IEEE Access* **2019**, *7*, 124737–124749. [CrossRef]
26. Peng, H.; Song, N.; Li, F.; Tang, S. A Mechanistic-Based Data-Driven Approach for General Friction Modeling in Complex Mechanical System. *ASME J. Appl. Mech.* **2022**, *89*, 071005. [CrossRef]
27. Peng, H.; Song, N.; Kan, Z. Data-Driven Model Order Reduction with Proper Symplectic Decomposition for Flexible Multibody System. *Nonlinear Dyn.* **2022**, *107*, 173–203. [CrossRef]
28. Kang, B.; Lucia, S. Learning-based Approximation of Robust Nonlinear Predictive Control with State Estimation Applied to a Towing Kite. In Proceedings of the 18th European Control Conference (ECC), Naples, Italy, 25–28 June 2019; pp. 16–22.
29. Lee, T.; Kang, Y. Performance Analysis of Deep Neural Network Controller for Autonomous Driving Learning from a Nonlinear Model Predictive Control Method. *Electronics* **2021**, *10*, 767. [CrossRef]
30. Winkler, D.A.; Le, T.C. Performance of Deep and Shallow Neural Networks, the Universal Approximation Theorem, Activity Cliffs, and QSAR. *Mol. Inform.* **2017**, *36*, 1–2.
31. Lucia, S.; Karg, B. A Deep Learning-based Approach to Robust Nonlinear Model Predictive Control. *IFAC-PapersOnLine* **2018**, *51*, 511–516. [CrossRef]

Article

A Hybrid Asynchronous Brain–Computer Interface Based on SSVEP and Eye-Tracking for Threatening Pedestrian Identification in Driving

Jianxiang Sun and Yadong Liu *

College of Intelligence Science and Technology, National University of Defense Technology, Changsha 410073, China
* Correspondence: liuyadong@nudt.edu.cn

Abstract: A brain–computer interface (BCI) based on steady-state visual evoked potential (SSVEP) has achieved remarkable performance in the field of automatic driving. Prolonged SSVEP stimuli can cause driver fatigue and reduce the efficiency of interaction. In this paper, a multi-modal hybrid asynchronous BCI system combining eye-tracking and EEG signals is proposed for dynamic threatening pedestrian identification in driving. Stimuli arrows of different frequencies and directions are randomly superimposed on pedestrian targets. Subjects scan the stimuli according to the direction of arrows until the threatening pedestrian is selected. The thresholds determined by offline experiments are used to distinguish between working and idle states of the asynchronous online experiments. Subjects need to judge and select potentially threatening pedestrians in online experiments according to their own subjective experience. The three proposed decisions filter out the results with low confidence and effectively improve the selection accuracy of hybrid BCI. The experimental results of six subjects show that the proposed hybrid asynchronous BCI system achieves better performance compared with a single SSVEP-BCI, with an average selection time of 1.33 s, an average selection accuracy of 95.83%, and an average information transfer rate (ITR) of 67.50 bits/min. These results indicate that our hybrid asynchronous BCI has great application potential in dynamic threatening pedestrian identification in driving.

Keywords: brain–computer interface (BCI); steady-state visual evoked potential (SSVEP); electroencephalography (EEG); threatening pedestrians; eye-tracking

Citation: Sun, J.; Liu, Y. A Hybrid Asynchronous Brain–Computer Interface Based on SSVEP and Eye-Tracking for Threatening Pedestrian Identification in Driving. *Electronics* 2022, 11, 3171. https://doi.org/10.3390/electronics11193171

Academic Editor: Jose Eugenio Naranjo

Received: 3 September 2022
Accepted: 30 September 2022
Published: 2 October 2022

Publisher's Note: MDPI stays neutral with regard to jurisdictional claims in published maps and institutional affiliations.

Copyright: © 2022 by the authors. Licensee MDPI, Basel, Switzerland. This article is an open access article distributed under the terms and conditions of the Creative Commons Attribution (CC BY) license (https://creativecommons.org/licenses/by/4.0/).

1. Introduction

In recent years, the brain–computer interface (BCI) has become a research hotspot in the field of artificial intelligence, aiming at building communication between the human brain and external devices. Electroencephalography (EEG), reflecting brain activity, is the common signal source of BCI applications. As a non-invasive and low-cost signal, EEG has shown high levels of reliability [1,2]. As a new interactive mode, BCI has been widely used in the fields of medical assistance [3] automobile driving [4], robot control [5], etc.

As a complex BCI application, there is a direct control pathway between the brain and the vehicle in Brain-Controlled Vehicles (BCV). At present, the BCI paradigms adopted by BCV systems are mainly P300 [6], motor imagery (MI) [7], and steady-state visual evoked potential (SSVEP) [8]. P300, which is always evoked by a visual stimulus with poor real-time performance, can only be used for the control of static targets, such as switches, wipers, etc. The real-time performance of MI is also poor, and the degrees of freedom available is limited (generally less than four), which makes it impossible to complete the overall driving task. SSVEP, which is an electrophysiological response to a repetitive visual stimulus, has a high information transfer rate (ITR) and good real-time performance. When subjects focus their attention on a stimulus, the corresponding frequency appears in the representation of the EEG signals recorded mainly in occipital regions [9]. Studies [10,11] have shown that the human cerebral cortex will produce SSVEP characteristic components at the fundamental or multiplicative frequency of the target stimuli when exposed to a fixed-frequency visual stimulus. The target stimuli can be identified by detecting the dominant frequency of SSVEP. Based on its high applicability, simplicity, and high accuracy, BCI adopting SSVEP is conducive to the selection of threat targets in the process of automatic driving. However, it is easy to cause fatigue by long visual flicker.

With the maturity of eye-tracking technology and the continuous improvement of human requirements for interaction comfort, interaction based on eye-tracking has attracted more and more attention. In contrast to EEG, interaction based on eye-tracking is more natural, which can further reduce fatigue. In addition, eye-tracking interaction learning is inexpensive, and most users can operate it without special training [12]. However, there are still some drawbacks to eye-tracking. Some eye movements are not guided by volitional attention. If the system does not distinguish between these eye movements, it is likely to misunderstand human intentions and cause false triggering, which is called the "Midas Touch" problem [13]. In addition, eye-tracking technology is not completely reliable. In addition, some random instability factors can cause system errors. Several eye-movement interactions have been applied to text spelling [14] and robot control [15].

A hybrid BCI system is generally composed of one BCI and another system (which might be another BCI) and can perform better than a conventional BCI [16]. Some studies [17] adopt hybrid systems to recognize characters, combining EEG and EOG. In addition, eye-tracking, which is a popular technology in the field of computer vision, has been gradually adopted to combine with BCI to control games [18], robotic arms [19], and drones [20].

At present, the significant improvement of computer information fusion capability is constantly promoting the development of automatic driving. Autonomous driving is gradually moving from specific scenarios (such as highways, experimental parks) to complex urban traffic. Urban traffic conditions are relatively complex, with many dynamic pedestrian targets and variable trajectories. In such a complex road situation, the environment perception approach based on computer vision technology cannot predict a threatening pedestrian target quickly and accurately. Driver intention is integrated into the vehicle's environment perception through BCI, which can help to improve the comfort and safety of driving.

In this work, a multi-modal hybrid BCI combining SSVEP with eye-tracking is proposed for the selection of potentially threatening pedestrians. The arrows in different directions are randomly superimposed on pedestrian targets. SSVEP is evoked by the stimuli of the corresponding frequency while subjects scan the threatening pedestrian target according to the direction of arrows. I-VT filter is applied to process eye-movement tracks, and canonical correlation analysis (CCA) is adopted to detect EEG signals. The combination of eye-tracking and EEG can not only be used to distinguish between working and idle states, but also shorten target selection time and improve accuracy. The experimental results of six subjects show that the proposed hybrid asynchronous BCI system of eye-tracking and SSVEP achieves better performance compared with a single SSVEP-BCI, with an average selection time of 1.33 s, an average selection accuracy of 95.83%, and an average information transfer rate of 67.50 bits/min.

The remainder of this paper is presented as follows: Section 2 introduces a hybrid BCI system, target detection and tracking, graphical stimuli interface, participants, signal acquisition and preprocessing. Section 3 presents the process of experiments, evaluation metrics, and the results of experiments. Section 4 is the discussion of the hybrid BCI system, and Section 5 summarizes the main work of this paper.

2. Materials and Methods

Figure 1 shows the overall framework for threatening pedestrian identification. Yolov5 is introduced to detect pedestrian targets, and DeepSORT is used to track pedestrians. SSVEP stimuli of different frequencies are superimposed on the obtained pedestrian coordinates. Subjects scan pedestrians according to the direction of superimposed arrowhead stimuli. The three decisions effectively reduce the false positives and improve the reliability of threatening pedestrian identification.

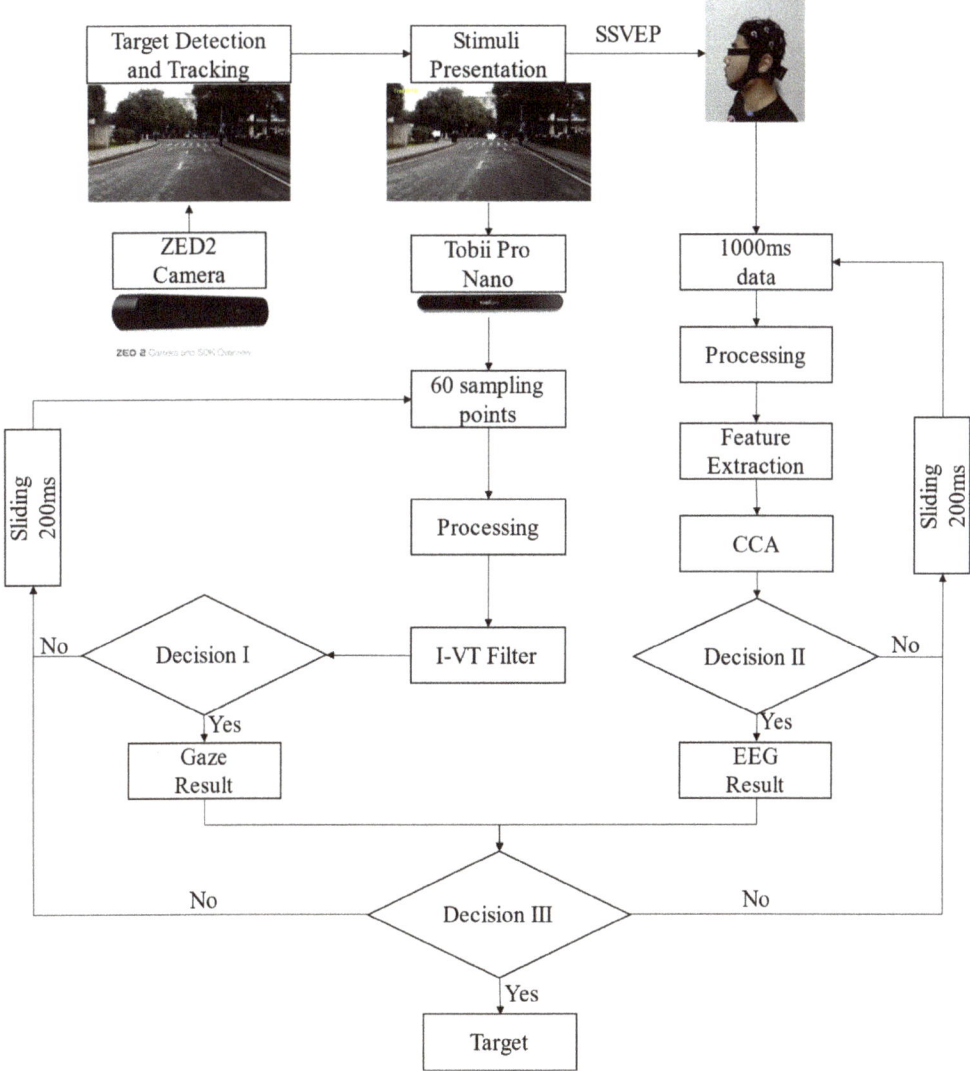

Figure 1. Hybrid asynchronous BCI system for dynamic pedestrian detection.

2.1. System Description

The purpose of this study is to evaluate the performance of a multi-modal BCI that combines eye-tracking and SSVEP for pedestrian tracking and selection. First, the ZED2 camera collects real-time video of driving foreground road conditions and performs multi-target detection and tracking. The coordinates and IDs of the pedestrians are transmitted to the remote computer through the LAN. Second, flashing arrows of different stimuli appear on the targets after receiving data, and follow their movements. The arrows point to a random distribution. After calibrating the eye tracker, participants gaze at the stimulation interface. The eye tracker and the EEG acquisition instrument begin to collect the corresponding signals at the same time. The flow of online signal processing is shown in Figure 1. The sampling frequency of the eye tracker is 60 Hz. In the processing of eye-movement data, I-VT filter is introduced to process visual trajectories. Decision I: When the confidence of the trajectory change over 60 consecutive sampling points exceeds 70%, the result $\{r1, r2 \ldots \}$ of eye-tracking is output. In the processing of EEG signals, the canonical correlation analysis (CCA) algorithm performs feature extraction on 1000 ms of EEG data and outputs the maximum correlation coefficient (ρ). Decision II: Output the result $\{s1\}$ of EEG selection when ρ exceeds the pre-set threshold. Decision III: Output selection target when $\{r1, r2 \ldots \} \cap \{s1\} \neq \emptyset$. Otherwise, no result output is considered an idle state. Slide the window forward 200 ms to acquire eye-tracking data and EEG data for the next 1 s and process until the target result is output.

2.2. Target Detection and Tracking

Detecting and tracking pedestrians in the driving environment can reduce the cognitive load of drivers to a certain extent, and assist the vehicle intelligence system in making decisions, which plays a very important role in improving the safety of intelligent vehicles, and is a hot research topic in intelligent driving and computer vision.

In recent years, with the development of big data and the improvement of computer performance, deep learning has been widely applied in the field of computer vision and has achieved good performance. As a representative target-detection algorithm at the present stage, YOLO algorithms have excellent performance in both detection speed and accuracy, which can achieve end-to-end training. YOLO takes the whole image as the input of the network, and directly outputs the coordinates and IDs of the objects after inference. Compared with other algorithms, yolov5s [21] has higher detection accuracy, faster detection speed, and lower consumption of computation, which can better meet the real-time requirements and be easier to apply in the actual systems. However, detecting the position of pedestrians is not enough. Each object must be tracked before being chosen.

Pedestrian detection determines the position and ID of the object in a particular frame, and pedestrian tracking locks the target in consecutive frames. Most application scenarios involve the tracking of multi-targets. DeepSORT [22] extracts the appearance characteristics of targets, and adopts recursive Kalman filtering and frame-by-frame correlation [23] to match the trajectory of multi-objects, which can effectively reduce the number of target IDs transitions. In this study, we use yolov5 and DeepSORT to process the driving foreground video, which realizes multi-object detection and tracking accurately and quickly, and obtains the position coordinates and IDs of pedestrians in real time.

2.3. Graphical Stimuli Interface

According to the object positions and IDs obtained by the object-detection and tracking module, flicker stimuli of different frequencies are superimposed on each pedestrian in Figure 2, and participants can achieve their selection by staring at stimuli. The length of arrows that flash alternately in black and white is 60 pixels. The frequency list is set to meet a variable number of pedestrians. Studies [24] have shown that a frequency band of 8~15 Hz can induce a relatively strong SSVEP response. Moreover, each frequency should satisfy that there is no overlap between the fundamental frequency and the frequency doubling. The interval between frequencies is set as large as possible to ensure the distinguishability of signals. Considering the above factors, the frequency list is set to 6.10 Hz, 8.18 Hz, 15.87 Hz, 12.85 Hz, 10.50 Hz, 8.97 Hz, 13.78 Hz, 9.98 Hz, 11.23 Hz, 7.08 Hz, 14.99 Hz, and 11.88 Hz. The frequencies of the superimposed stimuli are sequentially selected from the frequency list according to the coding order of each pedestrian ID. During the experiment, participants find the threatening target and follow his movement until the flicker of the target stack stops and turn yellow.

Figure 2. Stimuli presentation interface of hybrid BCI system on trial 6 of a block. (**a**) Arrows in different directions are randomly superimposed on pedestrians according to the frequency list corresponding to the ID order; (**b**) Subjects select the threatening object, and the arrow stops flashing and turns yellow.

2.4. Participants

Six healthy subjects (22–25 years; four men, two women) with an average age of 23.2 years were recruited from the campus and participated in the study. No one was left-handed. In addition, each subject reported no history of any psychiatric deficits. Following the Declaration of Helsinki, all subjects signed a letter of commitment after receiving a detailed description of the procedure.

2.5. Signal Acquisition and Processing

As shown in Figure 3, the eye-movement data were collected by Tobii Pro Nano at a frequency of 60 Hz and an operating distance of 80 cm. Subjects were required to calibrate their eye trackers before participating in experiments. An LCD screen (LEGION Y27gq-25, 1920 × 1080 pixels) was used to present stimuli with a refresh rate of 240 Hz.

Figure 3. Placement of data acquisition equipment.

A 64-channel extended international 10/20 system was used to record the EEG signals in this experiment. Figure 4 shows the placement of 9 electrodes for EEG collection, which were placed in Pz, PO7, PO3, POz, PO4, PO8, O1, Oz and O2. The reference electrode was placed behind the right ear, and the ground electrode was placed on the forehead. Before data acquisition with BrainAmp DC amplifier (Brain Products GmbH, Germany), the impedance of each electrode was reduced to less than 10 kΩ. The sampling frequency was 200 Hz and was filtered by a 4–35 Hz bandpass and notch at 50 Hz. BCI2000 [25] served as the control platform to collect EEG signals, the PyGame [26], a Python expansion package, presented the stimuli interface, and MATLAB was responsible for real-time signal processing. The display interface and the control platform were connected through the TCP/IP protocol.

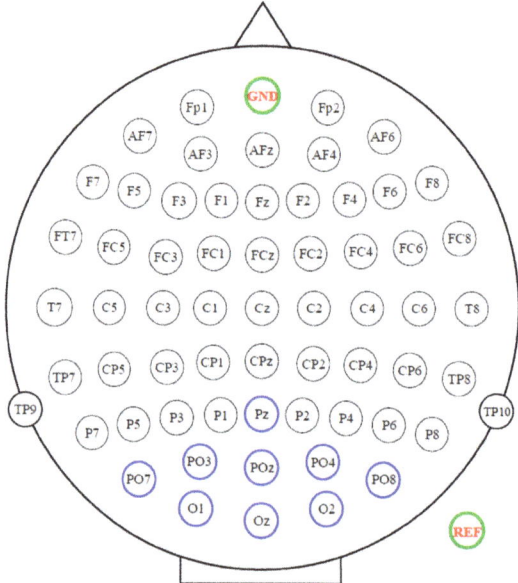

Figure 4. Placement of electrodes. The blue circles are the placements of the sampling electrode. The reference electrode is placed on the green circle behind the right ear, and the ground electrode is placed on the forehead.

Canonical correlation analysis (CCA) [27] was applied to extract features of preprocessed EEG signals, which fuses multi-channel data and identifies the target by calculating the correlation coefficient between multi-channel EEG signal and stimuli frequency. The target was the option corresponding to the maximum SSVEP response score. Periodic stimuli were represented as square-wave periodic signals that could be decomposed into Fourier harmonic series:

$$Y_f(t) = \begin{bmatrix} \sin(2\pi ft) \\ \cos(2\pi ft) \\ \sin(2\pi * 2ft) \\ \cos(2\pi * 2ft) \\ \cdots \\ \sin(2\pi * Nft) \\ \cos(2\pi * Nft) \end{bmatrix}, t = \frac{1}{S}, \frac{2}{S}, \cdots, \frac{L}{S}, \qquad (1)$$

where N is the number of harmonics, t is the current time, L is the number of sampling points of the original signals, and S is the sampling rate of EEG. CCA is a multivariate statistical analysis method, which calculates the maximum correlation coefficient (ρ) of the linear combination of variables ($x = X^T W_x$, $y = Y^T W_y$) in two data sets (X, Y), to reflect the correlation of the two groups of signals. The calculation formula for ρ is as follows:

$$\rho(x,y) = \max_{\omega_x, \omega_y} \frac{E[x^T y]}{\sqrt{E[x^T x] E[y^T y]}} = \max_{\omega_x, \omega_y} \frac{E[\omega_x^T X Y^T \omega]_y}{\sqrt{E[\omega_x^T X X^T \omega_x] E[\omega_y^T Y Y^T \omega_y]}}, \quad (2)$$

The velocity threshold recognition (I-VT) filter is a popular speed-based eye-tracking method [28], which realizes the classification of eye tracks by analyzing the speed of eye movement. As shown in Formula (3), the eye-movement velocity can be obtained by the ratio of the distance between the two sampling points to the corresponding sampling time. Speed is commonly expressed in visual degrees per second (°/s). When the speed is higher than the set threshold, the sample associated with the speed is determined to be a saccade, and below the threshold is fixation.

$$v_x = \frac{x_2 - x_1}{t_2 - t_1}, v_y = \frac{y_2 - y_1}{t_2 - t_1}, \quad (3)$$

where v_x represents the velocity in the x direction, v_y represents the velocity in the y direction, and (x_1, y_1) is the coordinate of the eyeball's position at the moment of t_1. Similarly, (x_2, y_2) is the coordinate of the eyeball position at t_2 moment.

3. Results

3.1. Evaluation Metrics

The performance of hybrid BCI selection is evaluated by accuracy and information transfer rate (ITR). In addition, ITR is calculated as follows (bits per minute):

$$ITR = \frac{\left(\log_2 N + P \log_2 P + (1-P) \log_2\left(\frac{1-P}{N-1}\right)\right) * 60}{T}, \quad (4)$$

$$T = t_s + t_b, \quad (5)$$

where N represents the total number of targets, P is the target selection accuracy, and T represents the time of target selection, including the stimuli flicker time of the target (t_s) and flicker interval time (t_b). It can be seen that the ITR is not only related to the classification accuracy, but also related to the number of selected targets.

3.2. Performance of the Offline Experiment

The threshold is set for the output of the SSVEP to distinguish between idle and working states in online experiments. If the maximum correlation coefficient is higher than the threshold, it is considered to be the working state; otherwise, it is considered to be the idle state. The goal is that the results are not output when the subjects are not staring at the target. In one trial of the offline experiment, participants tend to select threatening pedestrians by staring at flickering stimuli blocks according to cues. Each trial consists of an interval time of 2 s and a stimuli time of 4 s. Each participant participates in the experiment with 2 blocks, and each block contains 10 trials. After a block, participants are given a 5-min break.

Since the SSVEP responses of the participants are individually different, a specific threshold is set for each participant. Ten correct choices of each subject in the offline experiment are randomly selected to calculate the SSVEP response score, and the minimum value is taken as the threshold. As shown in Figure 5, the SSVEP response scores of S5 in 10 correct selection tasks are 0.6387, 0.5647, 0.7696, 0.7065, 0.5630, 0.6896, 0.7323, 0.5981, 0.4541, and 0.6721, respectively. In addition, the minimum response score (0.45) is set as the threshold for Subject 5. The SSVEP response scores of the 6 subjects in the random correct selection tasks for ten times are shown in Table 1, and the statistical thresholds are 0.56, 0.62, 0.51, 0.47, 0.45, and 0.39, respectively.

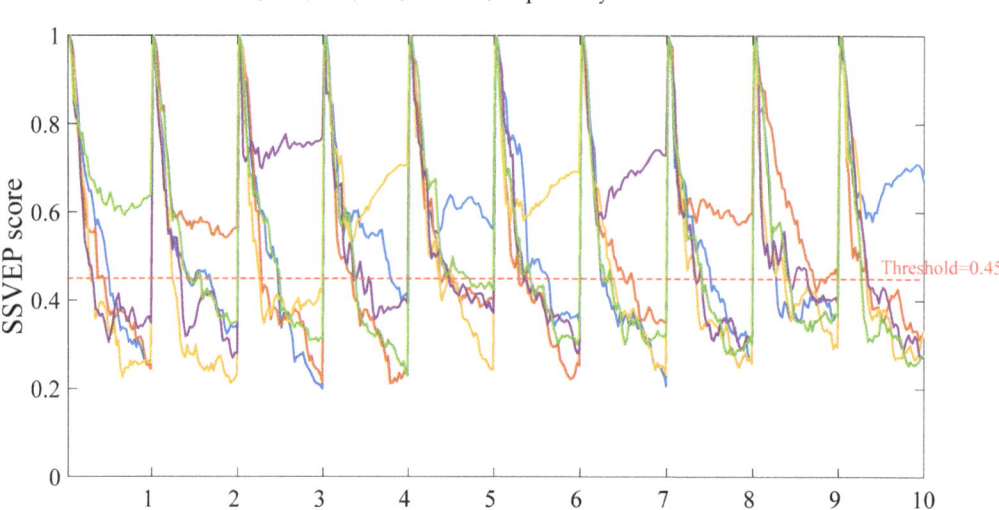

Figure 5. The SSVEP response scores of S5 for selecting correctly for 10 trials. Multi-colored lines represent different stimuli frequencies. The minimum response score of 10 trials is 0.4541, which is set as the threshold for Subject 5.

Table 1. SSVEP response threshold of 6 subjects.

Subject	S1	S2	S3	S4	S5	S6
Threshold	0.56	0.62	0.51	0.47	0.45	0.39

3.3. Performance of Asynchronous Online Experiment

Thresholds obtained from offline experiments are used for online experiments. In the online experiment, the subjects choose threatening pedestrians according to their subjective cognition instead of prompts. There is no time limit for the subjects to complete the experiment. The system continuously outputs control commands to realize the relative real-time selection of threatening pedestrians. The other settings are the same as those for offline experiments. EEG collection and eye-tracking acquisition are performed simultaneously. At the beginning of the experiment, the subjects saccade the stimuli according to the direction of arrows until the color of the stimuli flicker turns yellow and lasts for 0.5 s. If there is a result output, the subjects proceed to the next trial.

Figure 6 shows the change in sight of the subject scanning arrows in the hybrid BCI system. In the one-second time window of the selected target, one of the coordinates of the saccade points remains basically unchanged, and the absolute value of the other coordinate change is about equal to the length of the arrows (60 pixels).

SSVEP-BCI is introduced to verify the effectiveness and availability of the hybrid BCI structure. In Table 2, several evaluation metrics such as the accuracy, target selection time, and ITR are shown to evaluate the performance of two models in which six subjects select dynamic threatening pedestrians. Hybrid BCI achieves a higher selection accuracy (95.83%), shorter selection time (1.33 s), and higher ITR (67.5 bits/min). Compared to SSVEP-BCI, the selection time is shortened by 0.69 s, the accuracy is improved by 5%, and the ITR is increased by 25.2 bits/min. Subject 2 performs perfectly in both SSVEP-BCI and hybrid BCI, with a selection accuracy of 100%. It is worth mentioning that Subject 2 selects threatening pedestrians within 1 s, and the ITR reaches 92.88 bits/min in hybrid BCI. Subject 5 performs poorly in SSVEP-BCI with an accuracy of 80% and an ITR of 25.71 bits/min. By combining eye tracks with EEG data, the accuracy of target selection is significantly improved to 90% and the selection time is shortened from 2.3 s to 1.6 s. These results show that in the hybrid SSVEP architecture, the selection time and accuracy of subjects selecting dynamic threatening pedestrians meet the requirements of online experimental tasks. The advantage of hybrid BCI lies in the addition of eye tracks, which effectively avoids the wrong results caused by inattention. At the same time, the multi-modal fusion of eye movements and EEG enables subjects to make choices in a shorter time. The single eye-tracking system is not stable, and the phenomenon of "Midas Touch" often occurs. In actual traffic scenarios, the wrong choice of threatening targets will lead to traffic accidents caused by the inaccurate operation of self-driving vehicles. The stability and robustness of the hybrid BCI can ensure that the drivers can make the judgment and choose the threatening targets quickly and accurately in assisted driving.

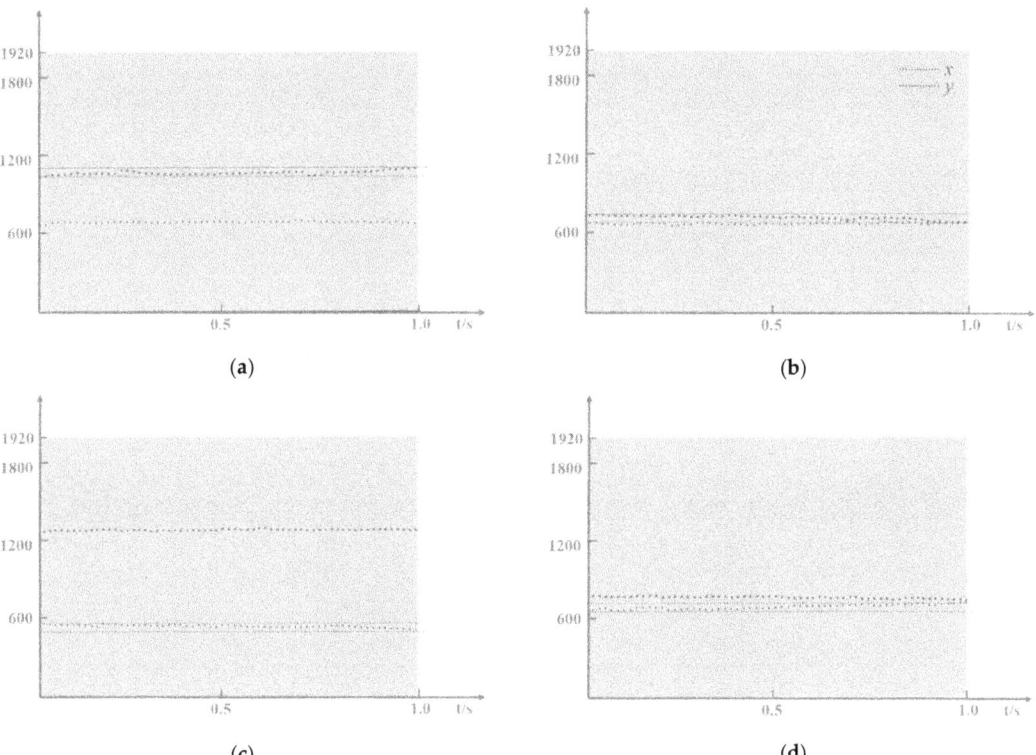

Figure 6. The change of eye movements during saccade in the direction of arrows. (**a**) Top-to-bottom saccade; (**b**) Bottom-to-top saccade; (**c**) Right-to-left scanning; (**d**) Left-to-right saccade. The blue line represents the change in the x direction and the orange line represents the change in the y direction. The distance between the two gray lines is 60 pixels.

Table 2. Results of asynchronous online selection of threatening pedestrians by SSVEP-BCI and hybrid BCI.

Subject	SSVEP			Hybrid		
	Mean Time (s)	Accuracy (%)	ITR (bits/min)	Mean Time (s)	Accuracy (%)	ITR (bits/min)
S1	1.9 + 0.5	95	48.39	1.2 + 0.5	95	68.31
S2	1.8 + 0.5	100	60.57	1.0 + 0.5	100	92.88
S3	2.0 + 0.5	95	46.45	1.3 + 0.5	100	77.39
S4	1.9 + 0.5	90	41.32	1.4 + 0.5	95	61.12
S5	2.3 + 0.5	80	25.71	1.6 + 0.5	90	47.23
S6	2.2 + 0.5	85	31.38	1.5 + 0.5	95	58.07
Mean	2.02 + 0.5	90.83	42.30	1.33 + 0.5	95.83	67.50
Std	0.18	6.72	11.43	0.20	3.44	14.62

4. Discussion

In complex road environments, pedestrians have a great impact on the safety of vehicle driving. The threat to driving safety is usually only a few pedestrians with special locations or trajectories. However, they significantly interfere with the driving route, and even directly determine whether the vehicle can pass safely. Therefore, marking potential threats from many pedestrian targets and feedbacking the location information of these pedestrians to the computer can help vehicles make safer decisions in subsequent control.

This paper proposes a hybrid BCI paradigm for threatening pedestrian selection based on object detection and tracking. The object-detection and tracking method based on deep learning obtains the coordinates and IDs of pedestrian targets, providing initial information for hybrid BCI. This study takes the traffic scenes as the background and combines computer vision with hybrid BCI, aiming at the judgment of dynamic threatening pedestrians. Participants need to judge and select pedestrians who pose a threat to driving safety according to their own subjective experience. Six subjects participated in offline experiments and asynchronous online experiments. The thresholds determined by offline experiments are used to distinguish between the working and idle states of the online experiments. In asynchronous online experiments, the average selection time is 1.33 s, average accuracy reaches 95.83%, and an average ITR reaches 67.5 bits/min. These results show that hybrid BCI has great application potential in dynamic threatening pedestrian selection.

5. Conclusions

This paper designs a hybrid BCI that combines eye-tracking and EEG for threatening pedestrian recognition in the driving environment. The experimental results of six subjects show that hybrid BCI achieves better performance compared with a single SSVEP-BCI, with an average selection time of 1.33 s, an average selection accuracy of 95.83%, and an average information transfer rate (ITR) of 67.50 bits/min. The three proposed decisions filter out the results with low confidence, which effectively improves the selection accuracy of hybrid BCI. The driver's understanding of the environment is fed back to the machine, and human–machine collaborative driving is realized to a certain extent. Compared with methods that rely solely on computer vision, this method has more advanced environmental semantic understanding ability and is safer and more reliable in driving. The system has been verified online in several specific experimental scenarios, but its applicability needs to be further enhanced in scenarios where multiple threatening pedestrians exist or threatening pedestrians suddenly appear. In future work, we will develop more rapid and accurate signal-processing methods to analyze SSVEP, and combine Bayesian probability to decide on threatening pedestrians in different scenarios.

Author Contributions: Conceptualization, Y.L.; methodology, J.S.; software, J.S.; validation, J.S.; writing—original draft preparation, J.S.; writing—review and editing, J.S. and Y.L.; supervision, Y.L.; funding acquisition, Y.L. All authors have read and agreed to the published version of the manuscript.

Funding: This research was supported by National Natural Science Foundation of China, grant number U19A2083.

Conflicts of Interest: The authors declare no conflict of interest. The funders had no role in the design of the study; in the collection, analyses, or interpretation of data; in the writing of the manuscript; or in the decision to publish the results.

References

1. Sánchez-Reyes, L.M.; Rodríguez-Reséndiz, J.; Avecilla-Ramírez, G.N.; García-Gomar, M.L.; Robles-Ocampo, J.B. Impact of eeg parameters detecting dementia diseases: A systematic review. *IEEE Access* **2021**, *9*, 78060–78074. [CrossRef]
2. Ortiz-Echeverri, C.; Paredes, O.; Salazar-Colores, J.S.; Rodríguez-Reséndiz, J.; Romo-Vázquez, R. A Comparative Study of Time and Frequency Features for EEG Classification. *Lat. Am. Conf. Biomed. Eng.* **2019**, *75*, 91–97.
3. Liu, Y.; Liu, Y.; Tang, J.; Yin, E.; Hu, D.; Zhou, Z. A self-paced BCI prototype system based on the in-corporation of an intelligent environment-understanding approach for rehabilitation hospital environmental. *Comput. Biol. Med.* **2020**, *118*, 103618. [CrossRef] [PubMed]
4. Li, W.; Duan, F.; Sheng, S.; Xu, C.; Liu, R.; Zhang, Z.; Jiang, X. A human-vehicle collaborative sim-ulated driving system based on hybrid brain–computer interfaces and computer vision. *IEEE Trans. Cogn. Dev. Syst.* **2017**, *10*, 810–822. [CrossRef]
5. Leeb, R.; Tonin, L.; Rohm, M.; Desideri, L.; Carlson, T.; Millan, J.D.R. Towards Independence: A BCI Telepresence Robot for People With Severe Motor Disabilities. *Proc. IEEE* **2015**, *103*, 969–982. [CrossRef]
6. Bi, L.; Fan, X.-A.; Luo, N.; Jie, K.; Li, Y.; Liu, Y. A Head-Up Display-Based P300 Brain–Computer Interface for Destination Selection. *IEEE Trans. Intell. Transp. Syst.* **2013**, *14*, 1996–2001. [CrossRef]
7. Zhuang, J.; Yin, G. Motion control of a four-wheel-independent-drive electric vehicle by motor imagery EEG based BCI system. In Proceedings of the 2017 36th Chinese Control Conference (CCC), Dalian, China, 26–28 July 2017; pp. 5449–5454.
8. Stawicki, P.; Gembler, F.; Volosyak, I. Driving a Semiautonomous Mobile Robotic Car Controlled by an SSVEP-Based BCI. *Comput. Intell. Neurosci.* **2016**, *2016*, 4909685. [CrossRef]
9. Fernandez-Fraga, S.M.; Aceves-Fernandez, M.A.; Rodríguez-Resendíz, J.; Pedraza-Ortega, J.C.; Ramos-Arreguín, J.M. Steady-state visual evoked potential (SSEVP) from EEG signal modeling based upon recurrence plots. *Evol. Syst.* **2019**, *10*, 97–109. [CrossRef]
10. Ortner, R.; Allison, B.Z.; Korisek, G.; Gaggl, H.; Pfurtscheller, G. An SSVEP BCI to Control a Hand Orthosis for Persons With Tetraplegia. *IEEE Trans. Neural Syst. Rehabil. Eng.* **2010**, *19*, 784–796. [CrossRef]
11. Wu, Z.; Lai, Y.; Xia, Y.; Wu, D.; Yao, D. Stimulator selection in SSVEP-based BCI. *Med. Eng. Phys.* **2008**, *30*, 1079–1088. [CrossRef]
12. Pavan Kumar, B.N.; Balasubramanyam, A.; Patil, A.K.; Chethana, B.; Chai, Y.H. GazeGuide: An Eye-Gaze-Guided Active Immersive UAV Camera. *Appl. Sci.* **2020**, *10*, 1668.
13. Jacob, R.J.K. Eye Tracking in Advanced Interface Design. In *Virtual Environments and Advanced Interface Design*; Oxford University Press: Oxford, UK, 1995; Volume 258, p. 288.
14. Stawicki, P.; Gembler, F.; Rezeika, A.; Volosyak, I. A Novel Hybrid Mental Spelling Application Based on Eye Tracking and SSVEP-Based BCI. *Brain Sci.* **2017**, *7*, 35. [CrossRef]
15. Kishore, S.; González-Franco, M.; Hintemüller, C.; Kapeller, C.; Guger, C.; Slater, M.; Blom, K.J. Comparison of SSVEP BCI and eye tracking for controlling a humanoid robot in a social environment. *Presence Teleoper. Virtual Environ.* **2014**, *23*, 242–252. [CrossRef]
16. Pfurtscheller, G.; Allison, B.Z.; Bauernfeind, G.; Brunner, C.; Solis Escalante, T.; Scherer, R.; Zander, T.O.; Mueller-Putz, G.; Neuper, C.; Birbaumer, N. The hybrid BCI. *Front. Neurosci.* **2010**, *4*, 3. [CrossRef] [PubMed]
17. Zhou, Y.; He, S.; Huang, Q.; Li, Y. A Hybrid Asynchronous Brain-Computer Interface Combining SSVEP and EOG Signals. *IEEE Trans. Biomed. Eng.* **2020**, *67*, 2881–2892. [CrossRef]
18. Myna, K.N.; Tarpin-Bernard, F. Evaluation and comparison of a multimodal combination of BCI paradigms and eye tracking with affordable consumer-grade hardware in a gaming context. *IEEE Trans. Comput. Intell. AI Games* **2013**, *5*, 150–154. [CrossRef]
19. McMullen, D.P.; Hotson, G.; Katyal, K.D.; Wester, B.A.; Fifer, M.S.; McGee, T.G.; Harris, A.; Johannes, M.S.; Vogelstein, R.J.; Ravitz, A.D.; et al. Demonstration of a semi-autonomous hybrid brain–machine interface using human intracranial EEG, eye tracking, and computer vision to control a robotic upper limb prosthetic. *IEEE Trans. Neural Syst. Rehabil. Eng.* **2013**, *22*, 784–796. [CrossRef]
20. Kim, B.H.; Kim, M.; Jo, S. Quadcopter flight control using a low-cost hybrid interface with EEG-based classification and eye tracking. *Comput. Biol. Med.* **2014**, *51*, 82–92. [CrossRef]
21. Thuan, D. *Evolution of Yolo Algorithm and Yolov5: The State-of-the-Art Object Detention Algorithm*; Oulu University of Applied Sciences: Oulu, Finland, 2021.
22. Wojke, N.; Bewley, A.; Paulus, D. Simple online and realtime tracking with a deep association metric. In Proceedings of the 2017 IEEE International Conference on Image Processing (ICIP), Beijing, China, 17–20 September 2017; pp. 3645–3649.
23. Bewley, A.; Ge, Z.; Ott, L.; Ramos, F.; Upcroft, B. Simple online and realtime tracking. In Proceedings of the 2016 IEEE International Conference on Image Processing (ICIP), Phoenix, AZ, USA, 25–28 September 2016; pp. 3464–3468.
24. Zhang, N.; Zhou, Z.; Liu, Y.; Yin, E.; Jiang, J.; Hu, D. A Novel Single-Character Visual BCI Paradigm With Multiple Active Cognitive Tasks. *IEEE Trans. Biomed. Eng.* **2019**, *66*, 3119–3128. [CrossRef]

25. Schalk, G.; McFarland, D.; Hinterberger, T.; Birbaumer, N.; Wolpaw, J. BCI2000: A General-Purpose Brain-Computer Interface (BCI) System. *IEEE Trans. Biomed. Eng.* **2004**, *51*, 1034–1043. [CrossRef]
26. Kelly, S. Basic introduction to pygame. In *Python, PyGame and Raspberry Pi Game Development*; Springer: Berlin/Heidelberg, Germany, 2016; pp. 59–65.
27. Lin, Z.; Zhang, C.; Wu, W.; Gao, X. Frequency Recognition Based on Canonical Correlation Analysis for SSVEP-Based BCIs. *IEEE Trans. Biomed. Eng.* **2006**, *53*, 2610–2614. [CrossRef] [PubMed]
28. Olsen, A. *The Tobii I-VT Fixation Filter*; Tobii Technology: Danderyd Municipality, Sweden, 2012; Volume 21, pp. 4–19.

Article

Space Discretization-Based Optimal Trajectory Planning for Automated Vehicles in Narrow Corridor Scenes

Biao Xu [1,2], Shijie Yuan [1], Xuerong Lin [1], Manjiang Hu [1,2,*], Yougang Bian [1,2] and Zhaobo Qin [1,2]

1. State Key Laboratory of Advanced Design and Manufacturing for Vehicle Body, College of Mechanical and Vehicle Engineering, Hunan University, Changsha 410082, China
2. Wuxi Intelligent Control Research Institute, Hunan University, Wuxi 214115, China
* Correspondence: manjiang_h@hnu.edu.cn

Abstract: The narrow corridor is a common working scene for automated vehicles, where it is pretty challenging to plan a safe, feasible, and smooth trajectory due to the narrow passable area constraints. This paper presents a space discretization-based optimal trajectory planning method for automated vehicles in a narrow corridor scene with the consideration of travel time minimization and boundary collision avoidance. In this method, we first design a mathematically-described driving corridor model. Then, we build a space discretization-based trajectory optimization model in which the objective function is travel efficiency, and the vehicle-kinematics constraints, collision avoidance constraints, and several other constraints are proposed to ensure the feasibility and comfortability of the planned trajectory. Finally, the proposed method is verified with both simulations and field tests. The experimental results demonstrate the trajectory planned by the proposed method is smoother and more computationally efficient compared with the baseline methods while significantly reducing the tracking error indicating the proposed method has huge application potential in trajectory planning in the narrow corridor scenario for automated vehicles.

Keywords: automated vehicle; trajectory planning; narrow corridor scene; space discretization strategy

Citation: Xu, B.; Yuan, S.; Lin, X.; Hu, M.; Bian, Y.; Qin, Z. Space Discretization-Based Optimal Trajectory Planning for Automated Vehicles in Narrow Corridor Scenes. *Electronics* **2022**, *11*, 4239. https://doi.org/10.3390/electronics11244239

Academic Editors: Bai Li, Youmin Zhang, Xiaohui Li and Tankut Acarman

Received: 16 November 2022
Accepted: 15 December 2022
Published: 19 December 2022

Publisher's Note: MDPI stays neutral with regard to jurisdictional claims in published maps and institutional affiliations.

Copyright: © 2022 by the authors. Licensee MDPI, Basel, Switzerland. This article is an open access article distributed under the terms and conditions of the Creative Commons Attribution (CC BY) license (https://creativecommons.org/licenses/by/4.0/).

1. Introduction

Automated vehicles have drawn a huge amount of attention from academia and industry in recent years because of the foreseen potential to improve driving safety and efficiency [1–3]. With researchers' continuous efforts, autonomous driving technologies have made great progress over the past few decades [4–7]. However, as one of the core modules in an autonomous vehicle system, trajectory planning remains to be challenging, especially in complicated environments. Narrow corridors are typically among the most complicated scenes, in which generating a safe, feasible, and smooth trajectory is difficult due to the exterior and interior restrictions.

1.1. Related Work

Numerous studies have been devoted to vehicle trajectory planning [8–11]. The mainstream methods in these studies can be classified into four groups: curve interpolation methods, sampling methods, graph search methods, and numerical optimal control methods [12]. The curve interpolation methods generate a trajectory by interpolating the waypoints in a known set with curves such as the RS curve, clothoid, polynomial, Bezier, B-spline, and so on [13]. For example, in [14], Bae et al. adopted a quintic Bezier curve to generate candidate paths in the lane change maneuver while using lateral acceleration as the path judgment index. Sampling methods try to search for the connectivity of the configuration space by randomly sampling knots in it [15]. Rapid-exploring Random Tree (RRT) is the most typical sampling-based method. For example, in [16], Zheng et al. applied RRT to autonomous parking with vehicle nonholonomic constraints considered. The aforementioned two categories of methods are easy to implement, yet they are not applicable in complicated environments. In detail, paths generated by curve interpolation are deeply influenced by the preset waypoints, while paths generated by randomly sampling are not consistent. Moreover, paths generated by these two methods are suboptimal. Instead, graph search methods could construct the optimal trajectory by traversing the entire state space to find paths with the minimum cost [17]. One famous algorithm of graph search-based methods is the A* algorithm. For example, in [18], Min et al. proposed an improved A* algorithm in an unstructured environment, in which profile collision avoidance was realized by simply setting a redundant security space. Graph search-based methods are inefficient while being used in high-dimensional spaces. To improve this method, some researchers combined the graph search methods with the former curve interpolation methods or sampling methods. In [19], Li et al. used the RS curve and the hybrid A* algorithm to generate paths for automatic parking, yet the path is not curvature continuous. In [20], Dirik and Kocamaz proposed an RRT-Dijkstra algorithm to plan paths with discontinuous curvature. Although graph search methods could construct the optimal trajectories, they could not directly deal with complicated constraints such as obstacle avoidance and actuator limitation. Conversely, optimization-based methods could construct an optimal trajectory while well handling complicated constraints by creating accurate mathematical models [21]. In [22], Li et al. designed a moving trend function in a framework of nonlinear model predictive control, using a risk index to realize collision avoidance. In [23], Dixit et al. used an MPC controller to generate feasible and collision free trajectories by combing the artificial potential field method. In [24], Zhu et al. constructed an optimization problem of parameterized curvature control to realize trajectory generation in dynamic on-road driving environments. Though the trajectories generated by these optimization-based methods are feasible, smooth, and continuous, even in complicated environments, yet they are computational complex.

Narrow corridors are a special but common working scene for vehicles, especially for special purpose vehicles working in a fixed route. These vehicles include logistics vehicles in parks, forklifts in warehouses, underground scrapers in mines, and so on. Passable areas of those vehicles in narrow corridors are strictly limited by the corridor boundaries. Thus, vehicle motion in narrow corridors should be as precise as possible, otherwise, vehicles could collide with the corridor boundaries or could not pass the turning areas. Existing studies on trajectory planning for narrow corridors are concentrated on mobile robots such as unmanned aerial vehicles, and sampling-based methods are the most popular [25]. However, unlike those mobile robots with holonomic constraints who could perform in situ steering, vehicle turning maneuvers are restricted by the non-holonomic vehicle kinematics constraints. Thus, sampling-based methods, whose trajectory curvature is not continuous, are not suitable for vehicle trajectory planning in narrow corridors. There are a few studies on vehicle trajectory planning in narrow passable areas. In [26], Kim et al. used Dubins curves to generate paths for narrow parking lots in a predefined collision-free space, in which curvature of the generated path is discontinuous and quite a part of accessible areas is sacrificed. In [27], considering the parking environment with uncertainty, Li et al. proposed a parallel stitching strategy to replan the trajectory for avoiding the new appeared obstacles utilizing the accessible areas. In [28], Do et al. proposed a method based on the support vector machine (SVM) and fast marching method (FMM) to plan paths for narrow passage, in which obstacle avoidance and vehicle kinematics were considered yet other constraints such as vehicle actuator limitation and terminal postures were ignored. In [29], Tian et al. explored a method about how to turn around in narrow environments, in which RS curves and Bezier curves were applied. With RS curves, a lot of free space would be occupied in the place where the forward and back segments meet. Therefore, this method would not be applicable in a strictly restricted narrow corridor. In [30], Li et al. proposed a progressively constrained strategy that solves a sequence of easier planning problems with shrunk obstacles before handling their nominal sizes. Although the finally derived trajectories are curvature-continuous and optimal, the runtime is usually long and the algorithm performance relies highly on the initial guess. In [31], with the aim of reducing computational time, Li et al. proposed a lightweight iterative framework to generate an optimal trajectory for autonomous parking. In [32], Lin et al. proposed a trajectory planning method for the mine scene but the experiments were not convincing due to the lack of field tests. Methods mentioned above do generate collision-free paths in narrow area scenes. However, the studies above either only consider the trajectory property of collision avoidance or aim at totally different subject models and vehicle operation space, which restricts their application in the narrow corridor scene.

1.2. Contributions

In this paper, we extend our previous work [32] in terms of successfully applying the space discretization-based optimal trajectory planning method (hereinafter called the SOTP method) for automated vehicles in multi-corner narrow corridor scenes, wherein the trajectory curvature is guaranteed to be continuous, every inch of the precious drivable space is sufficiently utilized, and exterior/interior constraints are strictly satisfied. In the proposed SOTP method, we first design a mathematically-described driving corridor and discretize its centerline to generate reference waypoints. Based on these derived reference waypoints, we thereafter formulate a trajectory optimization model in the spatial domain with the consideration of travel time minimization, boundary collision avoidance, and constraint satisfaction in terms of vehicle kinematics, actuator range limitation, side force, etc. Finally, the constructed trajectory optimization model is verified with both simulations and field tests. The main contributions of this paper are as follows:

(1) A novel space discretization-based optimization method is proposed to solve the challenging trajectory planning problem in narrow corridor scenes with very limited passable areas. Compared with [32], more complicated constraints, e.g., boundary avoidance is considered and processed with the accurately established mathematical models of the vehicle and the narrow corridor being embedded into the trajectory generation process.

(2) A space discretization strategy is designed for the construction of the trajectory optimization model. In this strategy, we consider the target trajectory to be described by several discrete waypoints with velocity information. The simulation is designed to demonstrate the enhanced smoothness and computational efficiency of the trajectory planned by the proposed method compared with the baseline algorithm. A sensitivity analysis of the key parameter, e.g., the safety margin is conducted to show the performance of the proposed method.

(3) The field tests related to the trajectory generation ability and the quality of the generated trajectory are conducted to illustrate the advantages of the proposed method over a popular method in the application of the narrow corridor scenes.

1.3. Organization

The rest of this paper is organized as follows. Section 2 formulates the problem. Section 3 introduces the methodology. Section 4 and Section 5 present the simulation and field test results respectively. Finally, Section 6 concludes the paper.

2. Problem Statement

Narrow corridors concerned in this paper refer to one-way roads with the ratio of vehicle width to road width exceeding 0.5. The workspace for vehicles operating in such corridors is usually closed and fixed, which brings a challenge for trajectory generation. In this case, how to generate such a reference trajectory is exactly what we explore below. We assume that the boundary and the centerline of the narrow corridor could be accessible by the perception technology or high definition map.

We consider a typical narrow corridor with specified left and right boundaries. Since the twisted narrow corridor makes it difficult for a vehicle to traverse, we simplify the boundaries via line segments and assume the corridor to be straight (Figure 1). By linear interpolation of the boundary position, we describe the corridor mathematically by using the formulas as follows:

$$A_i^l x + B_i^l y + C_i^l = 0, \quad (1)$$
$$A_i^r x + B_i^r y + C_i^r = 0, \quad (2)$$

where A_i^l, B_i^l, C_i^l are expression parameters of the left boundary in segment i and A_i^r, B_i^r, C_i^r are the right boundary. Suppose the corridor is comprised of N_s segments, so i ranges from 1 to N_s.

Figure 1. A typical narrow corridor with definite boundaries. The corridor is comprised of three segments. Hence, this corridor can be represented by the parameter groups A_i^l, B_i^l, C_i^l and A_i^r, B_i^r, C_i^r, $i \in \{1,2,3\}$.

In the process of passing the narrow corridor, the vehicle is expected to run smoothly and efficiently without any collisions. Therefore, the reference trajectory which the automated vehicle tries to track should be smooth, efficient, and collision-free under the premise of feasibility. To generate such a trajectory, several factors are considered. In detail,

(1) For smoothness, large side-force related to high speed and large curvature is avoided, and the vehicle posture at the corridor terminal is restricted and the motion parameters are limited.

(2) For efficiency, the travel time is minimized.

(3) For collision avoidance, obstacle avoidance conditions based on the circle-fitting strategy are designed.

(4) For feasibility, trajectory generation is set up upon vehicle kinematics and the trajectory is designed as per the actuator range.

3. Methodology

To generate a time-optimal trajectory for a narrow corridor, we explore the trajectory optimization model as follows.

3.1. Vehicle Kinematics Modeling

Vehicles in a narrow corridor generally travel with a low or medium speed because of safety concerns, so the vehicle kinematic model [33], i.e., the bicycle model (Figure 2) that satisfies Ackerman's steering principles, is capable here.

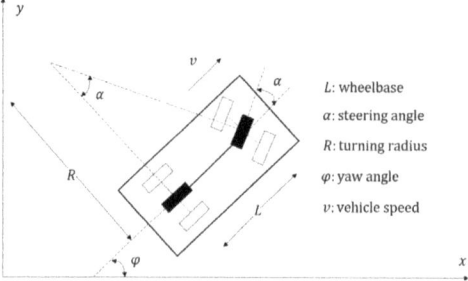

Figure 2. The kinematic model with front-wheel steering.

Similar to [32], we use the coordinate of the rear axle center to represent the vehicle position. With the above-mentioned kinematic model, the vehicle kinematic equations generally expressed in the time domain are as follows:

$$\dot{x} = v\cos(\varphi), \quad (3)$$

$$\dot{y} = v\sin(\varphi), \quad (4)$$

$$\dot{\varphi} = \frac{v\tan(\alpha)}{L}, \quad (5)$$

where x and y are the vehicle position in the global coordinate system, φ is the vehicle yaw angle, v is the velocity, α is the steering angle and L is the wheelbase.

In this study, we discuss the trajectory planning problem in the space domain to make better use of the position information of the corridor boundaries. Therefore, we transform the kinematic relationship from the time domain to the space domain with the chain rule $\frac{d(\cdot)}{dt} = \frac{d(\cdot)}{ds}\frac{ds}{dt} = \frac{vd(\cdot)}{ds}$. The transformed kinematic equations are as follows:

$$x' = \frac{dx}{ds} = \frac{\dot{x}}{v} = \cos(\varphi), \quad (6)$$

$$y' = \frac{dy}{ds} = \frac{\dot{y}}{v} = \sin(\varphi), \quad (7)$$

$$\varphi' = \frac{d\varphi}{ds} = \frac{\dot{\varphi}}{v} = \frac{\tan(\alpha)}{L}, \quad (8)$$

where x' and y' are the first derivatives of the vehicle position versus space, and φ' is the first derivative of the vehicle yaw angle versus space.

3.2. Space Discretization Strategy

In narrow corridors, the total travel time is unpredictable due to the unknown velocity. On the contrary, the total travel distance can be approximated by the corridor centerline due to the limited drivable area. Particularly, the approximation error is tiny between the corridor centerline discrete interval and the target trajectory discrete interval. Therefore, we design a space discretization strategy here to construct the trajectory optimization model. In detail, we take the discrete waypoints on the corridor centerline as a reference and use its discrete interval to approximate the target trajectory discrete interval. In this way, the mathematical relationships in the target trajectory can be explicitly described and the target trajectory can be directly solved.

In the space discretization strategy, we discretize the vehicle trajectory in the space domain and describe the trajectory by the discrete waypoints with velocity information. To provide reference waypoints for this discrete trajectory, we discretize the corridor centerline with certain rules. For the reason that vehicle operation in corner areas is more difficult than that in other areas, we deal with these two situations differently by dividing the corridor into turning areas and straight areas. In this process, the turning area is decided by the unique tangent points of the corridor boundaries and corridor turning arcs with a fixed corridor turning center, see Figure 3. Then, we discretize the centerline in the turning areas and the straight areas separately, see Figure 4. The discrete points in the turning areas are expected to be denser than the discrete points in the straight areas because of the much harder driving environment in the turning areas. Thus, the waypoint interval in the turning areas should be smaller than that in the straight areas. With the rules mentioned above, we discretize the corridor centerline into N waypoints and then record their position information. Here $x_o(k), y_o(k), k = 1, 2, \cdots, N$ is used to represent the waypoint position in the global system and $\Delta s(k), k = 1, 2, \cdots, N-1$ is used to represent the waypoint interval.

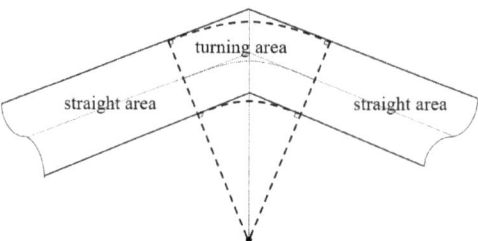

Figure 3. Corridor division. The turning area at the corridor is determined by the fixed road turning center and corridor turning angle.

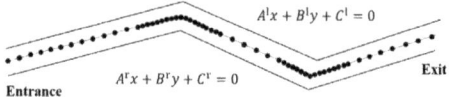

Figure 4. Centerline discretization. The upper formula represents the left boundary while the lower formula represents the right boundary.

Now that the reference centerline has been discretized into N waypoints, the target trajectory can be approximated by these N points in the form of $\xi(s(k)) = \xi(s(1)) + \sum_{j=1}^{k-1} \Delta s(j), k = 2, 3, \cdots, N$, in which target trajectory waypoints interval are replaced by the centerline waypoint interval $\Delta s(k)$. We choose the vehicle position, vehicle yaw angle, and vehicle velocity as the trajectory states. Thus, the final target trajectory $\xi(s(k))$ can be represented by a sequence $X(k), k = 1, 2, \cdots, N$ as follows:

$$X(k) = [x(k), y(k), \varphi(k), v(k)]. \tag{9}$$

What should be noted here is that the final trajectory would be approximate because the actual intervals of the target trajectory waypoints are approximated by the centerline waypoint intervals to express mathematical relationships explicitly and to directly plan the target trajectory. Moreover, the approximation strategy used here is reasonable and workable for the reason that the error caused by this strategy is very small in the case of a narrow corridor.

3.3. Vehicle Trajectory Optimization

This subsection presents the construction of the trajectory optimization model based on the space discretization strategy, considering the constraints related to safety, feasibility, smoothness, and travel time.

3.3.1. Objective Function

Concerns about vehicle trajectory planning are mainly concentrated on safety, feasibility, comfort, and efficiency. The first three ones are hard constraints, while the efficiency requirement is a soft one. Thus, we choose efficiency as the objective here, and an objective function based on the corridor travel time is designed. By taking the discrete waypoints as a reference, the total travel time can be approximately expressed as:

$$J = \sum_{k=1}^{N-1} \frac{\Delta s(k)}{v(k)}. \tag{10}$$

3.3.2. Terminal Posture Constraints

The vehicle positions in the corridor entrance and exit are expected to be on the corridor centerline, while the vehicle yaw angles are expected to be parallel to the corridor heading orientations, which makes it quite difficult to find an exact solution. Thus, we constrain the vehicle's final state to be near the expected posture. By taking the discrete waypoints as a reference, the terminal constraint can be expressed as:

$$x(1) = x_o(1), x(N) = x_o(N) \pm \Delta x, \tag{11}$$
$$y(1) = y_o(1), y(N) = y_o(N) \pm \Delta y, \tag{12}$$
$$\varphi(1) = \theta_{enter}, \varphi(N) = \theta_{exit} \pm \Delta \theta, \tag{13}$$

where θ_{enter} and θ_{exit} are the orientations of the corridor entrance and exit. The sizes of $\Delta x, \Delta y, \Delta \theta$ are expected to be appropriate to guarantee that the vehicle is within the corridor.

3.3.3. Vehicle Kinematics Constraints

Vehicle operation follows the kinematics relationship. Hence, the vehicle kinematics constraints are considered here. By taking the discrete waypoints as a reference, the kinematics constraints in the space domain can be approximately expressed as follows:

$$x(k+1) = x(k) + \Delta s(k) \cos(\varphi(k)), \tag{14}$$
$$y(k+1) = y(k) + \Delta s(k) \sin(\varphi(k)), \tag{15}$$
$$\varphi(k+1) = \varphi(k) + \frac{\Delta s(k) \tan(\varphi(k))}{L}. \tag{16}$$

3.3.4. Vehicle Speed Constraints

We consider the constraint of vehicle speed here. By taking the discrete waypoints as a reference, the vehicle speed constraint is as follows:

$$v_{min} \leq v(k) \leq v_{max}. \tag{17}$$

3.3.5. Actuator Range Constraints

There are both longitudinal and lateral actuator constraints in a vehicle's operations. We ignore the limitation of longitudinal actuators here for the fact that a vehicle's normal operation in a narrow corridor would never reach its acceleration/deceleration limits. Though the longitudinal actuator limitation is ignored here, we would restrict the longitudinal motion parameters for comfort. This would be discussed in a later subsection. For the later actuator, its range could not be ignored since an adequate steering angle is necessary in a narrow turning area. By taking the discrete waypoints as a reference, the constraint of steering is as follows:

$$-\alpha_{max} \leq \alpha(k) \leq \alpha_{max}. \tag{18}$$

3.3.6. Tire Side-Force Constraints

A big tire slip angle caused by a large side force would lead to vehicle model invalidation. To avoid this situation, the side force related to the vehicle speed and path curvature is constrained by the formula $\frac{v^2}{R} \leq \mu g$, where R is the curvature radius, μ is the side-force coefficient, and g is the gravity coefficient. Since the vehicle follows the Ackerman steering principle, R is dependent on the formula $R = \frac{L}{\tan(\alpha)}$. By taking the discrete waypoints as a reference, the final expression of the side-force avoidance constraint is as follows:

$$\frac{v(k)^2 \tan(\alpha(k))}{L} \leq \mu g. \tag{19}$$

3.3.7. Acceleration and Angular Velocity Constraints

Motion change has a great influence on the smoothness and comfort of vehicle operation. The vehicle motion is realized by speed control in the longitudinal direction and front steering control in the lateral direction. Thus, we consider constraining the change of the speed and steer angle here. That is to say, we constrain the acceleration and steer angular velocity. Since we take the discrete waypoints as a reference, the constraints can be approximately expressed as:

$$\left|\frac{v^2(k+1) - v^2(k)}{2\Delta s(k)}\right| \leq a_{max}, \quad (20)$$

$$\left|\frac{v(k)(\alpha(k+1) - \alpha(k))}{\Delta s(k)}\right| \leq \omega_{max}, \quad (21)$$

where a_{max} represents the limit value of accleration and ω_{max} represents the limit value of the steering angle velocity.

3.3.8. Collision Avoidance Constraints

Safety is the most important consideration in vehicle trajectory planning. For narrow corridor scenarios, safety means no border collision happens. For border-collision detection, we adopt a strategy of approximating a vehicle body shape with circles here to describe the collision condition intuitively and leave some safety margin at the same time [34], as shown in Figure 5.

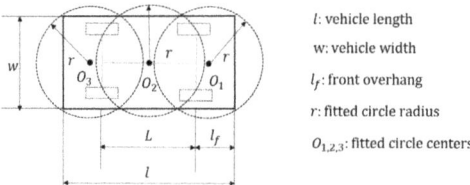

Figure 5. Diagram of the circle fitting strategy.

In the circle fitting strategy, the circle parameters are decided by the circle number and vehicle body parameters. Suppose the vehicle is fitted by N_c circles, and then the circle's parameters are calculated as follows:

$$x_j^c(k) = x(k) + (l_f + L + \frac{l}{N_c}(0.5 - j))\cos(\varphi(k)), \quad (22)$$

$$y_j^c(k) = y(k) + (l_f + L + \frac{l}{N_c}(0.5 - j))\sin(\varphi(k)), \quad (23)$$

$$r = 0.5\sqrt{(\frac{l}{N_c})^2 + w^2}, \quad (24)$$

where r is the circle radius, $x_j^c(k), y_j^c(k)$ are the circle center position of the j_{th} fitted circle, L is the wheel base, l and w are the length and width of the vehicle respectively.

Now that the circle centers and radius are quantitatively derived. Boundary-collision detection can be easily realized by comparing the value of the circle radius with all the distance from the circle centers to the corresponding boundary segments, as shown in Figure 6.

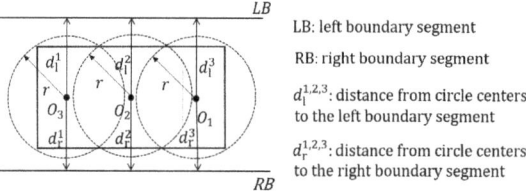

Figure 6. Corridor boundary-collision checking in trajectory planning.

If any distance is less than the circle radius, a collision between the vehicle and corridor boundary happens. Therefore, all the distances should be greater than the circle radius. In particular, the distance between a circle center and the boundary segments is as follows:

$$d_j^l(k) = \frac{\left|A_i^l x_j^c(k) + B_i^l y_j^c(k) + C_i^l\right|}{\sqrt{A_i^{l2} + B_i^{l2}}}, \tag{25}$$

$$d_j^r(k) = \frac{\left|A_i^r x_j^c(k) + B_i^r y_j^c(k) + C_i^r\right|}{\sqrt{A_i^{r2} + B_i^{r2}}}. \tag{26}$$

By taking the discrete waypoints as a reference, the collision avoidance constraints can be expressed as:

$$\left(A_i^l x_j^c(k) + B_i^l y_j^c(k) + C_i^l\right)^2 - r^2\left(A_i^{l2} + B_i^{l2}\right) > 0, \tag{27}$$

$$\left(A_i^r x_j^c(k) + B_i^r y_j^c(k) + C_i^r\right)^2 - r^2\left(A_i^{r2} + B_i^{r2}\right) > 0, \tag{28}$$

in which k range from 1 to N. On the basis of [32], the principle of the parameter and its influence on the results will be discussed in Section 4.3.

3.4. SOTP Method Design

With the objective function and the constraints mentioned above, the trajectory optimization model is finally constructed as follows:

$$\begin{aligned} &\min \ (10), \\ &\text{s.t.} \ (11) \text{ to } (21), (27) \text{ to } (28). \end{aligned} \tag{29}$$

The trajectory optimization model would lead to an optimal sequence of vehicle velocities and steering angles. By the integration of this sequence, a target trajectory with the characteristics of safety, feasibility, comfort, and efficiency would finally be generated.

What should be noticed here is that the target trajectory is used for reference in a fixed scene. Thus, the optimization model is solved in a single time rather than the rolling horizon way. Moreover, if the solving process fails to find an optimal solution because of being trapped in a local infeasibility point, a trajectory recorded by an experienced driver could be used to replace the target trajectory.

4. Numerical Simulation

Numerical simulation is conducted to verify the proposed SOTP method. We first test the practicability in the single corner scenario and subsequently conduct statistical analysis in the narrow corridor scenario with multi-corners compared with the baseline methods.

4.1. Single Corner Scenario

Narrow corridors are hard for vehicles to pass because the corridor boundaries strictly limit the collision-free space, especially in the turning area. Therefore, in this subsection, we apply the proposed method SOTP to the single corner scenario for investigating two problems: (1) could the proposed method be feasible for the narrow corridor scene with single corner? (2) what is the minimum corridor turning angle for passing?

4.1.1. Simulation Setup

To explore the solving ultimate limitation of the proposed method SOTP, we here construct a narrow corridor cluster with a decreasing amplitude of 5°, which ranges from 180° to the minimum angle within their solving capability. Parameters of the narrow corridor cluster are shown in Table 1. As for the vehicle model and vehicle motion limitation, we adopt the same parameters as in Table A2 and Table A3 respectively. As we construct terminal posture constraints, we need to set the threshold value $\Delta x, \Delta y, \Delta \theta$ to ensure the vehicle's final posture to be near the expected posture. The key parameters of the SOTP method are all listed in Table 2.

Table 1. Narrow corridor cluster parameters.

Symbol	Meaning	Value
NC1	Corner angle of the first narrow corridor	180°
NC2	Corner angle of the second narrow corridor	175°
...	...	...
NCz	Corner angle of the z_{th} narrow corridor	$(185-5z)°$
W_{NC}	Corridor width of the narrow corridors	3.5 m

Table 2. The proposed method parameters.

Symbol	Meaning	Value
Δx	The allowable error in the x-axis	0.0625 m
Δy	The allowable error in the y-axis	0.0625 m
$\Delta \theta$	The allowable heading error	0.0685 rad
N_c	The number of fitted circles	3
N	The number of discrete waypoints	60

We use the platform of MATLAB to design the mathematically-described narrow corridors and establish the proposed trajectory generation model. Since the trajectory planning process discussed in this paper is converted to a nonlinear program (NLP), we here use the NLP solver, IPOPT [35], to find the solution. After realizing trajectory generation by the SOTP method, we subsequently set the trajectory tracking simulation on a co-simulation platform of Simulink and CarSim to prove the generated trajectory could be tracked in the practical application. For the established tracking controller, we take the famous Stanley algorithm [36] for lateral tracking and the classical PID algorithm for longitudinal tracking. All processes are executed on an Intel Core i5-11300H CPU with 16 GB RAM that runs at 3.1 Ghz.

4.1.2. Simulation Result

The minimum corner angle required to find a feasible solution is 120° for the proposed SOTP method. As for the study purpose of this subsection, we here choose five groups of data from the corridor clusters, within which the solvability of our method is guaranteed, to illustrate the proposed method's performance. These five groups of data belong to Narrow Corridor NC1, Narrow Corridor NC4, Narrow Corridor NC7, Narrow Corridor NC10 and Narrow Corridor NC13 of which the corner angles are 180°, 165°, 150°, 135° and 120°.

As shown in Figure 7, the planned trajectory is smooth with the guarantee of minimum travel time and collision free. With the decrease in the turning angle, the average velocity of the trajectory slow down as well while the controller could track the generated trajectory which indicates the proposed method has huge potential to solve the narrow corridor scenario.

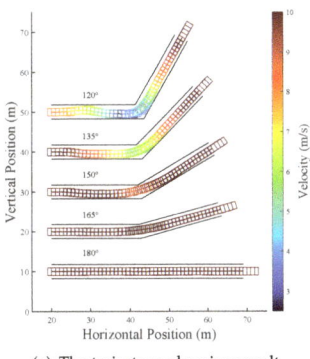
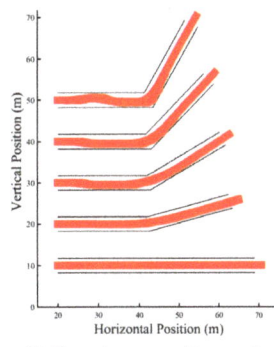

(a) The trajectory planning result   (b) The trajectory tracking result

Figure 7. The simulation results in the single corner scenario. (**a**) the trajectory planning result and (**b**) the trajectory tracking result. The minimum turning angle of the proposed method is 120° and the trajectory could be tracked in different corner angles.

4.2. Multi Corners Scenario

In this subsection, we apply the proposed method and the baseline methods to a more general scenario that has two corners and evaluate the quality of the generated trajectories according to the evaluation metrics to present the excellent performance of the proposed method. The narrow corridor scenario with limited collision-free space means the scope of the feasible solution is smaller compared with those scenes with a larger passable area. Hence, we only consider the narrow corridor scenario.

4.2.1. Baseline Methods

The baseline algorithm we choose for comparison is the hybrid state A* algorithm and dynamic window approach (DWA) method [37]. The hybrid state A* method is widely used in complicated scenes because of its advantage to find the optimal path satisfying the non-holonomic constraints while the DWA method is popular in mobile robot navigation by generating the control variables directly to avoid obstacles.

In the hybrid state A* algorithm, the path is formed by extending the nodes on the grid map with the lowest cost in the 3D kinematic state space. Since the extending nodes are influenced by the grip resolution, the target posture may never be reached. In this case, a termination condition is necessary so that the search process can be stopped once the extending node reaches a preset domain. In this study, we set the domain as: $\Delta x : \pm 0.25$ m, $\Delta y : \pm 0.25$ m, $\Delta \varphi : \pm 15°$ After finishing the search process, a rough path can be formed. Since the curvature of the original path is not continuous, a smoothing process is essential. We use the conjugate gradient descent algorithm introduced in [38] to smooth the original path here. Besides path planning, vehicle velocity also needs to be planned. Based on the path generated by the hybrid A* algorithm, we here adopt nonlinear programming to plan the vehicle velocity, considering the side force and acceleration constraint.

4.2.2. Simulation Setup

In this case study, the narrow corridor scenarios with more than two corners could be split into a scene with two corners. Therefore, we only consider two special narrow corridors. The first narrow corridor turns continuously in the same direction, while the second turns continuously in the opposite direction. The former situation can be continuous left to left turns or right to right turns. We name them the L2L mode and the R2R mode. The latter situation can be continuous right to left turns or left to right turns. We name them the R2L mode and the L2R mode. Because of the symmetry, we only need to simulate one mode in each situation. As shown in Figure 8, narrow corridors with the L2L and R2L modes are finally chosen to be the simulated scenes. More scenario details could be available in Table A1 and the parameter of the SOTP method is adopted in Table 2.

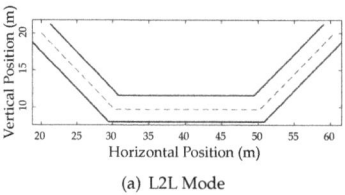
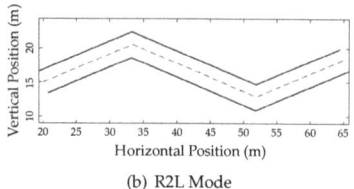

(a) L2L Mode (b) R2L Mode

Figure 8. Typical continuous turning situations in narrow corridors with (**a**) presents the L2L mode and (**b**) presents the R2L mode.

4.2.3. Evaluation Metrics

The assessment of the trajectory planning methods is concentrated on the ability to find the optimal solution and the quality of the generated trajectory. We consider computational performance, curvature, acceleration, and trajectory tracking error in the simulation assessment to evaluate the performance of the proposed method.

Computational Performance : we compute the trajectory generation time of each method on the same simulation platform to compare the computational performance. The shorter time presents a higher computational efficiency.

Curvature: although we have added a limitation for the curvature, the curvature variation stands for the smoothness of the trajectory and the smaller curvature variation shows a better trajectory performance.

Acceleration: acceleration changes sharply may cause the jerk to change violently which can lead to a terrible ride.

Travel Time: this point is also a kind of evaluation metric to present the travel efficiency. The shorter travel time shows a higher quality of trajectory.

4.2.4. Trajectory Generation Result

We draw the generated trajectories and the evaluation results of all methods in Figure 9 for the L2L mode while in Figure 10 for the R2L mode. We know that vehicle operation in a narrow corridor is very hard, especially at the corridor corner, because of the requirement of more free area for turning. As shown in Figures 9a and 10a, the trajectories planned by the SOTP method in both the L2L mode and R2L mode approach the outer boundaries before turning while to the inner boundaries in the other areas. This phenomenon conforms to the fact that the vehicle with the proposed method can find more free space before turning compared with other baseline methods and it needs to shorten the trajectory to save travel time in the other areas. Additionally, as the velocity heat distribution shown in Figures 9a and 10a, the planned velocity would slow down before and rise up after turning, which is in accordance with the characteristics of actual vehicle rides. Therefore, the trajectories generated by the SOTP method in both corridors with the L2L and R2L modes are qualitatively reasonable.

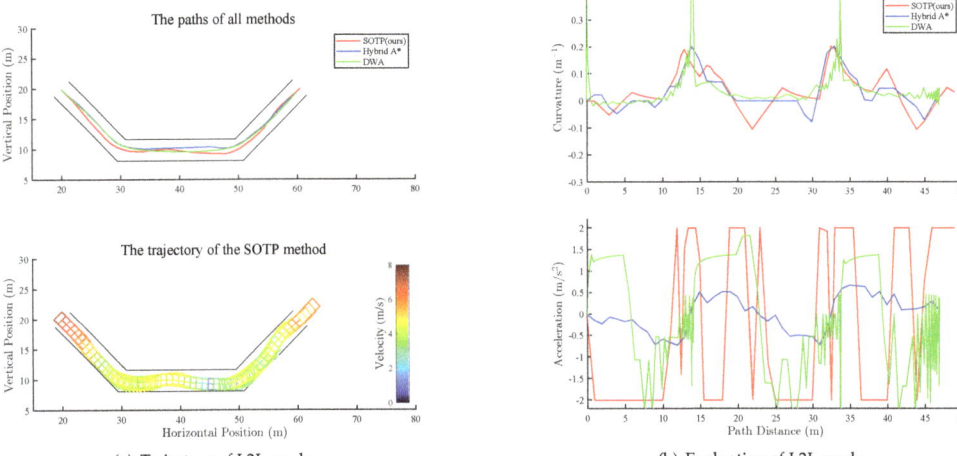

Figure 9. Trajectory generated by all methods in the narrow corridor scene with heat distribution indicating the velocity change. (**a**) the trajectory planned in the L2L mode. (**b**) the curvature and acceleration profile of the trajectories in the L2L mode.

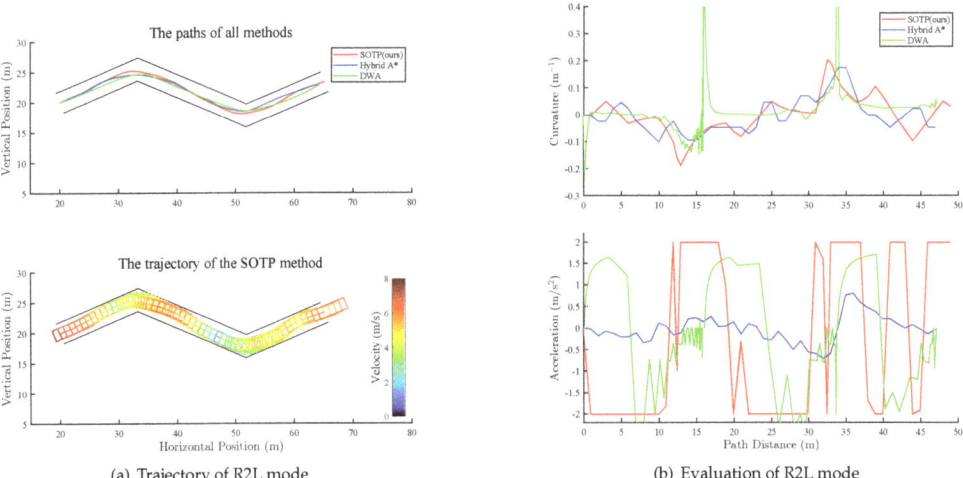

Figure 10. Trajectory generated by all methods in the narrow corridor scene with heat distribution indicating the velocity change. (**a**) the trajectory planned in the R2L mode. (**b**) the curvature and acceleration profile of the trajectories in the R2L mode.

The evaluation results including the curvature and the acceleration are shown in Figures 9b and 10b. More details concerning the maximum curvature, maximum acceleration, and average velocity are recorded in Table 3. From Figures 9b and 10b, the curvature of the DWA method changes sharply at each corner while the SOTP method and hybrid A* can availably restrain the curvature below 0.2 m^{-1}. As for the acceleration, both the SOTP method and the hybrid A* method can keep the acceleration in the allowable scope while the travel time of the SOTP method is shorter than the hybrid A* method. Furthermore, the computational time of the SOTP method is far less than that of the hybrid A* method which presents the proposed method has more powerful computational performance.

Table 3. The evaluation results.

Method	Mode	Max. Curvature (m^{-1})	Max. Acceleration (m/s^2)	Avg. Velocity (m/s)	Travel Time (s)	Computational Time (s)
Hybrid A star	L2L	0.20	0.70	3.96	12.21	83.83
	R2L	0.18	0.83	4.33	10.91	77.35
DWA	L2L	0.69	2.33	1.77	27.68	48.65
	R2L	0.62	2.42	1.47	33.33	82.62
SOTP(ours)	L2L	0.20	2.00	4.39	11.16	30.91
	R2L	0.20	2.00	4.84	10.11	9.37

It is concluded that the proposed method has the highest computational efficiency and shortest travel time with the highest average velocity compared with other baseline methods while its acceleration and curvature could also satisfy the motion limitation.

4.3. Sensitivity Analysis

As designed in Equation (24), we use the fitted circles to approximate the vehicle shape for collision avoidance constraints. In this case study, we utilize different fitted circle numbers to generate trajectory while the other parameters are the same as the previous setting. The safety margin r is 1.24 m, 1.05 m, and 0.99 m with the case of $N_c = 3$, $N_c = 5$, and $N_c = 7$ respectively. The planned trajectories are shown in the Figure 11 while the evaluation results are shown in the Table 4. With the fitted circle number increasing, the planned trajectory is smoother with a higher velocity in that the more fitted circles could reduce the redundant area, as a consequence of enlarging the feasible solution space. The trajectories with $N_c = 5$ and $N_c = 7$ have little difference indicating that the performance could not be improved by continuously increasing the fitted circles. The computational time of the L2L mode is totally greater than the R2L mode as a result of the limited space used to correct the vehicle posture for passing the next corner.

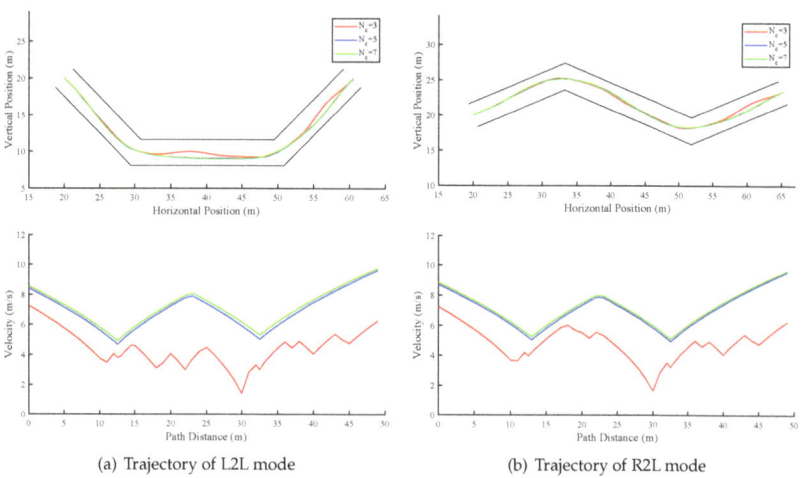

Figure 11. Trajectory generated by the SOTP method with different fitted circle numbers in the narrow corridor scene with the velocity curve. (**a**) the trajectory planned in the L2L mode. (**b**) the trajectory planned in the R2L mode.

Table 4. The result of sensitivity analysis.

Fitted Circle Number	Mode	Max. Curvature (m^{-1})	Max. Acceleration (m/s^2)	Avg. Velocity (m/s)	Travel Time (s)	Computational Time (s)
$N_c = 3$	L2L	0.20	2.00	4.39	11.16	30.91
	R2L	0.20	2.00	4.84	10.11	9.37
$N_c = 5$	L2L	0.13	2.00	6.75	7.27	29.29
	R2L	0.12	2.00	6.83	7.17	5.62
$N_c = 7$	L2L	0.12	2.00	6.94	7.05	33.03
	R2L	0.11	2.00	6.97	7.01	8.21

5. Field Experiment

Regarding the field test, we plan the trajectories by the SOTP method and the hybrid A* method separately and establish controller to track them in a real environment on a modified vehicle with an autopilot system.

5.1. Experiment Setup

In the field test, we test the trajectory planning and tracking in Narrow Corridor NC10, which is the most rigorous among the five narrow corridors introduced above. We ignore the generated velocity profiles and replace them with a constant speed of 10.8 km/h here for the safety concerns in real narrow corridor scenes. The vehicle parameters and the motion limitations are shown in Table A2 and Table A3 respectively.

The field test is conducted on a modified Lincoln MKZ platform as shown in Figure 12. The platform is equipped with an integrated Global Navigation Satellite System (GNSS) and an Inertial Measurement Unit (IMU). The vehicle also supports the by-wire control of the throttle, brake, steering and gear shifting system. The algorithms of motion planning and trajectory tracking control are implemented in C++ under Robot Operating System (ROS). Additionally, the parameters of the SOTP method and the hybrid A* method are the same as the previous setting. After generating the target trajectory, the vehicle is controlled by a nonlinear model predictive control algorithm introduced in [39] to track the planned trajectory at frequencies in excess of 100 Hz.

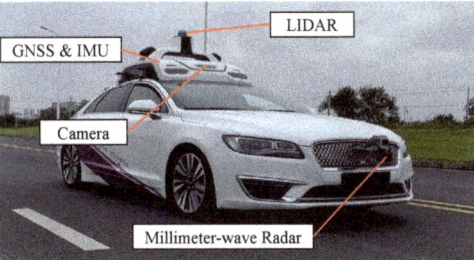

Figure 12. The modified Lincoln MKZ with an autopilot system used in the field experiment.

5.2. Experiment Result

The trajectories tracking results are shown in Figure 13, including the path tracking performance and the path tracking error. The lower tracking error shows that the trajectory could be tracked more easily and the corresponding method can be used in the real world more widely.

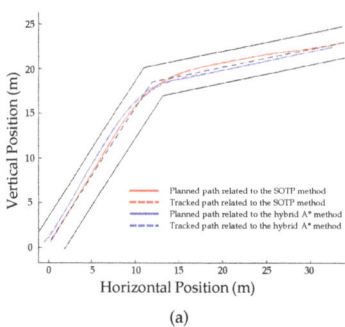 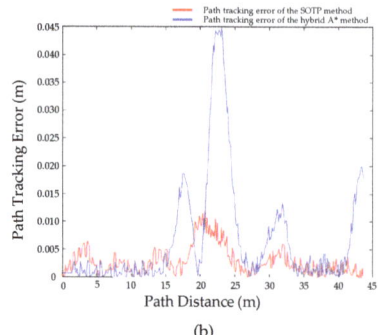

Figure 13. The trajectories tracking results in the field experiment. (**a**) the path tracking performance (**b**) the path tracking error.

The vehicle planned paths and the corresponding tracking paths in the field test are shown in Figure 13a. Intuitively, the dotted red path almost completely overlaps with the solid red path, while the dotted blue path has an obvious gap with the solid blue line. An objective illustration of the path tracking result is shown in the Figure 13b. The centimeter-level tracking error here is related to both the advanced tracking control algorithm and the low testing speed for safety concerns. Anyway, the peak value in red is only 0.01 m while the peak value in blue is almost 0.045 m, which means the tracking error in red is much less than that in blue. Moreover, the fluctuation frequency of the tracking error in red is much gentler than that in blue. Hence, it can be concluded that the SOTP method is superior indeed.

6. Conclusions

This paper presents a space discretization-based optimal trajectory planning method for automated vehicles in narrow corridor-related scenes, which we name the SOTP method. With the space discretization strategy, we take the pre-discretized centerline waypoints as a reference and construct the optimal trajectory generation model totally in the space domain. An objective function in the trajectory optimization model is designed considering the travel time, with the goal of high efficiency. For constraints, vehicle kinematics, boundary collision avoidance, side force, actuator range limitation, terminal states, and boundary collision-free constraints are considered to make sure that the generated trajectory is safe and feasible. The proposed SOTP method is verified with both simulations and field experiments. The results show that the SOTP method is capable of generating feasible, smooth, and collision-free trajectories in narrow corridor scenarios. Furthermore, compared to the popular hybrid state A* algorithm, the SOTP method owns higher efficiency to generate a trajectory in the narrow corridor scene and the generated trajectories are smoother and more efficient. Moreover, the tracking performance of the trajectories planned by the SOTP method is much better, which would lead to more stable conditions for vehicle rides. Consequently, the proposed method has ability to plan a feasible trajectory in a narrow corridor scenario regarded as a corner case in the autonomous driving, potentially overcoming the limitation of other search-based methods.

We consider a static scenario with the predefined passable corridor which may depend on the high definition map. Therefore, in future works, we are going to consider the narrow corridor scenario with dynamic obstacles and uncertainty such as the sudden appearance of a pedestrian shielded by other obstacles to explore a more universal method.

Author Contributions: Conceptualization, B.X. and M.H.; methodology, B.X.; writing—original draft preparation, X.L. and S.Y.; formal analysis, M.H., Y.B. and Z.Q. All authors have read and agreed to the published version of the manuscript.

Funding: This research was funded by National Key R&D Program of China, grant number 2021YFB2501800, National Natural Science Foundation of China, grant number 52102394, 52172384, and 52222216, and Hunan Provincial Natural Science Foundation of China, grant number 2021JJ40095 and 2021JJ40065.

Data Availability Statement: Not applicable.

Acknowledgments: The authors sincerely thanks to associate professor Bai Li of Hunan University for his critical discussion and reading during manuscript preparation.

Conflicts of Interest: The authors declare no conflict of interest.

Appendix A

Table A1. Narrow corridor parameters.

Symbol	Meaning	Value
W_{L2L}	Corridor width (L2L)	3.5 m
$T1_{L2L}$	The first corridor turning angle (L2L)	135°
$T2_{L2L}$	The second corridor turning angle (L2L)	135°
W_{R2L}	Corridor width (R2L)	3.5 m
$T1_{R2L}$	The first corridor turning angle (R2L)	135°
$T2_{R2L}$	The second corridor turning angle (R2L)	135°

Table A2. Vehicle size parameters.

Symbol	Meaning	Value
l	Vehicle length	4.925 m
L	Vehicle wheelbase	2.850 m
l_f	Vehicle front overhang	1.076 m
w	Vehicle width	1.864 m

Table A3. Vehicle motion parameters.

Symbol	Meaning	Value
α_{max}	The maximum steering angle	30°
ω_{max}	The limitation of wheel steering angular velocity	$30° \cdot s^{-1}$
v_{max}	The maximum allowable vehicle speed	$10 \text{ m} \cdot s^{-1}$
v_{min}	The minimum allowable vehicle speed	$1 \text{ m} \cdot s^{-1}$
a_{max}	The maximum limitation of vehicle acceleration	$2 \text{ m} \cdot s^{-2}$
a_{min}	The maximum limitation of vehicle deceleration	$-2 \text{ m} \cdot s^{-2}$
μ	The adhesive coefficient	0.3
g	The gravity coefficient	$9.8 \text{ m} \cdot s^{-2}$

References

1. Milakis, D.; Arem, B.V.; Wee, B.V. Policy and society related implications of automated driving: A review of literature and directions for future research. *J. Intell. Transp. Syst.* **2017**, *21*, 324–348. [CrossRef]
2. Chan, T.K.; Chin, C.S. Review of Autonomous Intelligent Vehicles for Urban Driving and Parking. *Electronics* **2021**, *10*, 1021. [CrossRef]
3. Guo, Z.; Huang, Y.; Hu, X.; Wei, H.; Zhao, B. A Survey on Deep Learning Based Approaches for Scene Understanding in Autonomous Driving. *Electronics* **2021**, *10*, 471. [CrossRef]

4. Li, B.; Tang, S.; Zhang, Y.; Zhong, X. Occlusion-Aware Path Planning to Promote Infrared Positioning Accuracy for Autonomous Driving in a Warehouse. *Electronics* **2021**, *10*, 3093. [CrossRef]
5. Ziegler, J.; Bender, P.; Schreiber, M.; Lategahn, H.; Strauss, T.; Stiller, C.; et al. Making bertha drive—an autonomous journey on a historic route. *IEEE Intell. Transp. Syst. Mag.* **2014**, *6*, 8–20. Available online: https://ieeexplore.ieee.org/document/6803933 (accessed on 15 November 2022). [CrossRef]
6. Vu, T.M.; Moezzi, R.; Cyrus, J.; Hlava, J. Model Predictive Control for Autonomous Driving Vehicles. *Electronics* **2021**, *10*, 2593. [CrossRef]
7. Sharma, O.; Sahoo, N.C.; Puhan, N.B. Recent advances in motion and behavior planning techniques for software architecture of autonomous vehicles: A state-of-the-art survey. *Eng. Appl. Artif. Intell.* **2021**, *101*, 104211. [CrossRef]
8. Paden, B.; Čáp, M.; Yong, S.Z.; Yershov, D.; Frazzoli, E. A survey of motion planning and control techniques for self-driving urban vehicles. *IEEE Trans. Intell. Veh.* **2016**, *1*, 33–55. Available online: https://ieeexplore.ieee.org/document/7490340 (accessed on 15 November 2022). [CrossRef]
9. Zhang, C.; Chu, D.; Liu, S.; Deng, Z.; Wu, C.; Su, X. Trajectory planning and tracking for autonomous vehicle based on state lattice and model predictive control. *IEEE Intell. Transp. Syst. Mag.* **2019**, *11*, 29–40. Available online: https://ieeexplore.ieee.org/abstract/document/8668708 (accessed on 15 November 2022). [CrossRef]
10. Jiang, B.; Li, X.; Zeng, Y.; Liu, D. Human-Machine Cooperative Trajectory Planning for Semi-Autonomous Driving Based on the Understanding of Behavioral Semantics. *Electronics* **2021**, *10*, 946. [CrossRef]
11. Li, B.; Ouyang, Y.; Li, L.; Zhang, Y. Autonomous Driving on Curvy Roads Without Reliance on Frenet Frame: A Cartesian-Based Trajectory Planning Method. *IEEE Trans. Intell. Transp. Syst.* **2022**, *23*, 15729–15741. [CrossRef]
12. González, D.; Pérez, J.; Milanés, V.; Nashashibi, F. A review of motion planning techniques for automated vehicles. *IEEE Intell. Trans. Intell. Transp. Syst.* **2015**, *17*, 1135–1145. Available online: https://ieeexplore.ieee.org/document/7339478 (accessed on 15 November 2022). [CrossRef]
13. Fraichard, T.; Scheuer, A. From Reeds and Shepp's to continuous-curvature paths. *IEEE Trans. Robot.* **2004**, *20*, 1025–1035. Available online: https://ieeexplore.ieee.org/document/1362698 (accessed on 15 November 2022). [CrossRef]
14. Bae, I.; Kim, J.H.; Moon, J.; Kim, S. Lane Change Maneuver based on Bezier Curve providing Comfort Experience for Autonomous Vehicle Users. In Proceedings of the 2019 IEEE Intelligent Transportation Systems Conference (ITSC), Auckland, New Zealand, 27–30 October 2019; pp. 2272–2277.
15. Elbanhawi, M.; Simic, M. Sampling-based robot motion planning: A review. *IEEE Access* **2014**, *2*, 56–77. Available online: https://ieeexplore.ieee.org/abstract/document/6722915/ (accessed on 15 November 2022). [CrossRef]
16. Zheng, K.; Liu, S. RRT based path planning for autonomous parking of vehicle. In Proceedings of the 2018 IEEE 7th Data Driven Control and Learning Systems Conference (DDCLS), Enshi, China, 25–27 May 2018; pp. 627–632.
17. Stentz, A. Optimal and efficient path planning for partially known environments. In Proceedings of the 1994 IEEE International Conference on Robotics and Automation (ICRA), San Diego, CA, USA, 8–13 May 1994; pp. 3310-3317.
18. Min, H.; Xiong, X.; Wang, P.; Yu, Y. Autonomous driving path planning algorithm based on improved A* algorithm in unstructured environment. *Proc. Inst. Mech. Eng. Part D J. Automob. Eng.* **2021**, *235*, 513–526. [CrossRef]
19. Li, C.; Du, J.; Liu, B.; Li, J. A Path Planning Method Based on Hybrid A-Star and RS Algorithm. In Proceedings of the 3rd International Forum on Connected Automated Vehicle Highway System, Jinan, China, 23–25 October 2020.
20. Dirik, M.; Kocamaz, F. RRT-Dijkstra: An Improved Path Planning Algorithm for Mobile Robots. *J. Soft Comput. Artif. Intell.* **2020**, *1*, 69–77. Available online: https://dergipark.org.tr/en/pub/jscai/issue/56697/784679 (accessed on 15 November 2022).
21. Kogan, D.; Murray, R. Optimization-based navigation for the DARPA Grand Challenge. In Proceedings of the 45th IEEE Conference on Decision and Control (CDC), San Diego, CA, USA, 13–15 December 2006.
22. Li, S.; Li, Z.; Yu, Z.; Zhang, B.; Zhang, N. Dynamic trajectory planning and tracking for autonomous vehicle with obstacle avoidance based on model predictive control. *IEEE Access* **2019**, *7*, 132074–132086. Available online: https://ieeexplore.ieee.org/document/8835033/ (accessed on 15 November 2022). [CrossRef]
23. Dixit, S.; Montanaro, U.; Dianati, M.; Oxtoby, D.; Mizutani, T.; Mouzakitis, A.; Fallah, S. Trajectory planning for autonomous high-speed overtaking in structured environments using robust MPC. *IEEE Intell. Trans. Intell. Transp. Syst.* **2019**, *21*, 2310–2323. Available online: https://ieeexplore.ieee.org/document/8734145 (accessed on 15 November 2022). [CrossRef]
24. Zhu, S.; Aksun-Guvenc, B. Trajectory Planning of Autonomous Vehicles Based on Parameterized Control Optimization in Dynamic on-Road Environments. *J. Intell. Robot. Syst.* **2020**, *100*, 1055–1067. Available online: https://link.springer.com/article/10.1007/s10846-020-01215-y (accessed on 15 November 2022). [CrossRef]
25. Szkandera, J.; Kolingerová, I.; Maňák, M. Narrow passage problem solution for motion planning. In Proceedings of International Conference on Computational Science (ICCS), Amsterdam, The Netherlands, 3–5 June 2020; pp. 459–470.
26. Kim, D.; Chung, W.; Park, S. Practical motion planning for car-parking control in narrow environment. *IET Control. Theory Appl.* **2010**, *4*, 129–139. [CrossRef]
27. Li, B.; Yin, Z.; Ouyang, Y.; Zhang, Y.; Zhong, X.; Tang, S. Online Trajectory Replanning for Sudden Environmental Changes During Automated Parking: A Parallel Stitching Method. *IEEE Trans. Intell. Veh.* **2022**, *7*, 748–757. [CrossRef]
28. Do, Q.H.; Mita, S.; Yoneda, K. Narrow passage path planning using fast marching method and support vector machine. In Proceedings of the 2014 IEEE Intelligent Vehicles Symposium Proceedings (IV), Ypsilanti, CA, USA, 8–11 June 2014; pp. 630–635.

29. Tian, X.; Fu, M.; Yang, Y.; Wang, M.; Liu, D. Local Smooth Path Planning for Turning Around in Narrow Environment. In Proceedings of the 28th IEEE International Symposium on Industrial Electronics (ISIE), Vancouver, BC, Canada, 12–14 June 2019; pp. 1924–1929.
30. Li, B.; Acarman, T.; Zhang, Y.; Zhang, L.; Yaman, C.; Kong, Q. Tractor-trailer vehicle trajectory planning in narrow environments with a progressively constrained optimal control approach. *IEEE Trans. Intell. Veh.* **2020**, *5*, 414–425. [CrossRef]
31. Li, B.; Acarman, T.; Zhang, Y.; Ouyang, Y.; Yaman, C.; Kong, Q.; Zhong, X.; Peng, X. Optimization-Based Trajectory Planning for Autonomous Parking With Irregularly Placed Obstacles: A Lightweight Iterative Framework. *IEEE Trans. Intell. Transp. Syst.* **2022**, *23*, 11970–11981. [CrossRef]
32. Lin, X.; Xie, H.; Xu, B.; Yuan, S.; Bian, Y.; Hu, M.; Qin, Z.; Hu, J. A Vehicle Trajectory Planning Method for Narrow Corridor in Mines Based on Optimal Control. *Control Inf. Technol.* **2022**, *05*, 23–29. Available online: https://kns.cnki.net/kcms/detail/detail.aspx?filename=BLJS202205004&dbname=cjfdtotal&dbcode=CJFD&v=MTE0NzhSNkRnOC96aFlVN3pzT1QzaVFyUmN6RnJDVVI3aWVaZWRyRmkza1Y3N0JKeUhCZmJHNEhOUE1xbzlGWUk= (accessed on 15 November 2022).
33. Rajamani, R. *Vehicle Dynamics and Control*, 2ed ed.; Springer: New York, NY, USA, 2012; pp. 24–25.
34. Ziegler, J.; Bender, P.; Dang, T.; Stiller, C. Trajectory planning for Bertha—A local, continuous method. In Proceedings of the 2014 IEEE Intelligent Vehicles Symposium Proceedings (IV), Ypsilanti, CA, USA, 8–11 June 2014; pp. 450–457.
35. Biegler, L.T.; Zavala, V.M. Large-scale nonlinear programming using IPOPT: An integrating framework for enterprise-wide dynamic optimization. *Comput. Chem. Eng.* **2009**, *33*, 575–582. [CrossRef]
36. Thrun, S.; Montemerlo, M.; Dahlkamp, H.; Stavens, D.; Aron, A.; Diebel, J.; Mahoney, P. Stanley: The robot that won the DARPA Grand Challenge. *J. Field Robot.* **2006**, *23*, 661–692. [CrossRef]
37. Fox, D.; Burgard, W.; Thrun, S. The dynamic window approach to collision avoidance. *IEEE Robot Autom. Mag.* **1997**, *4*, 23–33. [CrossRef]
38. Dolgov, D.; Thrun, S.; Montemerlo, M.; Diebel, J. Practical search techniques in path planning for autonomous driving. In Proceedings of the First International Symposium on Search Techniques in Artificial Intelligence and Robotics (STAIR-08), Chicago, IL, USA, 13–14 July 2008.
39. Yin, C.; Xu, B.; Chen, X.; Qin, Z.; Bian, Y.; Sun, N. Nonlinear Model Predictive Control for Path Tracking Using Discrete Previewed Points. In Proceedings of the 2020 IEEE 23rd International Conference on Intelligent Transportation Systems (ITSC), Rhodes, Greece, 20–23 September 2020; pp. 1–6.

Article

Trajectory Planning for an Articulated Tracked Vehicle and Tracking the Trajectory via an Adaptive Model Predictive Control

Kangle Hu * and Kai Cheng

School of Mechanical and Aerospace Engineering, Jilin University, Changchun 130025, China; chengkai@jlu.edu.cn
* Correspondence: hukl16@mails.jlu.edu.cn

Abstract: This paper focuses on the trajectory planning and trajectory tracking control of articulated tracked vehicles (ATVs). It utilizes the path planning method based on the Hybrid A-star and the minimum snap smoothing method to obtain the feasible kinematic trajectory. To overcome the highly non-linearity of ATVs, we proposed a linear-parameter-varying (LPV) kinematic tracking-error model. Then, the kinematic controller was formulated as the adaptive model predictive controller (AMPC). The simulation of the path planning algorithm showed that the proposed planning strategy could provide a feasible trajectory for the ATVs passing through the obstacles. Moreover, we compared the AMPC controller with the developed controller in four scenarios. The comparison showed that the AMPC controller achieved satisfactory tracking errors regarding the lateral position and orientation angle errors. The maximum lateral distance error by the AMPC controller has been reduced by 72.4% compared to the standard-MPC controller. The maximum orientation angle error has been reduced by 55.53%. The simulation results confirmed that the proposed trajectory planning and tracking control system could effectively perform the automated driving behaviors for ATVs.

Keywords: articulated tracked vehicle; adaptive model predictive control; Hybrid A-star; trajectory planning; trajectory tracking

1. Introduction

In articulated tracked vehicles, two double-tracked units are joined by an articulated mechanism. Unlike skid-steer vehicles, ATVs have articulated steering mechanisms driven by hydraulic actuators, which allow them to produce an articulation angle between the front and rear units of the ATVs and steer the front unit to the desired location [1]. ATVs have the advantage of a balanced driving force between both tracks, which results in minimal driving torque requirements in the steering maneuver compared to single and coupled tracked vehicles [2].

Research has been focused on off-road vehicles to enhance driving efficiency and ensure safety by using automated driving systems [3,4]. Several approaches have been applied to obtain a feasible kinematic trajectory for the off-ground vehicle [5]. As the articulated steering mechanism gives ATVs unique steering characteristics, the rear unit of ATVs contributes to the overall nonholonomic constraints. It is, therefore, impossible for traditional planning methods such as the RRT [6] or artificial potential field [7] to produce a feasible and smooth path for the ATVs. The Hybrid A-star algorithm could produce a smooth, kinematics feasible path for nonholonomic systems [8,9]. The Hybrid A-star is implemented in two stages, including the node search, to produce a kinematics-feasible trajectory. The second stage then locally improves the quality of the path using analytical expansion of the path.

For trajectory tracking, two types of modeling have been commonly used: kinematics-based modeling and dynamics-based modeling. Because those ATVs operate at low speeds, dynamic factors such as road-wheel load distribution and centrifugal force can be overlooked. The articulated steering vehicle (ASV) path-tracking deviation model has been extended from ordinary mobile robots in terms of speed deviation, lateral position deviation, and heading angle deviation. To simplify the automatic guidance of ASV, Nayl defined an improved path tracking model, considering the lateral displacement deviation, heading angle deviation, and curvature deviation [10]. A tracking error model, which includes both the position error and orientation error of both the front and rear units, has been applied in [11] to facilitate the control design for the rear unit of the ASV.

Based on the above path tracking error model, Ridley developed a full state feedback adaptive tracking method [12]. There are also numerous applications of complex controllers based on robust control [13], fuzzy control [14], and sliding mode control [15,16]. In addition, researchers proposed a linear switching control strategy that took advantage of the linearization of the ASV to overcome its nonlinear characteristics [17]. A simpler and more robust trajectory tracking controller, including the MPC controller, was suggested for the tracked vehicle [18]. A further advantage of the model predictive control (MPC) algorithm is that it takes constraints into account. Thus, the real-time status of the vehicle can be taken into account directly in the MPC controller. A switch controller consisting of multiple MPC controllers was proposed to account for the side angles of the ASV [19]. On the articulated vehicle platform, Kayacan implemented linear MPC, nonlinear MPC, and robust tube-based MPC algorithms for path tracking [20,21]. An articulated wheel loader achieved good tracking accuracy despite varying road curvature using the adaptive MPC method [22].

The steering characteristics of ATVs have been extensively studied in the recent research on ATVs in [23,24]. Furthermore, the design parameters regarding the steering characteristics of the ATVs were examined in [25,26]. A fuzzy-PID control system has been proposed to obtain the articulation angle of ATVs to track a predefined path [27]. After that, a closed-loop control of the steering torque of the ATV hydraulic-driven system was introduced to obtain the desired articulation angle [28]. Despite this, researchers have not studied motion control for ATVs during complex maneuvers in obstacle-filled environments. Using the trajectory provided by the path planner and optimization modules, this paper proposed a trajectory-tracking control framework based on an adaptive MPC control framework for tracking the trajectory of ATVs. The following works have been completed:

1. Using the Hybrid A-star path planning method to obtain a feasible kinematic trajectory.
2. Using the minimum snap method to optimize the planned trajectory and obtain the reference vehicle kinematic states.
3. Designing a kinematic controller based on the AMPC control scheme to achieve robust trajectory tracking control.

This paper structure is presented as follows. In Section 2, we establish the kinematic model as well as the trajectory-tracking-error model for the ATVs. Section 3 presents the trajectory planner for ATVs based on the Hybrid A-star method and the trajectory optimized method based on the minimum snap method. Section 4 describes the two-layer trajectory-tracking controller consisting of a forward control method and an adaptive model predictive method. We discuss the simulation results of the proposed trajectory planner and the control framework in Section 5. The conclusion of this work is presented in Section 6.

2. Autonomous Articulated Vehicle System

The geometry of the articulated tracked vehicle is shown in Figure 1. The ATVs comprise two vehicles connected by articulating mechanisms and hydraulic steering actuators. Changing the articulation angle allows the ATVs to perform steering maneuvers. Additionally, the front and rear vehicle's tracks are controlled to maintain the longitudinal speed. The motion control of ATVs is intended to guide the front and rear units of the ATVs to the reference trajectory determined by the trajectory planner. The reference trajectory is defined as $[x_r(t), y_r(t), \theta_r, \psi_r]$, where $[x_r(t), y_r(t), \theta_r]$ are the center of the front vehicle's gravity, and ψ_r denotes the orientation angle of the rear unit, respectively. In this work, the reference trajectory, namely, $q_r = [x_r(t), y_r(t), \theta_r(t), \psi_r(t)]^T$, and the derivatives of the reference trajectory are all continuous and bound. The longitudinal velocities of the front and rear vehicles are denoted by v. $\dot{\theta}$ and $\dot{\psi}$ are the yaw rates of the two ATV units, respectively. Then, the state vector $q = [x, y, \theta, \psi]^T$ denotes the ATVs' position and orientation.

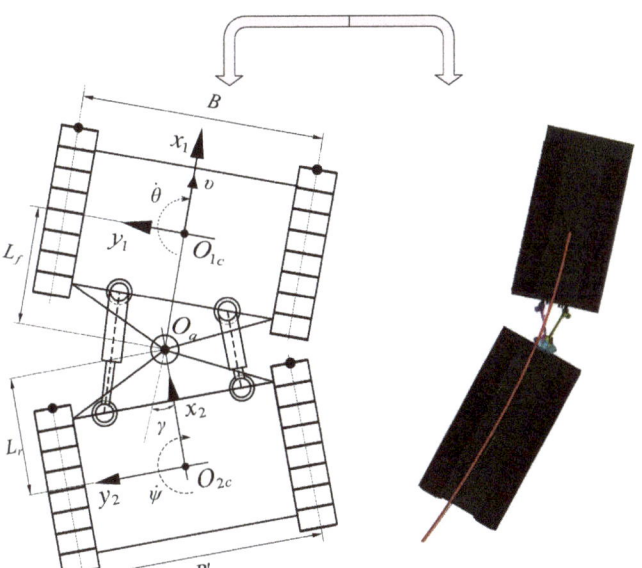

Figure 1. The modeling of an articulated tracked vehicle system in this work is divided into theoretical mathematical modeling and virtual multi-body dynamic modeling based on the real vehicle system. The modeling depicts the steering of the ATVs driven by the hydraulic cylinders, which results in the change in articulation angle γ, the articulation angular rate $\dot{\gamma}$, and the yaw-rate response of the front unit and rear unit $\dot{\theta}$ and $\dot{\psi}$.

In this paper, we consider the path planning and tracking of ATVs in a structured environment, including the boundaries and obstacles described by the rectangle. The problem of achieving a viable path and accurate control can be divided into two stages, trajectory planning, and trajectory-tracking control. Several goals must be fulfilled in the trajectory planning process, such as continuous driving velocity profiles and establishing a feasible path to avoid obstacles. This planner provides a discrete and smooth reference path for the trajectory-tracking controller. Finally, the controller produces accurate velocity and steering angle for the ATVs to travel to the destination safely. Figure 2 shows the scheme of the whole work, including the path planning and the path tracking control. In the control system, the trajectory planning module provides the reference positions $X_r, Y_r, \theta_r, \psi_r$ and the kinematic reference states $v_r, \dot{\theta}_r$ to the trajectory tracking controller as the external disturbance.

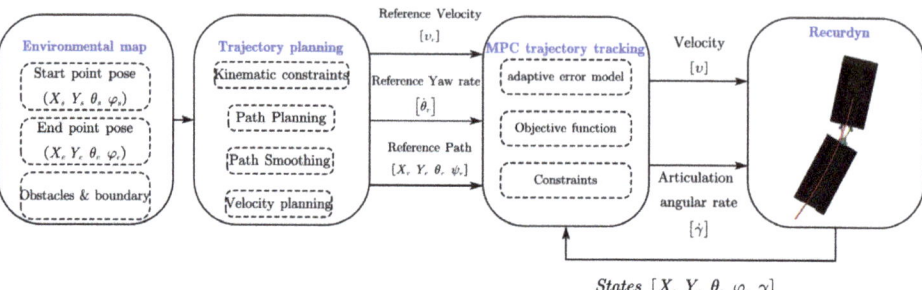

Figure 2. Overall scheme of the path planning and tracking modules for the articulated tracked vehicle system.

2.1. Kinematic Vehicle Models

In this section, the developed kinematic model is used to capture the main feature of the kinematics of the ATVs [15], which can be expressed as follows:

$$\begin{aligned}
\dot{x} &= \bar{v} \cos \theta \\
\dot{y} &= \bar{v} \sin \theta \\
\dot{\theta} &= \frac{\bar{v} \sin \gamma + L_r \dot{\gamma}}{L_f \cos \gamma + L_r} \\
\dot{\psi} &= \frac{\bar{v} \sin \gamma - L_f \dot{\gamma}}{L_f \cos \gamma + L_r}
\end{aligned} \quad (1)$$

where the variables x and y denote the coordinate of the geometry center of the front unit of the ATVs; θ and ψ denote the orientation angle of two parts of ATVs, respectively. γ and $\dot{\gamma}$ denote the articulate angle and the articulate angle rate. The difference between the orientation angle of two units is defined as the articulation angle γ. To maintain the safety of the ATVs in the steering process, the steering control action, namely articulation angle and articulation angle rate, should be less than the maximum value.

2.2. Tracking Error Dynamics Model

To obtain the tracking error between the vehicle and the reference trajectory, we define the variable of tracking error:

$$q_e = q - q_r \quad (2)$$

where q denotes the position and orientation of the ATVs, and q_r indicates the position and direction of the reference point on the desired path. Both q and q_r are expressed in the earth-fixed frame. The tracking error q_e should, however, be expressed in the vehicle-fixed frame to benefit from the computation of the kinematic controller. As a result, we use an orthogonal rotation matrix to translate the vehicle motion from the earth-fixed frame to a vehicle-fixed frame. Based on the kinematic parameter of ATVs, the transformation can be expressed as follows:

$$q_e = \begin{bmatrix} e_x \\ e_y \\ e_\theta \\ e_\psi \end{bmatrix} = \begin{bmatrix} \cos \theta_r & \sin \theta_r & 0 & 0 \\ -\sin \theta_r & \cos \theta_r & 0 & 0 \\ 0 & 0 & 1 & 0 \\ 0 & 0 & 0 & 1 \end{bmatrix} \begin{bmatrix} x - x_r \\ y - y_r \\ \theta - \theta_r \\ \psi - \psi_r \end{bmatrix} \quad (3)$$

where e_x and e_y denote the position distance deviation projecting on the longitudinal and lateral directions, respectively. e_θ and e_ψ denote the orientation error of two units, respectively. This work presents a kinematic controller to propel the ATVs to follow a predefined or planned trajectory while maintaining the vehicle states within physical limits. In terms of the position deviation and orientation deviation, we propose the following differential equations as follows:

$$\begin{cases} \dot{e}_x = \dot{\theta} e_y + v\cos e_\theta - \varphi_r \\ \dot{e}_y = -\dot{\theta} e_x - v\sin e_\theta \\ \dot{e}_\theta = \dot{\theta} - \dot{\theta}_r \\ \dot{e}_\psi = \dot{\psi} - \dot{\psi}_r \end{cases} \quad (4)$$

2.3. Kinematic LPV Modelling

To construct the linear parameter varying (LPV) tracking error model, we define a scheduling variable including the reference velocity and yaw-rate, expressed as $\rho(k) := [v_r, \dot{\theta}_r]$. In the LPV model of the tracking error system, the state, control, and output variables are defined as follows:

$$x = \begin{bmatrix} e_x \\ e_y \\ e_\theta \\ e_\psi \\ \gamma \end{bmatrix} \quad u = \begin{bmatrix} v \\ \dot{\gamma} \end{bmatrix} \quad y = \begin{bmatrix} e_x \\ e_y \\ e_\theta \\ e_\psi \end{bmatrix} \quad r = \begin{bmatrix} v_r \\ \dot{\theta}_r \end{bmatrix} \quad (5)$$

Then the tracking error model is transformed into the formulation of the LPV model:

$$\dot{x} = A(\rho(k))x + B_u(\rho(k))u + B_r r \quad (6)$$

where

$$A(\rho(k)) = \begin{bmatrix} 0 & \Xi_1 & 0 & 0 & 0 \\ -\Xi_1 & 0 & 0 & 0 & 0 \\ 0 & 0 & 0 & 0 & \Xi_2 \\ 0 & 0 & 0 & 0 & 0 \end{bmatrix}$$

$$B_u(\rho(k)) = \begin{bmatrix} 1 & e_y \frac{L_r}{D} \\ \sin e_\theta & -e_x \frac{L_r}{D} \\ 0 & \frac{L_r}{D} \\ 0 & -\frac{L_f}{D} \\ 0 & 1 \end{bmatrix} \quad B_r = \begin{bmatrix} 1 & 0 \\ 0 & 0 \\ 0 & -1 \\ 0 & -1 \\ 0 & 0 \end{bmatrix}$$

where $\Xi_1 = \frac{v_r \sin \gamma_r}{D}$ and $\Xi_2 = \frac{v_r}{D}$, $D = L_f + L_r$. From this LPV formulation of the tracking-error model, a polytopic representation for the error model can be expressed as

$$x(k+1) = (\sum_{i=1}^{2^{r_c}} \mu_i A_i)x(k) + (\sum_{i=1}^{2^{r_c}} \mu_i B_i)u + B_r r \quad (7)$$

where r_c is the number of scheduling component in $\rho(k)$. The matrix A_i and B_i denotes the each polytopic vertex of the matrix $A(\rho(k))$, $B_u(\rho(k))$, defined by the extreme realization of scheduling components in $\rho(k)$. The weighting coefficient μ_i is used to comprehensively describe the scheduling variables $\rho(k)$, which therefore determines the realization of $A(\rho(k))$, $B_u(\rho(k))$ in the control system. Using the available information of scheduling variable $\rho(k)$, we utilize the Takagi-Sugeno fuzzy method to obtain the state-space matrices of the LPV formulation [29]. The membership function μ_i can be expressed as follows:

$$\begin{cases} M_{v_r} = \frac{\bar{v}_r - \hat{v}}{\bar{v}_r - \underline{v}_r} \quad M_{\dot{\theta}_r} = \frac{\bar{\dot{\theta}}_r - \hat{\dot{\theta}}}{\bar{\dot{\theta}}_r - \underline{\dot{\theta}}_r} \\ \mu_1(v, \dot{\theta}) = M_{v_r} M_{\dot{\theta}_r} \\ \mu_2(v, \dot{\theta}) = M_{v_r}(1 - M_{\dot{\theta}_r}) \\ \mu_3(v, \dot{\theta}) = (1 - M_{v_r}) M_{\dot{\theta}_r} \\ \mu_4(v, \dot{\theta}) = (1 - M_{v_r})(1 - M_{\dot{\theta}_r}) \end{cases} \quad (8)$$

where $\bar{v}_r$ and $\underline{v}_r$ denote the upper bound and lower bound of the longitudinal speed of reference. $\bar{\dot{\theta}}_r$ and $\underline{\dot{\theta}}_r$ denote the upper bound and lower bound of the yaw rates of the reference. $\hat{v}_r$ denotes the measured longitudinal speeds of the vehicle; $\hat{\dot{\theta}}$ denotes the measured yaw rate of the front unit of the ATVs.

3. Trajectory Planning

This section defines the path planner of ATVs in two stages, namely path searching and trajectory optimization, as shown in Figure 3. The environment map is divided by the grids at first. In the simulation, the ATVs are shrunk into two coupled rectangles moving on a two-dimensional map. Obstacles, $\Lambda_i (i = 1, 2, \dots, n)$, are marked with a cyan color that the ATVs are not permitted to cross. The simulation has predefined the start point $(X, Y, \theta)_S$ and goal point $(X, Y, \theta)_T$. For the ATVs to reach the goal point and avoid obstacles, a Hybrid A-star algorithm is proposed to generate a feasible kinematic trajectory. The planned trajectory is optimized with the minimum snap algorithm for smoothness and continuous acceleration. By interpolating the kinematic states concerning the time, the optimal trajectory could be expressed by polynomial functions. It will check whether its velocity and acceleration meet the physical limits. At the final step, the reference path is summarized in terms of a series of points attached with the kinematic states of ATVs $(x_r, y_r, \theta_r, \psi_r)$.

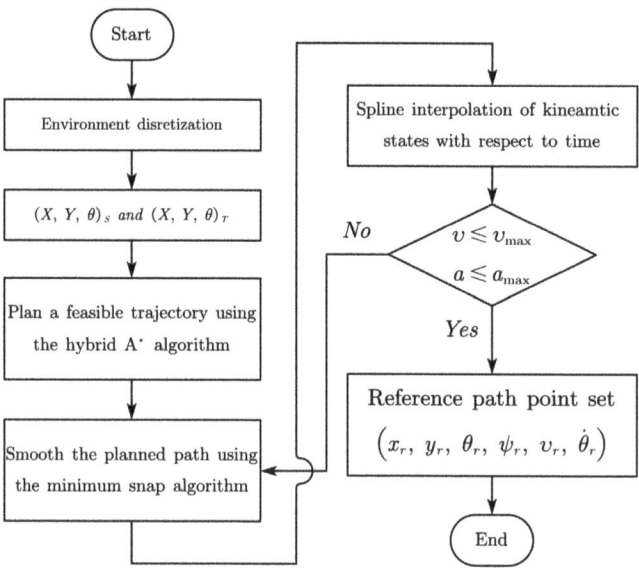

Figure 3. Flow chart of the trajectory planning algorithm of ATVs.

3.1. Node Expansion

Path search is used to construct a collision-free route for the ATVs from the initial position S to the goal position T. In this stage, we propose the Hybrid A-star algorithm, which expands path nodes in the map's continuous space to maintain the kinematic feasibility of the planned path. The continuous moving state is defined as (x, y, θ, γ), which refers to the position, orientation, and steering angle of the moving object. t is the expanded node, and its cost can be evaluated by the function $f(t)$. Based on the lowest value of the evaluation function $f(t)$, the Hybrid A-star algorithm expands the path node. It updates the evaluation function of sub-nodes associated with the current node t until all nodes have been traversed or the current node t is near the goal node T. Using the function $g(t)$, we could obtain the travel cost from the start node S to the current node t. In the meantime, a heuristic function $h(t)$ can predict the heuristic cost between the current position and the goal. The overall cost can be calculated using $f(t)$ as follows:

$$f(t) = g(t) + C_h h(t) \tag{9}$$

where C_h denotes the weighting coefficient of the heuristic cost. The list L denotes the collection of nodes for the next expansion, and the list C refers to the expanded node collection. The actual cost considers the moving distance from S to t. It includes additional penalties for the steering angle and steering angle increments to avoid unreasonable steering maneuvers. The actual cost $g(t)$ for the current node can be expressed as follows

$$g(t) = C_{distance} + \gamma * C_{steer} + (\gamma_{prev} - \gamma) * C_{steer\_diff} + C_{prev}. \tag{10}$$

$C_{distance}$ denotes the distance from t to its parent node; C_{prev} denotes the actual cost for the parent node. γ denotes the articulation steering angle for the current node while γ_{prev} denotes the steering action for the parent node. C_{steer} and C_{steer_diff} denote the weighting coefficients for the steering input and the steering input change.

3.2. Heuristics Cost

As shown in Equation (10), the actual cost from S to current node t has been evaluated by the function $g(t)$. Since the minimum value of $g(t)$ represents the optimal trajectory and the corresponding steering action, the steering action to the goal can be derived from the heuristic function $h(t)$. The classical Hybrid A-star algorithm usually adopts two kinds of heuristics. The first heuristic function is the nonholonomic-without-obstacles, which only considers the kinematics of a moving object without considering the obstacles. The second approach is holonomic-with-obstacles, which ignores the kinematics of moving objects and produces a trajectory with the minimum Euclidean distance between the current position and the destination in the presence of obstacles. To obtain the optimal guidance for the current node, the algorithm adopts the maximum value of the two heuristics as the heuristic function $h(t)$ for the current node. To expand the node from t to T, the first heuristics use the Reeds-Shepp (RS) curve, which takes into account the ATVs' kinematics. Using the grid A-star cost map as the second heuristic function enables the ATV to avoid inefficient path searches by providing the map's information to the vehicle. The maximum of both heuristics has been calculated. The value of function $h(t)$ will be determined by the maximum of the two heuristics.

3.3. Analytical Expansion

The discretized control actions achieve the expansion of the node during the forward node search. While using the optimal steering action, the moving object may not be able to precisely reach the predefined position and orientation of the goal node T. As a result, analytic expansions based on the RS curve would be used to guide the node search near the goal. The RS curve would also be checked for collision with the obstacles along the way. When the analytical expansion is employed, the algorithms perform an analysis that looks for an RS curve to the goal node T before conducting the node search from the node t. The node search would be terminated if the algorithm could find a collision-free path to the destination based on the RS curve. The RS curve would be augmented to the searched path, and the overall path planning from t to T is accomplished.

3.4. Trajectory Optimized

In the stage of the node expansion, the discretized steering angles would select the steering angle from a set of twenty steering angles from -35 degrees to 35 degrees. This would result in the unnatural swerve steering of the articulated steering vehicle by non-continuous steering actions. Moreover, the Hybrid A-star algorithm aims to produce the shortest path, causing the planned route would be very close to the obstacles. Therefore, the planned courses need to be improved regarding smoothness and safety.

Polynomial equations are often used to describe the trajectory of mobile transportation using the fifth-order and seventh-order polynomials. The polynomial trajectory is expressed by the n-order polynomial as follows:

$$
\begin{aligned}
p(t) &= p_0 + p_1 \times t + p_2 \times t^2 + \cdots + p_n \times t^n \\
&= \sum_{i=0}^{n} p^i \times t^i
\end{aligned}
\tag{11}
$$

where $p_0, p_1, \ldots, p_n$ are trajectory parameters. As the single polynomial cannot describe the complex trajectory, the entire trajectory could be divided into k-segment polynomials, and each segment is allocated with a certain time step as follows:

$$
p(t) = \begin{cases}
[1, t, t^2, \cdots, t^n] \cdot p_1 & t_0 \leq t \leq t_1 \\
[1, t, t^2, \cdots, t^n] \cdot p_2 & t_1 \leq t \leq t_2 \\
\cdots \\
[1, t, t^2, \cdots, t^n] \cdot p_k & t_{k-1} \leq t \leq t_k
\end{cases}
\tag{12}
$$

where k denotes the number of segments, and $p_i = [p_{i0}, p_{i1}, \ldots, p_{in}]^T$ denotes the polynomial coefficients of the ith segment. Then, the trajectory optimization process can be transformed into an optimization problem for obtaining the feasible coefficient $p_1, p_2, \ldots, p_k$ that minimizes the integration of the square of the fourth derivative of position, namely, snap. The optimal problem needs to consider the constraints, such as the continuity at the junction of adjacent segments and the limits on the velocity and acceleration. In all, this optimal problem has been formulated as a constrained quadratic problem with equality constraints and inequality constraints in this work as follows:

$$
\begin{cases}
\min & \mathcal{J}(p) = p^T Q p \\
\text{s.t.} & A_{eq} q = b_{eq} \\
& A_{ineq} q \leq b_{ineq}
\end{cases}
\tag{13}
$$

where the matrices Q, A_{eq} and A_{ineq} are functions of the time allocation $\delta_t \triangleq [\delta_{t_1}, \ldots, \delta_{t_k}]$. The equality constraint limits the states of movement, including the position, velocity, and acceleration within the segments, and ensures the continuity between the segments. Meanwhile, the inequality constraints form a trajectory corridor, which keeps a distance from the obstacles. To assign the appropriate time for each trajectory segment, the Euclidean distance between way-points is used to allocate time proportionally.

4. Control Design

In this section, we describe the design of the ATV trajectory tracking control scheme. We have divided the control framework into two parts. The first part deals with the longitudinal speed control of ATVs by adjusting the rotating speeds of the tracks. The second one is the steering motion of the ATVs by adjusting the articulation angular rate.

The control system produces feedback control actions by using an adaptive MPC algorithm. This system regulates the states of the tracking deviation system through feedback control. An error model for the tracking system has been proposed in Equation (6). To predict the system state, the MPC systems need future information about the planned trajectory to analyze the evolution of the tracking error over time. The path planning module could provide the control system with the possible disturbance on the ATVs. The path planner determines the disturbance vector r by the longitudinal velocity and yaw rate. u is a feedback control action aimed at minimizing the tracking error in the presence of disturbance.

4.1. Reference Trajectory

A path-planning method has been proposed in Section 3 to derive a smooth trajectory for ATVs. The reference states x, y, θ, ψ will be used to determine the ATVs' kinematic control. Based on the planned trajectory, we obtain the reference states v_r and $\dot{\theta}_r$. Planner subsystems formulate time-based reference trajectories by evaluating a given reference trajectory $(x_r^t, y_r^t, \theta_r, \psi_r)$ and its derivatives. The reference speeds v_r and the reference yaw rates $\dot{\theta}_r$ could be calculated as follows

$$\begin{aligned} v_r &= \sqrt{(\dot{x}_r^t)^2 + (\dot{y}_r^t)^2} \\ \dot{\theta}_r &= \frac{\ddot{y}_r^t \dot{x}_r^t - \ddot{x}_r^t \dot{y}_r^t}{(\dot{x}_r^t)^2 + (\dot{y}_r^t)^2} \end{aligned} \tag{14}$$

4.2. Adaptive MPC Controller

The path tracking of ATVs is controlled using the adaptive MPC algorithm, which adopts the LPV formulation instead of the non-linear model of the tracking error. Based on the information about scheduling variables, the LPV model in Equation (6) updates its state-space models. As defined in Section 3, $\rho = [v_r\ \dot{\theta}_r]$ is used as the scheduling variable based on the reference path given by the path planner. Using the Equations (7) and (8), the system matrices A_k and B_{uk} can be calculated at any instant k. Throughout the prediction horizon, the adaptive MPC controller could accurately predict the evolution of vehicle states. Thus, the LPV tracking error model in Equation (6) enables the MPC controller to balance computing complexity and efficiency.

To obtain a feasible control action u, the MPC controller must solve a quadratic optimization problem at each instant k. The values of past states x_k, the past control action u_{k-1} and the disturbance vector r_k are available to predict the system states x_{k+1} based on the related matrix coefficients A_k and B_{uk}. The optimization problem can be formulated as follows:

$$\min_{\Delta U} J_k = \sum_{i=1}^{H_p - 1} (x_{k+i}^T Q x_{k+i} + \Delta u_k^T R \Delta u_k)$$

s.t.

$$\begin{aligned} x_{k+i+1} &= x_{k+i} + (A_k x_{k+i} + B_{uk} u_{k+i} + B_r r_{k+i}) dt \\ u_{k+i} &= u_{k+i-1} + \Delta u_{k+i} \\ \Delta U &\in \Delta \Pi \\ U &\in \Pi \\ \bar{x} &= \hat{x}_k \end{aligned} \tag{15}$$

where $x = [e_x\ e_y\ e_\theta\ e_\psi\ \gamma]^T$ is the state vector for the nominal system; x_k denotes the estimated state vector at the time instant k. $r = [v_r\ \dot{\theta}_r]^T$ is the disturbance vector given by the proposed path planner. H_p denotes the prediction horizon. $Q \in \mathcal{R}^{5 \times 5}$ and $R \in \mathcal{R}^{2 \times 2}$ are semi-positive diagonal weighting coefficients for the state and control action increments to obtain a convex cost function, respectively. $u = [v\ \dot{\gamma}]$ denotes the vector of the control action. Δu denotes the vector of the increment of the control actions. U and ΔU denote the sequence of the control actions and their increments through the prediction horizon. Π and $\Delta\Pi$ denote the physical limit on the articulation angular rate and its increments, respectively.

If the optimization problem in Equation (15) could be successfully solved, then a sequence of control input increments can be obtained as $\Delta U = [\Delta u_k \; \Delta u_{k+1} \; \ldots \; \Delta u_{t+H_p-1}]^T$. Based on the past control action u_{k-1}, the control action at the current time instant k is the summation of the past control action u_{k-1} and the control increment Δu_k from the sequence ΔU as follows:

$$u_k = u_{k-1} + \Delta u_k \tag{16}$$

4.3. Track-Speed Control

Based on the control action u_k obtained by Equation (16), the longitudinal control of ATVs is performed by adjusting the tracks of both front and rear unit of ATVs. To achieve the desired longitudinal speed and avoid the excessive track slip, the speeds difference of different tracks, $[v^r_{fl}, v^r_{fr}, v^r_{rl}, v^r_{rr}]$ can be obtained by the calculation method developed in [15,30]:

$$\begin{aligned}
v^r_{fl} &= v + \frac{v\sin\gamma}{2(L_f + L_r\cos\gamma)} - \frac{v\dot\gamma B}{2(L_f + L_r\cos\gamma)} \\
v^r_{fr} &= v - \frac{v\sin\gamma}{2(L_f + L_r\cos\gamma)} + \frac{v\dot\gamma B}{2(L_f + L_r\cos\gamma)} \\
v^r_{rl} &= v\cos\gamma + \frac{vL_r\sin\gamma^2}{2(L_f + L_r\cos\gamma)} + \frac{vB\sin\gamma}{2(L_f + L_r\cos\gamma)} + \frac{\dot\gamma L_r(2L_f\sin\gamma - B\cos\gamma)}{2(L_f + L_r\cos\gamma)} \\
v^r_{rr} &= v\cos\gamma + \frac{vL_r\sin\gamma^2}{2(L_f + L_r\cos\gamma)} - \frac{vB\sin\gamma}{2(L_f + L_r\cos\gamma)} + \frac{\dot\gamma L_r(2L_f\sin\gamma + B\cos\gamma)}{2(L_f + L_r\cos\gamma)}
\end{aligned} \tag{17}$$

where γ and $\dot\gamma$ denote the articulation angle and articulation angular rate. B denotes the width of the front unit of ATVs. v^r_{fl} and v^r_{fr} denote the left and right tracks longitudinal speeds of the front vehicle, and v^r_{rl} and v^r_{rr} denote the linear speeds of tracks of the rear vehicle. The velocities of four tracks of the articulated vehicle are obtained from Equation (17), which ensures that the longitudinal speed of the front unit of ATVs is equal to the desired longitudinal velocity given by the adaptive MPC controller. It is worth noting that the track speeds given by Equation (17) could not deal with the lateral slippage of the tracks well. Therefore, the lateral motion of the vehicle may occur in the steering maneuver that causes the lateral tracking error.

5. Simulation and Discussion

5.1. Simulation Setup

This section evaluates the ATVs' path planning and path tracking algorithms on a simulation platform by MATLAB/Simulink and Recurdyn. Multi-body dynamics software Recurdyn features a high-speed track module. The virtual ATV model in Recurdyn is shown in Figure 4. The parameters of this virtual model are based on reality. This virtual model uses parameters derived from reality. The proposed AMPC controller is evaluated for its tracking performance on this virtual model. The simulation platform's structure is illustrated in Figure, whose parameters are given in Table 1.

We conducted the simulation on the MATLAB 2022a platform to verify the proposed path planning method. The algorithm is implemented in Matlab programming language. We designed two maps with different obstacles. The ATV model was configured with two rectangles, whose sizes are 2.5 × 2 m and 2 × 2 m. The driving speed of the ATVs was set to 2.5 m/s. The parameters of the Hybrid A-star planner and the trajectory optimization are listed in Table 2. The proposed method (Method 2) was compared to the original Hybrid A-star method (Method 1). Method 1 also considers the kinematic characteristics of ATVs and implements the node search by the discretized steer angles. At the same time, the Reed-Shepp

(RS) curve is not adopted by Method 1 in the path search. Moreover, the heuristic function of Method 1 is the Euclidean distance from the current node and the destination.

Figure 4. The virtual model of the articulated tracked vehicle constructed on the multi-body dynamics software Recurdyn.

The simulation results of the proposed AMPC algorithm were compared with the fuzzy control and MPC control published in [27,31,32]. The performance of the path-tracking controller could be mainly determined by the lateral position error and the orientation angle error of the front unit with respect to the reference path.

Four conditions were considered in the co-simulation of the path-tracking controllers. In the first condition, the ATV was controlled to track an arc of 25 m radius with a longitudinal velocity of 0.56 m/s. The second condition was to follow a path consisting of three circles with 30 m, 20 m, and 40 m radius, respectively. The longitudinal speed was set as 3 m/s. In the third condition, the reference path is a mixed path with three straight lines and two arcs of 20 m radius. The vehicle was set to move at a speed of 4 m/s. Finally, the fourth condition was to track a path generated by the Hybrid A-star planning method proposed in Section 3. The proposed AMPC controller was compared with the standard—MPC controller in this condition. In the above simulations, the parameters of the proposed controller were fixed and presented in Table 3.

Table 1. Kinematic parameters of the articulated tracked vehicle model.

Symbol	Description	Value	Unit
B	Width of ATVs	2.1	[m]
D	Length of ATVs	4.8	[m]
L_f	Distance from the hitch point to front unit	2.6	[m]
L_f	Distance from the hitch point to rear unit	2.2	[m]
γ	Articulation angle	[−0.75, 0.75]	[rad]
$\dot{\gamma}$	Articulation angular rate	[−0.18, 0.18]	[rad/s]
v	Vehicle longitudinal speed	[−1, 4]	[m/s]

Table 2. Parameters of the trajectory planning and optimization.

Description	Value	Unit
Minimum turn radius	10.4	[m]
Maximum velocity	5	[m/s]
Maximum acceleration	2	[m/s^2]
Maximum steering angle	0.5	[rad]
Maximum steering rate	0.15	[rad/s]
Grid resolution in distance	2	[m]
Grid resolution in yaw angle	15	[degree]
Motion step size	1	[m]
Number of steering angle candidate	20	
Steer angle change weighting coefficient	2	
Steer angle weighting coefficient	1	
Heuristic weighting coefficient	2	

Table 3. Parameters of the trajectory tracking AMPC controller.

Symbol	Description	Value
T_s	The sample time of controller	0.2 [s]
H_p	Length of the prediction horizon	10
H_c	Length of the control horizon	5
Q	Weighting coefficient for states	diag(0.5 0.5 1 0.1 0)
R	Weighting coefficient for control input	diag(0.1 0.2)
P	Terminal cost coefficient	diag(0.1 0.1 1 1 0)

5.2. Simulation of Path Planning

The comparisons of the two simulation results are illustrated in Figure 5a,b. The solid line denotes the results of Method 1, and the dash-dot line indicates the proposed Method 2. As shown in Figure 5a, both path planning methods could generate the path to the goal. A significant improvement can be observed as the path generated by Method 1 contained non-smooth segments, while the planned path of Method 2 is much smoother and contains fewer switches of turning direction. Moreover, in Method 1, the planning path is close to the obstacles when the ATVs try to turn between the obstacles, as shown in Figure 5b. On the contrary, the path by Method 2 could be in the middle of the obstacles to avoid unnecessary turns when crossing the corridor between the obstacles.

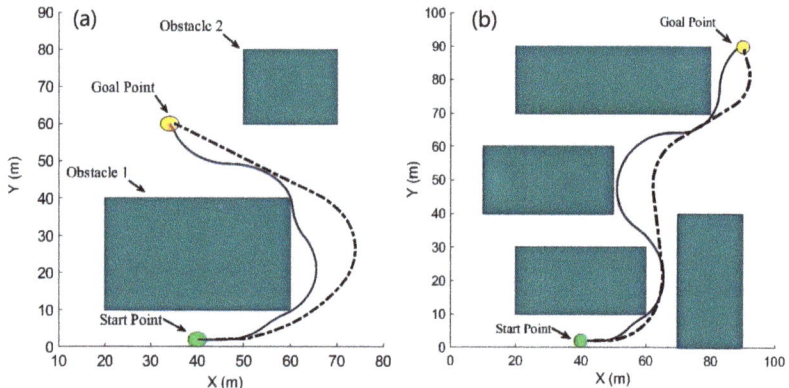

Figure 5. Original Hybrid A-star (solid line) and the proposed Hybrid A-star (dash-dot line) path planning in the simulation. (**a**) Map A; (**b**) map B.

Table 4 summarizes the comparison between Method 1 and Method 2 regarding the maximum curvature, path length, the number of steering direction changes, and execution time. The higher values indicate higher steering instability, a longer driving path, and a higher computation cost to find the goal. From the comparison, Method 2 generates a path with more minor curvatures than Method 1. Moreover, the path length of Method 2 is longer in map A, but shorter in map B, although the difference between the two methods is not significant (88.46 versus 96.06 and 128.94 versus 116.09, respectively). In both conditions, the proposed method's computation time are both longer than Method 1 because the candidate RS curves must be computed and checked for the collision in Method 2. Whereas the benefit of the RS curve is the much less number of the steering change in Method 2, as the RS curve could simplify the process of the path forward search.

Table 4. Comparison of original method and the proposed method.

Map	Method 1				Method 2			
	Curvature [1]	Number [2]	Length [3]	Time [4]	Curvature	Number	Length	Time
Map A	0.095	4	88.46	15.5	0.059	1	96.06	23.3
Map B	0.098	5	128.94	26.4	0.096	2	116.09	40.6

[1] **Curvature** denotes the maximum curvature of the planned path. [2] **Number** denotes the number of the steering direction change. [3] **Length** denotes the overall length of the planned path. [4] **Time** denotes the computation time of the planning method to obtain the planned trajectory.

5.3. Simulation of the Trajectory Tracking

5.3.1. Simulation Result of Case 1

In Case 1, the ATV is controlled to follow the curved path with a radius of 25 m, and the ATV is assumed to be positioned at the initial point. In the previous research [27], the fuzzy control system was utilized to guide the ATV to follow an arc. The simulation result of the fuzzy control and the proposed AMPC control have been presented in Figures 6–8. The maximum lateral error of the AMPC controller is almost the same as that of the fuzzy-PID controller, as shown in Figure 7. Moreover, the AMPC controller achieves a minor orientation deviation compared to the fuzzy-PID controller, as shown in Figure 8. The maximum orientation error in the AMPC controller has been reduced by 81.84% compared to the fuzzy-PID method. Thus, the trajectory generated by the AMPC controller is closer to the predefined path than the fuzzy-PID method, as shown in Figure 6.

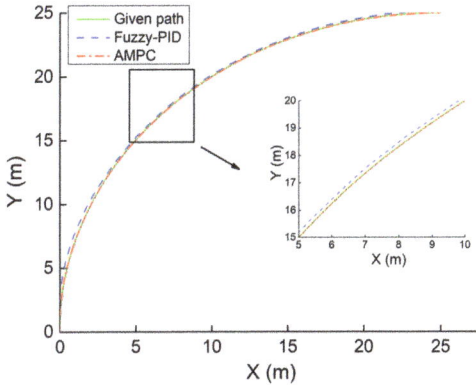

Figure 6. The trajectory of the AMPC controller and the fuzzy-PID controller in Case 1.

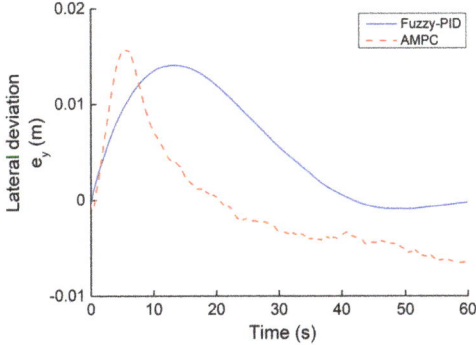

Figure 7. The lateral position error of the AMPC controller and the fuzzy-PID controller in Case 1.

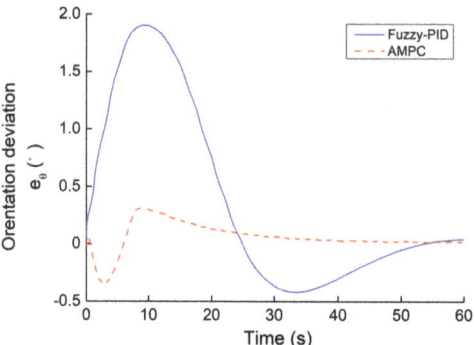

Figure 8. The orientation angle error of the AMPC controller and the fuzzy-PID controller in Case 1.

5.3.2. Simulation Result of Case 2

Case 2 consists of three circle paths to test the tracking performance of the ATV for the continuous steering mode. In the previous research [31], the standard-MPC controller was proposed for path tracking of autonomous articulated vehicles. The path-tracking error model of standard-MPC is based on lateral displacement, orientation, and curvature errors. The curvature error is dedicated to the circular path with a constant radius. As our work has not included the curvature error in the path-tracking model, we have compared the lateral position error and orientation error of the AMPC method with that of the standard-MPC.

The reference path and the trajectory produced by the standard-MPC and the AMPC are presented in Figure 9. The trajectory of the AMPC is closer to the defined path than the standard-MPC. The position and orientation errors produced by the standard-MPC and the AMPC have been illustrated in Figures 10 and 11. The AMPC method has achieved better performance of the lateral position error than that of the standard-MPC. The maximum position error of the standard-MPC is 2 m, while the maximum position error of the AMPC is 0.67 m. In addition, the position error response of the AMPC converges to zero at the final, while the standard-MPC retains a significant position error. The standard-MPC produces a more minor orientation angle error than the AMPC controller. The maximum orientation errors of the standard-MPC and the AMPC are 0.002 rad and 0.067 rad, respectively. The reason may be due to the unavoidable skid of the articulated tracked vehicles compared to the articulated vehicle with tires. Nevertheless, the orientation error caused by the AMPC is acceptable for the ATV in practical operation.

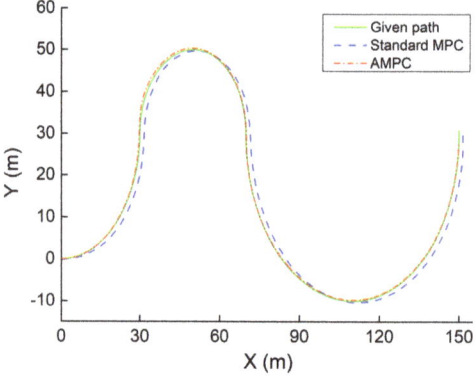

Figure 9. The trajectory of the AMPC controller and the standard-MPC controller in Case 2.

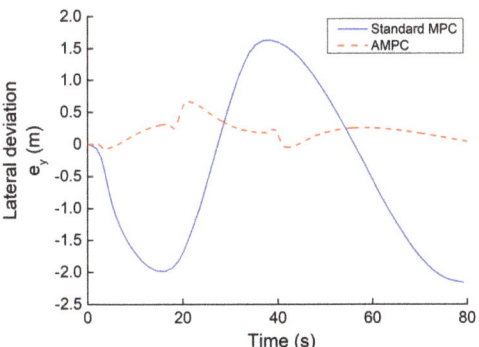

Figure 10. The lateral position error of the AMPC controller and the standard-MPC controller in Case 2.

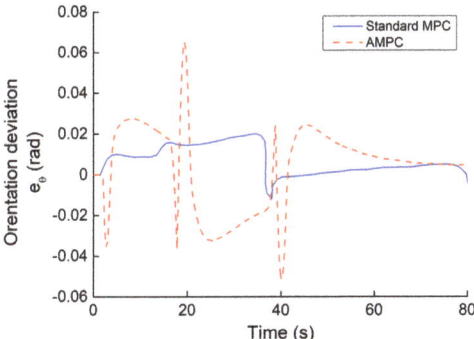

Figure 11. The orientation angle error of the AMPC controller and the standard-MPC controller in Case 2.

5.3.3. Simulation Result of Case 3

Case 3 consists of straight lines and arcs with a radius of 20 m. The previous research in [32] compared the tracking performance between the switch-MPC and the nonlinear MPC (NMPC) on the path tracking of the articulated mining vehicle with tires. Although the articulated vehicle of the research [32] differs from the ATV, the results in [32] have significant value and are worthy of reference.

Figure 12 illustrates the articulation angle response of three controllers. The maximum articulation angle response of the switch-MPC, the NMPC, and the AMPC controller is 0.5589 rad, 0.3940 rad, and 0.272 rad, respectively. Both the articulation angle response of the NMPC and the AMPC controllers exhibit less overshoot and change smoothly compared to the switch-MPC controller. Figure 13 illustrates the lateral position errors of the controllers. The maximum position errors by the switch-MPC, NMPC, and AMPC controller reached 0.7217 m, 0.0874 m, and 0.192 m, respectively. The ultimate position error generated by the AMPC controller was reduced by 73.4 % compared to the switch-MPC controller. Figure 14 presents the orientation errors of all controllers. The maximum orientation errors are 0.1458 rad, 0.0461 rad, and 0.0392 rad for the switch-MPC controller, the NMPC controller, and the AMPC controller, respectively. The maximum orientation error of the AMPC has been reduced by 73.11% compared to the switch-MPC controller and by 14.97 % with respect to the NMPC controller.

According to the research [32], the maximum computation times for the NMPC controller and the switch-MPC controller are 0.014 s, and 0.04 s, respectively. As the AMPC control system is constructed in the Matlab/Simulink, the profile report of the Simulink run-time indicates that the proposed controllers have been invoked 200 times during the whole simulation time of 40 s. The overall computation time of the controller is 0.399 s. The average computation time of the AMPC controller is approximately 0.002 s at each time step. The average computation time of the AMPC is much less than that of both the switch-MPC controller and the NMPC controller.

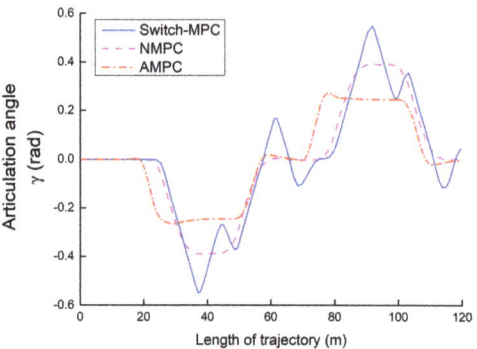

Figure 12. The articulation angle of the AMPC controller, the NMPC controller, and the switch-MPC controller in Case 3.

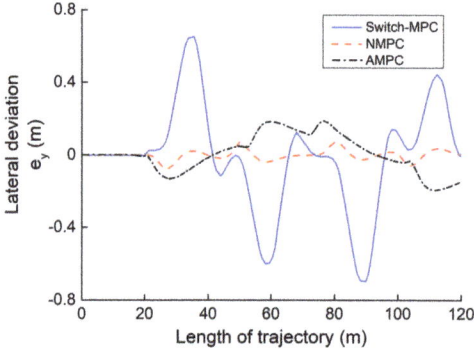

Figure 13. The lateral position error of the AMPC controller, the NMPC controller, and the switch-MPC controller in Case 3.

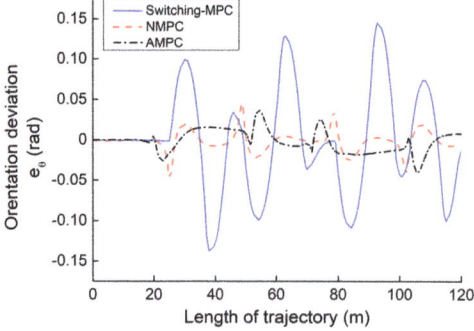

Figure 14. The orientation angle error of the AMPC controller, the NMPC controller, and the switch-MPC controller in Case 3.

5.3.4. Simulation Result of Case 4

To evaluate the reliability of the AMPC algorithm to track a non-uniform trajectory given by the proposed path planner, we conducted a simulation where the ATV is controlled to follow the non-uniform trajectory with varying curvature.

The simulation results are presented in Figures 15–19. The vehicle is set to move at 3 m/s. As shown in Figure 15, both the standard-MPC and the AMPC controllers could drive the ATV to follow the given path, while the AMPC controller causes less offset from the reference path compared to the standard-MPC controller. Figure 16 illustrates the articulation angle of both controllers. The ultimate articulation angle of the standard-MPC and the AMPC controller reached 0.611 rad and 0.532 rad, respectively. The maximum articulation angle of the standard-MPC is more than the limit on the articulation angle. Moreover, the articulation angle response of the AMPC controller presents less overshoot compared to the standard-MPC controller. Figure 17 illustrates the articulation angle rate of both controllers. The response of the AMPC controller exhibits more drastic changes when compared to the standard-MPC controller. At the same time, the AMPC controller outputs the articulation angle rate in advance to resist the disturbance, which could reduce the tracking error of the ATV. Figure 18 depicts the lateral tracking errors of the AMPC and the standard-MPC controller. The tracking error of the AMPC controller maintains much less than that of the MPC controller through the tracking process. Moreover, the maximum lateral tracking error of the ATV approached 0.832 m and 3.015 m by the AMPC and the standard-MPC controller, respectively. The number of maximum position errors by the AMPC controller was reduced by 72.4% compared to the standard-MPC controller. Figure 19 illustrates the orientation angle error of both controllers. The orientation error of the AMPC controller is also less than the standard-MPC controller along the time. The maximum orientation error of the AMPC controller and the standard-MPC controller is 0.125 rad and 0.269 rad, respectively. The maximum value of the orientation angle error by the AMPC controller was reduced by 53.53% compared to the standard-MPC controller.

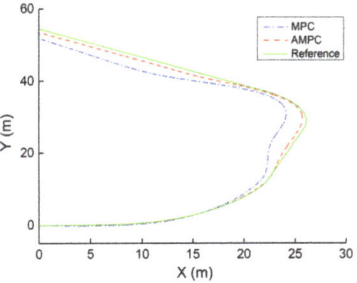

Figure 15. The trajectory of the AMPC controller and the standard-MPC controller in Case 4.

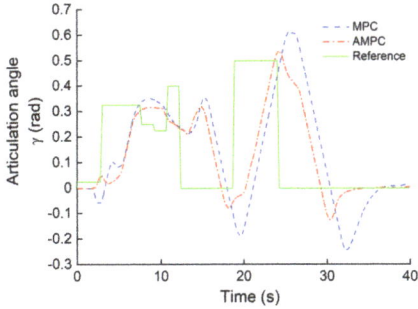

Figure 16. The articulation angles of the AMPC controller and the standard-MPC controller in Case 4.

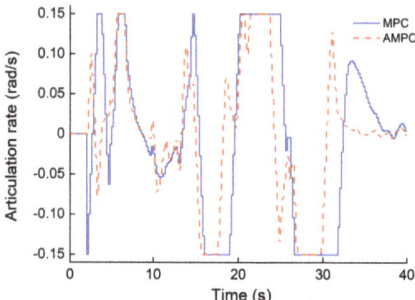

Figure 17. The articulation angle rates of the AMPC controller and the standard-MPC controller in Case 4.

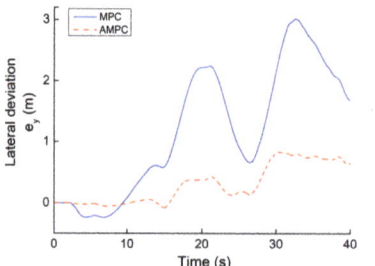

Figure 18. The lateral position errors of the AMPC controller and the standard-MPC controller in Case 4.

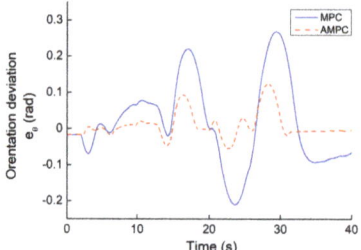

Figure 19. The orientation angle errors of the AMPC controller and the standard-MPC controller in Case 4.

6. Conclusions

To enhance the ability of ATVs to drive in a complex environment, we apply the Hybrid A-star method to plan a safe trajectory. Moreover, the planned path was optimized to ensure smoothness and continuity. Comparing the proposed path planner with the original Hybrid A-star method shows that the planner could generate a feasible trajectory with minimum steering direction change. Numerous studies have applied the MPC algorithm and verified its effectiveness for path-tracking control of the articulated vehicle. To achieve the trajectory tracking of the articulated tracked vehicle to follow the planned path, we propose an adaptive model predictive control (AMPC) control method that is based on the time-varying tracking error system. We obtain the following results and conclusions by comparing the AMPC controller with the previously developed fuzzy and MPC controller.

Firstly, the AMPC controller could rapidly track the reference path compared to the fuzzy-PID controller. The AMPC controller also achieves a minor orientation angle error. Secondly, the AMPC controller could achieve more minor tracking errors than the standard-

MPC method. Thirdly, the tracking accuracy of the AMPC method is still inferior to the NMPC method, while the AMPC method has the advantage of computation efficiency.

From the above analysis, the main contributions of this work could be summarized as follows:

1. Although the adaptive model predictive control algorithm has been applied in the path-tracking of the mobile robot, its application in the articulation vehicle is not mature. The MPC algorithm has yet to be applied in the path-tracking control of the articulated tracked vehicle. Thus, our work has extended the application of the MPC algorithm in the field of ATVs.
2. The ATVs have unique steering characteristics compared to the skid-steering tracked vehicles. The path tracking of the ATVs also needs to consider its kinematic characteristics, for example, the multi-input and multi-output for the ATV control system. Thus, it is challenging for the developed control methods to control the ATV in a complex maneuver accurately. To this end, our work provides a practical method for the path planning and path tracking of ATVs.
3. The simulation of several path-tracking cases has demonstrated that the standard-MPC controller cannot accurately control the ATV to follow a path with varying curvature. However, the proposed AMPC controller outperforms the standard-MPC controller, while the AMPC controller can achieve the same level of tracking performance compared to the nonlinear MPC controller.

However, the proposed Hybrid A-star planning method has the drawback of extensive computation time, which could be improved by the refined algorithm structure in further research. Moreover, the proposed AMPC method is applied in the kinematic control of the ATV, which could not deal with the high-speed driving condition. In a further study, we will focus on the dynamic control of ATVs and apply the AMPC method in the dynamic control of ATVs.

Author Contributions: The contributions of the K.H. incude the conceptualization, methodology, data curation and analysis; writing—original draft; writing—review and editing. The contributions of the K.C. include the supervision, project administration, and funding acquisition. All authors have read and agreed to the published version of the manuscript.

Funding: This research was funded by the National Key Research and Development Program of China (Grant 2016YFC0802703).

Data Availability Statement: The codes for this paper have been uploaded to the GitHub since the data of 24/03/2020. It is accessible for any visitors to this repository in GitHub https://github.com/HuKangle/Path_planing-and-Path_tracking-for-an-articulated-tracked-vehicle.

Conflicts of Interest: The authors declare no conflict of interest.

Abbreviations

The following abbreviations are used in this manuscript:

ASV	Articulated steering vehicle
ATV	Articulated tracked vehicle
AMPC	Adaptive model predictive control
NMPC	Nonlinear model predictive control
MPC	Model predictive control
RS	Reeds Shepp

References

1. Lopatka, M.J.; Rubiec, A. Concept and preliminary simulations of a driver-aid system for transport tasks of articulated vehicles with a hydrostatic steering system. *Appl. Sci.* **2020**, *10*, 5747. [CrossRef]
2. Watanabe, K.; Kitano, M. Study of Steerability of Articulated Tracked Vehicles—Part 1. Theoretical and Experimental Analysis. *J. Terramech.* **1986**, *23*, 69–83. [CrossRef]

3. Marshall, J.; Barfoott, T.; Larsson, J. Autonomous Underground Tramming for Center-Articulated Vehicles. *J. Field Robot.* **2008**, *25*, 400–421. [CrossRef]
4. Alshaer, B.J.; Darabseh, T.T.; Alhanouti, M.A. Path planning, modeling and simulation of an autonomous articulated heavy construction machine performing a loading cycle. *Appl. Math. Model.* **2013**, *37*, 5315–5325. [CrossRef]
5. Qing, G.; Li, L.; Bai, G. Longitudinal and Lateral Trajectory Planning for the Typical Duty Cycle of Autonomous Load Haul Dump. *IEEE Access* **2019**, *7*, 126679–126695.
6. Xu, T.; Xu, Y.; Wang, D.; Feng, L.H. Path Planning for Autonomous Articulated Vehicle Based on Improved Goal-Directed Rapid-Exploring Random Tree. *Math. Probl. Eng.* **2020**, 1–14. [CrossRef]
7. Sabiha, A.D.; Kamel, M.A.; Said, E.; Hussein, W.M. Trajectory Generation and Tracking Control of an Autonomous Vehicle Based on Artificial Potential Field and Optimized Backstepping Controller. In Proceedings of the 2020 12th International Conference on Electrical Engineering (ICEENG), Cairo, Egypt, 7–9 July 2020.
8. Chien, V.D.; Heungju, A.; Doo, S.L.; Sang, C.L. Improved Analytic Expansions in Hybrid A-Star Path Planning for Non-Holonomic Robots. *Appl. Sci.* **2022**, *12*, 5999.
9. Xu, Y.; Liu, R. Path planning for mobile articulated robots based on the improved A * algorithm. *Int. J. Adv. Robot. Syst* **2017** *14*, 1–10. [CrossRef]
10. Nayl, T.; Nikolakopoulos, G.; Gustafsson, T. A full error dynamics switching modeling and control scheme for an articulated vehicle. *Int. J. Control Autom. Syst.* **2015**, *13*, 1221–1232. [CrossRef]
11. Kayacan, E.; Ramon, H.; Saeys, W. Robust trajectory tracking error model-based predictive control for unmanned ground vehicles. *IEEE-ASME Trans. Mechatron.* **2015**, *21*, 806–814. [CrossRef]
12. Ridley, P.; Corke, P. Load haul dump vehicle kinematics and control. *J. Dyn. Syst. Meas. Control-Trans. ASME* **2003**, *125*, 54–59. [CrossRef]
13. Barbosa, F.M.; Marcos, L.B.; Grassi, V. Robust path-following control for articulated heavy-duty vehicles. *Control Eng. Practice* **2019**, *85*, 246–256. [CrossRef]
14. Oreh, S.T.; Kazemi, R.; Azadi, S. A new desired articulation angle for directional control of articulated vehicles. *Proc. Inst. Mech. Eng. Part K J.-Multi-Body Dyn.* **2012**, *226*, 298–314. [CrossRef]
15. Nayl, T.; Nikolakopoulos, G.; Gustafsson, T. Design and experimental evaluation of a novel sliding mode controller for an articulated vehicle. *Robot. Auton. Syst.* **2018**, *103*, 213–221. [CrossRef]
16. Astolfi, A.; Bolzern, P.; Locatelli, A. Path-tracking of a tractor-trailer vehicle along rectilinear and circular paths: A Lyapunov-based approach. *IEEE Trans. Robot. Autom.* **2004**, *20*, 154–160. [CrossRef]
17. Sampei, M.; Tamura, T.; Kobayashi, T.; Shibui, N. Arbitrary path tracking control of articulated vehicles using nonlinear control theory. *IEEE Trans. Control Syst. Technol.* **1995**, *3*, 125–131. [CrossRef]
18. Ruslan, N.A.I.; Amer, N.H.; Hudha, K.; Kadir, Z.A.; Ishak, S.A.F.M.; Dardin, S.M.F.S. Modelling and control strategies in path tracking control for autonomous tracked vehicles: A review of state of the art and challenges. *J. Terramech.* **2023**, *105*, 67–79. [CrossRef]
19. Nayl, T.; Nikolakopoulos, G.; Gustafsson, T. Effect of kinematic parameters on MPC based on-line motion planning for an articulated vehicle. *Robot. Auton. Syst.* **2015**, *70*, 16–24. [CrossRef]
20. Kayacan, E.; Saeys, W.; Ramon, H.; Belta, C.; Peschel, J.M. Experimental Validation of Linear and Nonlinear MPC on an Articulated Unmanned Ground Vehicle. *IEEE/ASME Trans. Mechatron.* **2018**, *23*, 2023–2030. [CrossRef]
21. Kayacan, E.; Kayacan, E.; Ramon, H.; Saeys, W. Robust tube-based decentralized nonlinear model predictive control of an autonomous tractor-trailer system. *IEEE/ASME Trans. Mechatron.* **2015**, *20*, 447–456. [CrossRef]
22. Shi, J.; Sun, D.; Qin, D.; Hu, M.; Kan, Y.; Ma, K.; Chen, R. Planning the trajectory of an autonomous wheel loader and tracking its trajectory via adaptive model predictive control. *Robot. Auton. Syst.* **2020**, *131*, 103570. [CrossRef]
23. Dong, C.; Cheng, K.; Hu, W.; Yao, Y. Dynamic modelling of the steering performance of an articulated tracked vehicle using shear stress analysis of the soil. *Proc. Inst. Mech. Eng. Part D-J. Automob. Eng.* **2016**, *231*, 653–683. [CrossRef]
24. Dong, C.; Cheng, K.; Hu, K.; Hu, W. Dynamic modeling study on the slope steering performance of articulated tracked vehicles. *Adv. Mech. Eng.* **2017**, *9*, 1–26. [CrossRef]
25. Wu, J.; Wang, G.Q.; Zhao, H.; Sun, K. Study on electromechanical performance of steering of the electric articulated tracked vehicles. *J. Mech. Sci. Technol.* **2019**, *33*, 3171–3185. [CrossRef]
26. Hu, K.; Cheng, K. Dynamic modelling and stability analysis of the articulated tracked vehicle considering transient track-terrain interaction. *J. Mech. Sci. Technol.* **2021**, *35*, 1343–1356. [CrossRef]
27. Cui, D.; Wang, G.Q.; Zhao, H.; Wang, S. Research on a path-tracking control system for articulated tracked vehicle. *Strojniski Vestn.-J. Mech. Eng.* **2020**, *66*, 311–324. [CrossRef]
28. Tota, A.; Galvagno, E.; Velardocchia, M. Analytical study on the cornering behavior of an articulated tracked vehicle. *Machines* **2021**, *9*, 38. [CrossRef]
29. Alcala, E.; Puig, V.; Quevedo, J. TS-MPC for Autonomous Vehicles Including a TS-MHE-UIO Estimator. *IEEE Trans. Veh. Technol.* **2019**, *68*, 6403–6413. [CrossRef]
30. DeSantis, R.M. Path-tracking for a tractor-trailer-like robot: Communication. *Int. J. Robot. Res.* **1994**, *13*, 533–544. [CrossRef]

31. Dou, F.; Huang, Y.; Liu, L.; Wang, H.; Meng, Y. Path planning and tracking for autonomous mining articulated vehicles. *Int. J. Heavy Vehicle Systems* **2019**, *26*, 315–333. [CrossRef]
32. Bai, G.; Liu, L.; Meng, Y.; Luo, W.; Gu, Q.; Ma, B. Path Tracking of Mining Vehicles Based on Nonlinear Model Predictive Control. *Appl. Sci.* **2019**, *9*, 1372. [CrossRef]

Disclaimer/Publisher's Note: The statements, opinions and data contained in all publications are solely those of the individual author(s) and contributor(s) and not of MDPI and/or the editor(s). MDPI and/or the editor(s) disclaim responsibility for any injury to people or property resulting from any ideas, methods, instructions or products referred to in the content.

Article

GIS-Data-Driven Efficient and Safe Path Planning for Autonomous Ships in Maritime Transportation

Xiao Hu [1], Kai Hu [1], Datian Tao [2], Yi Zhong [2,*] and Yi Han [2,*]

1. China Ship Development and Design Center, Wuhan 430064, China
2. School of Information Engineering, Wuhan University of Technology, Wuhan 430070, China; taodatian@whut.edu.cn
* Correspondence: zhongyi@whut.edu.cn (Y.Z.); hanyi@whut.edu.cn (Y.H.); Tel.: +86-27-8785-8005 (Y.Z.)

Abstract: Maritime transportation is vital to the global economy. With the increased operating and labor costs of maritime transportation, autonomous shipping has attracted much attention in both industry and academia. Autonomous shipping can not only reduce the marine accidents caused by human factors but also save labor costs. Path planning is one of the key technologies to enable the autonomy of ships. However, mainstream ship path planning focuses on searching for the shortest path and controlling the vehicle in order to track it. Such path planning methods may lead to a dynamically infeasible trajectory that fails to avoid obstacles or reduces fuel efficiency. This paper presents a data-driven, efficient, and safe path planning (ESP) method that considers ship dynamics to provide a real-time optimal trajectory generation. The optimization objectives include fuel consumption and trajectory smoothness. Furthermore, ESP is capable of fast replanning when encountering obstacles. ESP consists of three components: (1) A path search method that finds an optimal search path with the minimum number of sharp turns from the geographic data collected by the geographic information system (GIS); (2) a minimum-snap trajectory optimization formulation with dynamic ship constraints to provide a smooth and collision-free trajectory with minimal fuel consumption; (3) a local trajectory replanner based on B-spline to avoid unexpected obstacles in real time. We evaluate the performance of ESP by data-driven simulations. The geographical data have been collected and updated from GIS. The results show that ESP can plan a global trajectory with safety, minimal turning points, and minimal fuel consumption based on the maritime information provided by nautical charts. With the long-range perception of onboard radars, the ship can avoid unexpected obstacles in real time on the planned global course.

Keywords: kinematics; improved A* algorithm; path planning; GIS

Citation: Hu, X.; Hu, K.; Tao, D.; Zhong, Y.; Han, Y. GIS-Data-Driven Efficient and Safe Path Planning for Autonomous Ships in Maritime Transportation. *Electronics* **2023**, *12*, 2206. https://doi.org/10.3390/electronics12102206

Academic Editor: Felipe Jiménez

Received: 25 March 2023
Revised: 8 May 2023
Accepted: 10 May 2023
Published: 12 May 2023

Copyright: © 2023 by the authors. Licensee MDPI, Basel, Switzerland. This article is an open access article distributed under the terms and conditions of the Creative Commons Attribution (CC BY) license (https://creativecommons.org/licenses/by/4.0/).

1. Introduction

With the rapid development of the global economy, according to the survey of the Baltic and International Maritime Council/International Chamber Shipping (BIMCO/ICS), the maritime industry has accounted for 80% of the world's trade and transportation [1]. Thus, the safety and efficiency of maritime transportation are of paramount importance. Current issues of maritime transportation include: (1) about 75–96% marine vessel accidents being caused by humans; (2) a severe shortage of seafarers and management personnel; (3) more than 80% of shipping costs being from fuel and labor [2]. Autonomous navigation is vital to mitigate the above issues for ships in that it can be more vigilant than humans at avoiding accidents by perceptions from heterogeneous sensors such as a camera, laser scanner, and mmWave radar [3]. Ship autonomy not only saves human labor costs but also utilizes intelligent path planning methods to achieve optimized fuel consumptions.

Realizing autonomous ships requires localization and path planning. Currently, the global positioning system (GPS) and compass have been commonly available to provide reliable location services. In contrast, path planning for ships still poses several challenges.

Existing works mainly focus on the path planning for lightweight surface vessels, which are agile and applicable to harbor patrol, marine resource exploration, etc. [4]. Ships, however, exhibit high inertia and thus a significant delay in motion control. When encountering sudden situations, e.g., encountering a large iceberg, its dynamic nature prevents agile avoidance [5]. In addition to the "shortest path", dynamical feasibility is crucial for ship path planning [6].

Existing ship path planners are typically optimal path searching methods based on A* and its modifications [7]. Although they provide optimal paths in a given map [8], the optimality is limited in ideal maps including the occupancy grid map, Voronoi-visibility roadmap [9], risk contour map [10], etc. Their path searching is conducted by discretized heading directions without considering the ships' dynamical constraints, making the planned path dynamically infeasible. More recent works [11,12] have taken the dynamical constraints into consideration. However, their iterative methods have high computational complexities, failing to plan in real time to avoid expected sudden risks. Researchers also proposed hybrid approaches that fuse the artificial potential field (APF) algorithm with velocity odometry and path optimization [13–15] to achieve real-time obstacle avoidance in complex maritime environments. However, the APF causes oscillations when searching for paths through narrow areas, causing frequent turning and increasing fuel consumptions and navigation risks. In addition to optimization-based methods, researchers incorporate reinforcement learning into path planning [16,17]. However, learning-based methods suffer from a trade-off between generality and accuracy. Their stochastic results cannot guarantee the safety and efficiency of ship navigation. In summary, none of the existing path planning methods meet the safety and efficiency needs when considering ship dynamics.

This paper presents ESP, a combinatorial optimized path planning approach that generates a safe, smooth, and dynamically feasible trajectory while minimizing the shipping cost. Realizing such an elegant approach poses several challenges: (1) to quantify the turning cost in optimal and dynamically feasible path searching; (2) to minimize the shipping cost in terms of fuel by formulating a minimum-snap problem, which is non-trivial in combining the dynamic model of ships; (3) to cope with sudden risks, e.g., avoid expected obstacles or enemy vessels, which requires replanning a smooth, safe, feasible and optimized path in real time.

To address the above challenges, ESP consists of three components. First, we propose A-turning, a path searching algorithm that quantifies the turning cost in order to obtain the optimal path with fewer turns. Then, we formulate the minimum-snap optimization problem subject to the dynamic constraints of ships to achieve the minimum shipping cost in terms of fuel. Finally, we propose a real-time path replanning algorithm using quasi-uniform cubic B-spline, achieving millisecond-level path replanning to cope with sudden risks.

In summary, the contributions of this paper include: (1) quantifying the turning cost and incorporating it into an optimal global path search through a modified A* algorithm; (2) formulating a minimum-snap optimization problem to generate a smooth trajectory that consumes the least fuel and satisfies the ship's dynamic constraints; (3) enabling real-time obstacle avoidance for ships through a B-spline-based local trajectory replanner.

ESP is evaluated in a data-driven simulator implemented by MATLAB and our developed geographic information system (GIS). The simulation results demonstrate the effectiveness of ESP in generating a safe, smooth, and feasible path with minimal turns and fuel consumption. Moreover, ESP enables a quick reaction for ships to smoothly avoid unexpected obstacles by path replanning in less than 48 ms.

The rest of this paper is organized as follows. Section 2 reviews related works. Then, we elaborate on the design of ESP in Section 3. The performance evaluation in Section 4 demonstrates the effectiveness of ESP. Section 5 concludes this paper.

2. Related Works

Path planning can be divided into two steps: path searching and path optimization. Path searching involves searching for an obstacle-free path from the start to the end. Path optimization involves optimizing the searched path to meet users' specific objectives, e.g., the shortest sailing distance, minimum fuel consumption, and minimum shipping cost.

Path searching has been well-studied for decades. It can be categorized by graph-search-based and random-sampling-based path searching approaches. Graph-search-based path planning methods follow a set of steps to generate unique navigation paths. The classic algorithm, Dijkstra, expands a large number of irrelevant nodes during searching, which greatly slows down the searching process. In order to improve the searching efficiency, A*-family algorithms have been proposed. They make the searching process more purposeful to the destination by introducing heuristic functions [18–20]. These heuristic functions treat vessels as a mass point with unlimited turning and sailing speeds. Their results may have large-angle steers between consecutive path segments. However, to the best of our knowledge, the maximum speed of a ship (displacement > 320 t) is 15 knots, the maximum acceleration is 1, and the turning radius is three times the ship length. Simply considering the ship as a mass point leads to infeasible path planning, making the above heuristic solutions impractical. In addition, Yu et al. [21] proposed an A* algorithm with velocity variation and global optimization (A*-VVGO), which achieves the purpose of obstacle avoidance by changing the speed of the ship, and combines the artificial potential field method to ensure the smoothness of the path. Sang et al. [22] proposed a hybrid algorithm of an artificial potential field based on A* and local programming, which is often combined with many algorithms, such as the genetic algorithm (GA) [23], Fuzzy artificial potential field (FAPF) [24], etc. These hybrid algorithms contain various advantages. However, these methods do not consider vehicles' dynamic constraints. Tracking the paths cannot guarantee safety and smoothness. Moreover, graph-search-based methods cannot work efficiently in large environments due to the searching space being exponential to the size of the occupancy grid maps.

To address the searching efficiency problem with respect to the occupancy grid maps, random-sampling-based algorithms have been proposed to incrementally build maps by sampling. They can work in the planning of the ocean. Zhang et al. [25] proposed the adaptive hybrid dynamic step size and target attractive force–RRT (AHDSTAF–RRT), imposing the dynamic constraints of unmanned surface vehicles (USVs) to allow USVs to navigate complex aquatic environments. Webb et al. [26] proposed Kinodynamic RRT*, achieving asymptotically optimal motion planning for robots. However, these approaches suffer from slow convergence and inflexible settings of step size. Thus, Strub et al. [27] designed a heuristic function in the exploitation of random sampling with the aim that the new samples would be more likely to be closer to the destination. Xu et al. [28] proposed a simplified map-based regional sampling RRT* (SMRS–RRT*) algorithm to achieve path planning in complex environments. Dong et al. [29] proposed a path planning method based on improved RRT*–Smart, which optimizes the node sampling method by sampling in the polar coordinate system with the origin of USV, improves the search efficiency, and ensures that the navigation path follows the International Regulation for Preventing Collision at Sea. This design does not only improve the convergence speed but also improves the quality of the solution. Nevertheless, random-sampling-based methods cannot provide optimal solutions. Their results are not unique. The searched path usually contains many sharp turns, which is especially evident in open water.

Based on the path searching from graph-search-based and random-sampling-based methods, researchers tried to generate smooth trajectories. A strawman option is to use interpolation. Liang et al. [30] interpolated the trajectory with the Dobbins curve to ensure the smoothness and reduce the number of sharp turns, but the trajectory curvature was discontinuous. To solve this problem, Candeloro et al. [31] used the Fermat spiral to connect the straight line segment with the curved segment, generating the trajectory with a continuous curvature. Wang et al. [32] used B-spline interpolation to construct smooth trajectories with sparse waypoints. It, however, does not impose the vehicle's dynamic constraints, making trajectories infeasible to be executed. To generate dynamic feasible trajectories, control-space sampling approaches [5,6,33] are simple and effective. However, such approaches lack purpose, so their sampling process could take too much time and fail to plan paths in real time. MahmoudZadeh et al. [34] combined a novel B-spline data frame and the particle swarm optimization (PSO) algorithm to establish a continuous route planning system to achieve route planning for USV ocean sampling missions. Zheng et al. [35] proposed a ship collision avoidance decision method based on improved cultural particle swarm to achieve the steering collision avoidance of a ship, but without considering the speed constraint of the ship.

3. Methods

3.1. Problem Formulation

The obstacles considered in this paper are the static obstacles in the chart and the unexpected static obstacles that appear within the detection range of the ship's radar during the actual navigation of the ship. One primary objective of our path planning is to be collision-free. Additionally, we optimize two more objectives: best stability and minimum fuel consumption. The specific objective function and constraints are given in the following subsections.

3.2. Kinematic Model

The previous studies often ignored the influence of marine environments on the ship's motion state for the ease of modeling. In order to make the planned path fit the actual sailing situation, this paper establishes the kinematics model of ships considering the ocean current.

Figure 1 illustrates the kinematic model of a ship. $O_e X_e Y_e$ denotes the world frame, which refers to the coordinate system with respect to the earth. The earth's gravity points to the positive direction of the z-axis. The x–y–z axes follow the right-hand rule. The origin of the world frame O_e is the geometrical center's initial position, the positive direction of the $O_e X_e$ axis points to east, and the positive direction of $O_e Y_e$ points to north. $O_b X_b Y_b$ denotes the local frame, which refers to the ship's body frame, O_b is used as the center of gravity of the ship, the positive direction of the $O_b X_b$ axis points to the bow, and the positive direction of the $O_b Y_b$ axis points to the port side. Ψ denotes the yaw angle, u the surge velocity, v the sway velocity, and δ the rudder angle. According to Newton's second law, considering surge, sway, and yaw, the force at the center of gravity of the ship is

$$\begin{cases} X_e = m\ddot{x} \\ Y_e = m\ddot{y} \\ N_r = I_Z \ddot{\Psi} \end{cases} \quad (1)$$

$$I_Z = \int_V \left(x^2 + y^2\right) \rho_m dV \quad (2)$$

where X_e denotes the force along the x-axis, Y_e the force along the y-axis, x, y the position of the ship's center of gravity in the world frame, m the mass of the ship, N_r the force along the z-axis, $\ddot{\Psi}$ the angular acceleration, and I_Z the moment of inertia around the z-axis. As shown in Equation (2), it depends on the volume of the ship V and the mass density ρ_m. With the yaw Ψ, we express the transformation between the world frame and the local frame as

$$\begin{bmatrix} X_b \\ Y_b \end{bmatrix} = \begin{bmatrix} \cos\Psi & -\sin\Psi \\ \sin\Psi & \cos\Psi \end{bmatrix} \begin{bmatrix} X_e \\ Y_e \end{bmatrix} \quad (3)$$

Then, the forces on the surge and sway directions can be expressed as

$$\begin{cases} X_b = m(\dot{u} - vr) \\ Y_b = m(\dot{v} + ur) \end{cases} \quad (4)$$

where r denotes the yaw rate, and $\dot{u}$ and $\dot{v}$ denote the acceleration on the surge and sway directions, respectively. From Equations (3) and (4), we obtain the kinematic model as follows.

$$\begin{bmatrix} \dot{x} \\ \dot{y} \\ \dot{\Psi} \end{bmatrix} = \begin{bmatrix} \cos\Psi & -\sin\Psi & 0 \\ -\sin\Psi & \cos\Psi & 0 \\ 0 & 0 & 1 \end{bmatrix} \begin{bmatrix} u \\ v \\ r \end{bmatrix} \quad (5)$$

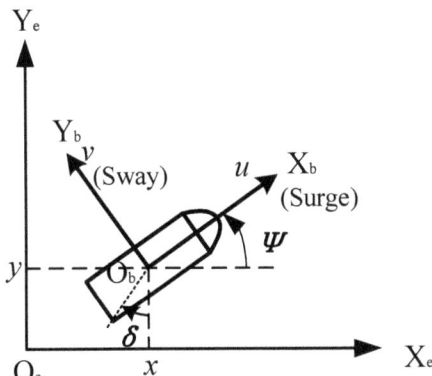

Figure 1. The kinematic model of a ship.

3.3. Dynamic Model

This paper uses the first-order K–T model to represent the hydrodynamic model of the ship, assuming that the port and starboard sides of the ship are symmetrical and the ship mass is uniformly distributed. The hydrodynamic equation can be expressed as:

$$M\dot{v} + Cv + Dv = \tau \quad (6)$$

where $v = [u, v, r]^T$,

$$M = \begin{bmatrix} m_{11} & 0 & 0 \\ 0 & m_{22} & 0 \\ 0 & 0 & m_{33} \end{bmatrix} = \begin{bmatrix} m - X_{\dot{u}} & 0 & 0 \\ 0 & m - Y_{\dot{v}} & 0 \\ 0 & 0 & I_z - N_{\dot{r}} \end{bmatrix} \quad (7)$$

$$C = \begin{bmatrix} 0 & 0 & -(m - Y_{\dot{v}})v \\ 0 & 0 & (m - X_{\dot{u}})u \\ (m - Y_{\dot{v}})v & -(m - X_{\dot{u}})u & 0 \end{bmatrix} \quad (8)$$

$$D = \begin{bmatrix} d_{11} & 0 & 0 \\ 0 & d_{22} & 0 \\ 0 & 0 & d_{33} \end{bmatrix} = \begin{bmatrix} -X_u & 0 & 0 \\ 0 & -Y_v & 0 \\ 0 & 0 & -N_r \end{bmatrix} \quad (9)$$

$$\tau = \tau_E + \tau_r \quad (10)$$

where M denotes the inertial mass matrix, C the Coriolis centripetal force matrix, and D the drag coefficient matrix. X_u and Y_v denote the derivatives for the hydrodynamic, $X_{\dot{u}} = \frac{\partial X}{\partial u}$, $Y_{\dot{v}} = \frac{\partial Y}{\partial v}$, and $N_{\dot{r}} = \frac{\partial N}{\partial r}$. τ_E denotes the force imposed by the environment and τ_r denotes the thrust of the propeller.

3.4. Ocean Circulation Model

Affected by the environment such as sea wind and ocean currents, a ship easily deviates from its course or even capsizes during sailing, resulting in property damage and even casualties. Therefore, we consider the influence of ocean currents on the ship's motion state when planning its path.

Ocean currents are formed when seawater flows in a certain direction at a regular, relatively steady speed. It is a large-scale, aperiodic form of seawater movement. According to the characteristics of its location and time, ocean currents can be divided into uniform currents, non-uniform currents, steady currents, and unsteady currents. In offshore or seabed areas with irregular topography, the model of ocean currents is more complicated. To simplify the modeling, we assume that ocean currents are constant and uniform. Let V_c denote the ocean current speed and Ψ_c the direction of the current. Then, the velocity of the ocean currents can be expressed as

$$v_c = \begin{bmatrix} V_c \cos \Psi_c & V_c \sin \Psi_c \end{bmatrix}^T \tag{11}$$

Affected by ocean currents, the actual velocity of the ship is different from its velocity in still water. At this time $v_r = v - v_c$, where v_r is the velocity of the ship relative to the ocean current.

3.5. Optimization Objectives

When a ship sails along a trajectory, the collision-free cost function is

$$f_c = -\sum_{i=0}^{n} Dis(Obstacle(p_i)) \tag{12}$$

where $Dis(Obstacle(p_i))$ denotes the minimum distance from a waypoint p_i to the obstacles, which can be obtained by the Euclidean signed distance field (ESDF) [36]. The distance will be negative if a waypoint is within an obstacle.

The smoothness is determined by the sum of snaps along the trajectory. The smoothness cost can be defined as

$$f_s = \int_0^T (p^{(4)}(t))^2 dt \tag{13}$$

where $p^{(4)}(t)$ denotes the fourth-order derivative, i.e., jerk, at time t.

The fuel consumption depends on the sailing speed. We use the exponential distribution model proposed in [37] as follows to describe the relationship between fuel consumption and speed.

$$FCPH = 0.128e^{0.243V} \tag{14}$$

where $V = \sqrt{v^2 + u^2}$. Thus, we define the cost function of fuel consumption as

$$f_o = \sum_{i=1}^{n-1} t_i \cdot FCPH_i \tag{15}$$

where t_i denotes the time duration between waypoint p_i and p_{i+1} and $FCPH_i$ the fuel consumption per hour between waypoint p_i and p_{i+1}.

Combining the collision-free cost, the smoothness cost, and the fuel consumption cost, we obtain the overall optimization objective function

$$F = \min\{f_c + f_s + f_o\} \tag{16}$$

subject to

$$|u| \leq u_{max} \tag{17}$$

$$|v| \leq v_{max} \tag{18}$$

$$|r| \leq r_{max} \tag{19}$$

where u_{max}, v_{max}, and r_{max} are the maximum surge velocity, the maximum sway velocity, and the maximum yaw rate, respectively.

3.6. Occupancy Grid Map Construction

In order to apply the environmental information provided by the electronic chart for path planning, it is necessary to process the chart into a binary occupancy grid map as shown in Figure 2. In this process, we first set an appropriate binarization threshold that converts an RGB image into a binary image. Such a threshold [38] is vital to constructing an accurate grid map that ensures the feasibility of path planning. If the threshold is inappropriate, as shown in Figure 2c, then a shoal is identified as a passable area, greatly increasing the navigation risk. Second, the grid size determines the resolution of path planning. Too large a grid cannot capture the subtle details of environments, e.g., small obstacles, resulting in unsafe path searching. On the other hand, too small a grid greatly increases the search space, reducing the computation speed.

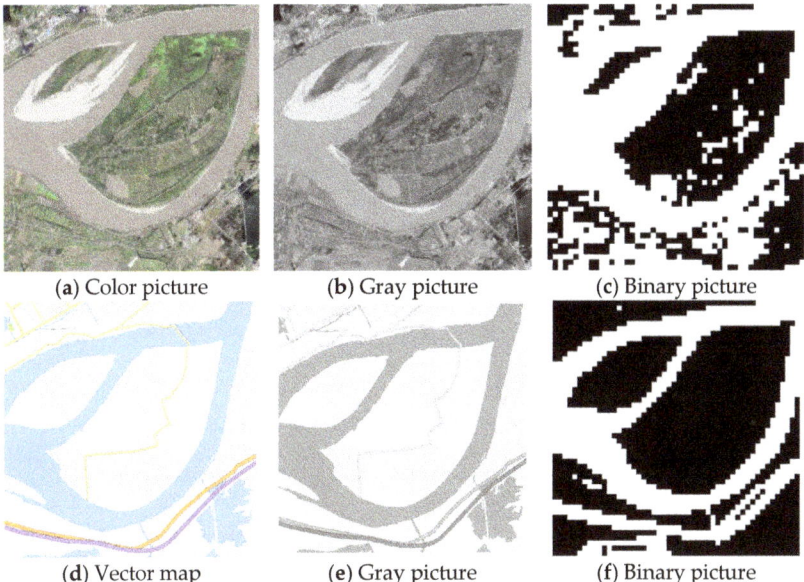

Figure 2. Converting an electronic chart into an occupancy grid map, where the black areas in the grid map represent obstacles and the white areas represent passable areas.

Considering the maneuverability of a ship, this paper chooses the minimum turning radius as the criterion to measure the grid size. The ship's minimum turning radius can be measured through the ship's maneuverability experiment. The turning radius depends on the sailing speed and water flow velocity. According to [39], in our simulation, we set the minimum turning radius of the ship as three times the length of the ship. Finally, we map the geographic chart represented in terms of longitude and latitude to the grid map using the following equation:

$$\begin{cases} lon_{(i,j)} = tlLon + \frac{|brLon-tlLon|}{w} \cdot (i-1) \\ lat_{(i,j)} = brLat + \frac{|tlLat-brLat|}{h} \cdot (j-1) \end{cases} \quad (20)$$

where $lon_{(i,j)}$ and $lat_{(i,j)}$ denote the longitude and latitude of position (i,j). $tlLon$ and $tlLat$ denote the longitude and latitude at the top-left corner of the selected area. $brLon$ and

$brLat$ denote the longitude and latitude at the bottom-right corner of the selected area. w and h denote the width and height of the electronic chart.

In our simulated forward exploration, we use eight discretized directions in the grid map to search for paths as shown in Figure 3.

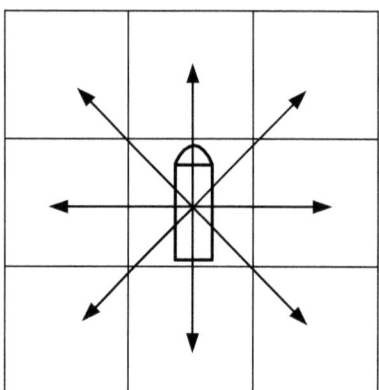

Figure 3. 8 discretized directions in path searching for autonomous ships.

3.7. Quantification of Turning Cost

The A* algorithm uses Equation (18) as the evaluation function to obtain a path with the shortest distance.

$$f(N_i) = g(N_i) + h(N_i) \tag{21}$$

where $f(N_i)$ denotes the estimated cost from the starting point to the target point, $g(N_i)$ the actual cost from the initial node to node N_i, and $h(N_i)$ the estimated cost of the best path from node N_i to the target node.

However, in path searching for ships, the turning is much more difficult than for cars or aerial vehicles. Thus, we have to add the cost of turning in order to better evaluate the path. Here we use the diagonal distance to compute the turning cost as shown in Figure 4. The yaw ϕ_i between waypoint p_i and p_{i+1} can be computed as follows.

$$\phi_i = arctan\left(\left|\frac{p_{i+1}(y) - p_i(y)}{p_{i+1}(x) - p_i(x)}\right|\right) \tag{22}$$

where $p_i(x), p_i(y)$ denote the coordinates of waypoint p_i, $p_{i+1}(x), p_{i+1}(y)$ the coordinates of waypoint p_{i+1}, $p_{i-1}(x), p_{i-1}(y)$ the coordinates of waypoint p_{i-1}. Preventing collisions is still of the highest priority in path planning. Thus, it is not reasonable to simply pursue the minimum turning cost in planning. We add a penalty to the turning cost:

$$c(N_i) = \varepsilon \cdot max(0, \Delta\phi_i - \phi) \tag{23}$$

where ϕ is the penalty threshold of the yaw and ε the penalty coefficient. Empirically, we set $\phi = 30°$ and $\varepsilon = 0.8$. $\Delta\phi_i$ is computed as follows.

$$\Delta\phi_i = \left|arctan\left(\frac{p_{i+1}(y) - p_i(y)}{p_{i+1}(x) - p_i(x)}\right) - arctan\left(\frac{p_i(y) - p_{i-1}(y)}{p_i(x) - p_{i-1}(x)}\right)\right| \tag{24}$$

Finally, the new evaluation function of our path searching algorithm is defined as:

$$f(N_i) = g(N_i) + h(N_i) + c(N_i) \tag{25}$$

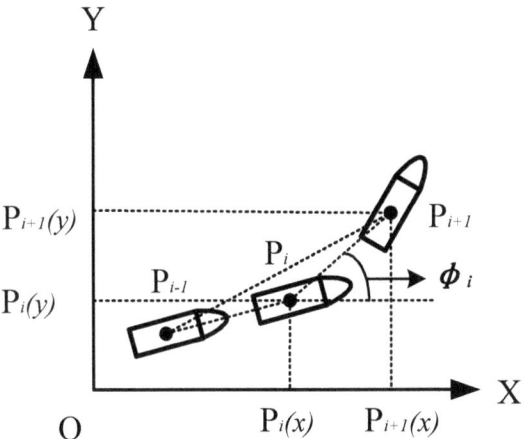

Figure 4. An illustration of how to compute the turning cost.

The improved A* algorithm is shown in Algorithm 1.

Algorithm 1. Improved A* algorithm

Input: Start node X_{start}, end node X_{end}
1: OPEN_list:= X_{start}, where $f(X_{start}) = h(X_{start})$
2: CLOSE_list:={ }
3: **while** OPEN_list is not empty **do**
4:     current node X_n:= the node in the OPEN_list with the lowest $f(X)$
5:     **if** $X_n = X_{end}$ **break**
6:     Remove X_n from OPEN_list and add it to CLOSE_list
7:     **for** each adjacent node, X_i of X_n **do**
8::   **if** $f(X_i) = 0$ || $X_i \in$ CLOSE_list **continue**
9:         **if** $X_i \notin$ OPEN_list
10:             add X_i into OPEN_list
11:             the parent node of X_i=, $X_{i\,>\,parent} = X_n$
12:             calculate $f(X_i), g(X_i), h(X_i)$ and $c(X_i)$
13:         **if** $X_i \in$ OPEN_list
14:             calculate $f(X_i)$ via (25)
15:     Resort and keep OPEN_list sorted by f value
16: $X_p = X_{end}$
17: Path_list:= X_p
18: **while** $X_p \neq X_{start}$ **do**
19:     $X_p = X_{p.parent}$
20:     Path_list = {Path_list, X_p}
Return: Path_list

3.8. Global Trajectory Optimization

The previous path searching gives us a discrete path with the minimum cost. However, the path does not consider the dynamic feasibility with respect to time, velocity, and acceleration. This part requires a further step to optimize the searched path into a smooth trajectory. The trajectory can be defined as a -order polynomial.

$$p(t) = p_0 + p_1 t + p_2 t^2 + \ldots + p_n t^n = \sum_{i=0}^{n} p_i t^i \qquad (26)$$

where $p_0, p_1, \ldots, p_n$ are the coefficients of this trajectory. We denote $P = [p_0, p_1, \ldots, p_n]^T$, and then Equation (23) can be rewritten as

$$p = \begin{bmatrix} 1, t, t^2, \ldots, t^n \end{bmatrix} \cdot P \tag{27}$$

Then, Equation (14) can be expressed as

$$\int_0^T \left(p^{(4)}(t)\right)^2 dt$$

$$= \sum_{i=1}^k \int_{t_{i-1}}^{t_i} \left(p^{(4)}(t)\right)^2 dt$$

$$= \sum_{i=1}^k \int_{t_{i-1}}^{t_i} \left(\left[0,0,0,0,24,\ldots,\frac{n!}{(n-4)!}t^{n-4}\right] \cdot p\right)^T \left[0,0,0,0,24,\ldots,\frac{n!}{(n-4)!}t^{n-4}\right] \cdot p dt$$

$$= \sum_{i=1}^k p^T \int_{t_{i-1}}^{t_i} \left[0,0,0,0,24,\ldots,\frac{n!}{(n-4)!}t^{n-4}\right]^T \left[0,0,0,0,24,\ldots,\frac{n!}{(n-4)!}t^{n-4}\right] dt \cdot p \tag{28}$$

Let

$$Q_i = \int_{t_{i-1}}^{t_i} \left[0,0,0,0,24,\ldots,\frac{n!}{(n-4)!}t^{n-4}\right]^T \left[0,0,0,0,24,\ldots,\frac{n!}{(n-4)!}t^{n-4}\right] dt \tag{29}$$

We have

$$\int_0^T \left(p^{(4)}(t)\right)^2 dt = \sum_{i=1}^k p^T Q_i p \tag{30}$$

However, the polynomial expression cannot explicitly control the shape of the trajectory. To gain better control, we choose the Bezier curve using Bernstein polynomials. The k-th segment of the trajectory can be expressed as

$$B_k(t) = \sum_{i=0}^n c_k^i b_n^k(t) \tag{31}$$

where $b_n^k(t) = \binom{n}{k} \cdot t^i \cdot (1-t)^{n-i}$, $t \in [0,1]$ c_k^i denotes the control point of the k-th segment of the Bezier curve.

Since the trajectory must pass through the first and last control points, it can satisfy the positional constraints of the initial and final states. In addition, based on the hodograph of the Bezier curve, we impose constraints on the velocity and acceleration of the trajectory, ensuring the multi-order continuity of the trajectory.

3.9. Real-Time Obstacle Avoidance

In a static chart, a ship can navigate safely along the aforementioned global trajectory. However, the marine environment is complex and changeable. Ships need to deal with unexpected risks when sailing, e.g., avoiding islands and reefs. As shown in Figure 5, if the ship maintains the planned global trajectory, it will collide with a temporary obstacle. To address this issue, we perform local path planning based on the B-spline curve. The advantage of the B-spline trajectory is that it can change the curve locally by adjusting few control points, while any control point of a Bezier curve will change the shape of the whole trajectory. Moreover, it guarantees that the locally replanned trajectory still satisfies the ship's kinematic and dynamic constraints. This not only achieves the goal of real-time obstacle avoidance, but also satisfies all optimization objectives.

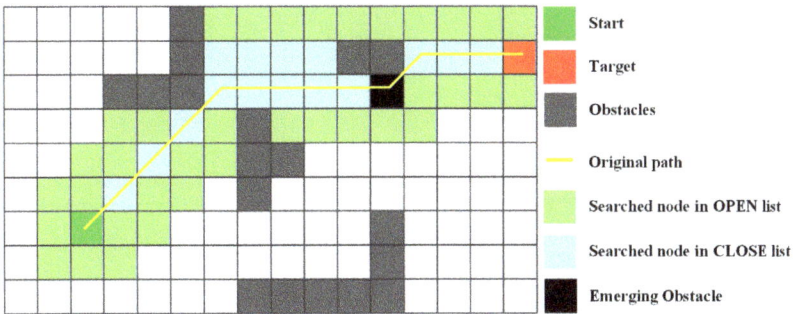

Figure 5. A ship can hit an expected obstacle by sailing along the generated global trajectory.

A B-spline can be expressed as

$$C_p(u) = \sum_{i=0}^{n} N_{i,p}(u) P_i \tag{32}$$

where P_i denotes the i-th control point and $N_{i,p}(u)$ is the B-spline basis function of degree p. $u = [u_0, u_1, \ldots, u_m]$ is the knot vector. Typically, a three-degree B-spline can ensure the smoothness of accelerations. Thus, we have

$$P_{(0,3)} = \frac{1}{6} \begin{bmatrix} 1 & t & t^2 & t^3 \end{bmatrix} \begin{bmatrix} 1 & 4 & 1 & 0 \\ -3 & 0 & 3 & 0 \\ 3 & -6 & 3 & 0 \\ -1 & 3 & -3 & 1 \end{bmatrix} \begin{bmatrix} P_0 \\ P_1 \\ P_2 \\ P_3 \end{bmatrix} \tag{33}$$

Figure 6 illustrates the collision avoidance algorithm. In this figure, $P_0 = [x_0, y_0]$ and $P_3 = [x_3, y_3]$ are the start and the end of the local planning. The gray area $ABCD$ represents an obstacle. Ψ_i denotes the yaw at position P_i. d_i denotes the distance between P_{i-1} and P_i. The geometrical relationship among these positions can be expressed as

$$\begin{cases} x_1 = x_0 + d_1 \cos \Psi_0 \\ y_1 = y_0 + d_1 \sin \Psi_0 \end{cases} \tag{34}$$

$$\begin{cases} x_2 = x_3 - d_3 \cos \Psi_2 \\ y_2 = y_3 - d_3 \sin \Psi_2 \end{cases} \tag{35}$$

First, based on random sampling [40] and collision detection [41], we obtain d_1, d_3 and Ψ_2 ($\Psi_0 = 0$), and then use the geometric relations in Equations (34) and (35) to solve for the position of $P_1 = [x_1, y_1]$ and $P_2 = [x_2, y_2]$. At last, the locally replanned B-spline can be generated by the control points P_0, P_1, P_2, and P_3.

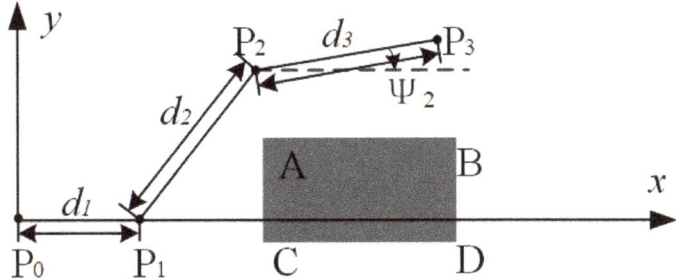

Figure 6. An illustration of the collision avoidance algorithm.

4. Simulation and Results

We conduct the evaluation of ESP by implementing a data-driven simulator in MAT-LAB. All simulation experiments are run on a quad-core 2.40 GHz Intel i5-1135G7 processor and 16 GB RAM. We input the data from the database of our developed GIS. The data include image and vector maps, longitude and latitude coordinates, and ship route information. During the simulation, we use environmental data such as shoals and whirlpools that are not currently marked in the GIS database to evaluate the effectiveness of ESP. The simulated settings are listed in Table 1. Specifically, the ship length is 30 m, the maximum turning radius is 3 times the ship length. The maximum velocity is 7.7 m/s. The maximum acceleration and jerk are 1 m/s^2 and 10 m/s^3 [42], respectively.

Table 1. Simulation parameters.

Parameter	Value
Length of the ship	30 m
Maximum velocity	7.7 m/s
Maximum acceleration	1 m/s^2
Maximum jerk	10 m/s^3

We simulate ESP in two scenes using the GCJ-02 coordinate system. In both scenes, the velocities and accelerations at the start and the end are 0. The simulated occupancy grid map is of size 50 × 50. We compare the path searching results of ESP with the A* algorithm as shown in Figure 7. The results show that ESP effectively reduces the number of turning points and the planned path is safe.

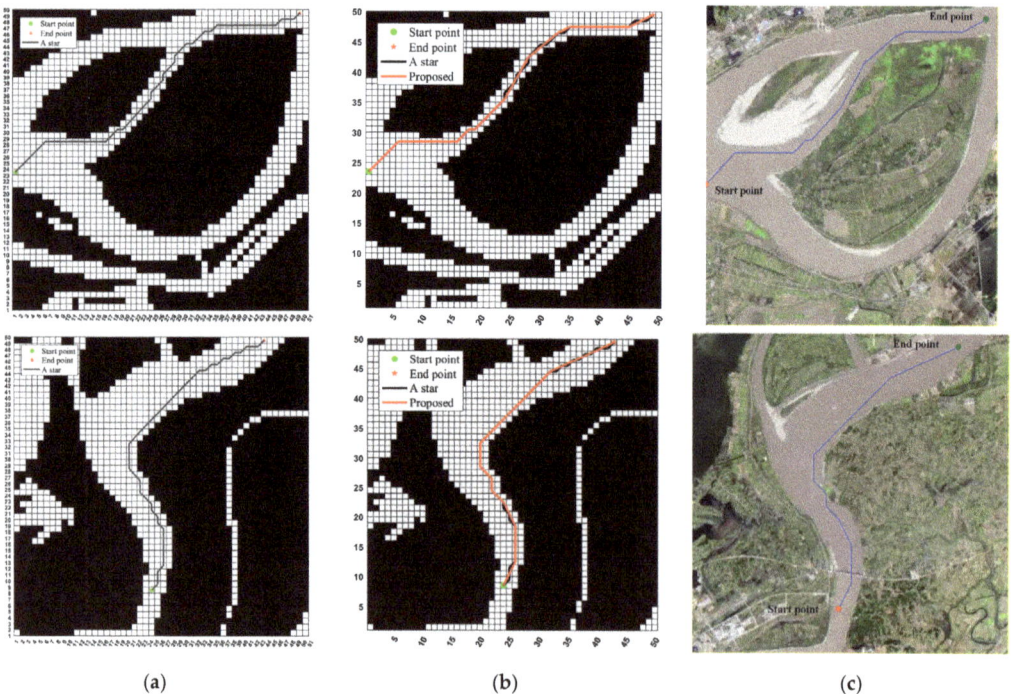

Figure 7. In (**a**), the black line represents the searched path by A*. The green dot denotes the start and the red star denotes the goal. In (**b**), the red line represents the searched path by ESP, considering the turning cost. (**c**) shows the searched path by ESP in the chart.

Figure 8 shows the performance of RRT. Due to the randomness of RRT, the planned paths have many unnecessary turning points, which is detrimental to the safe navigation of ships.

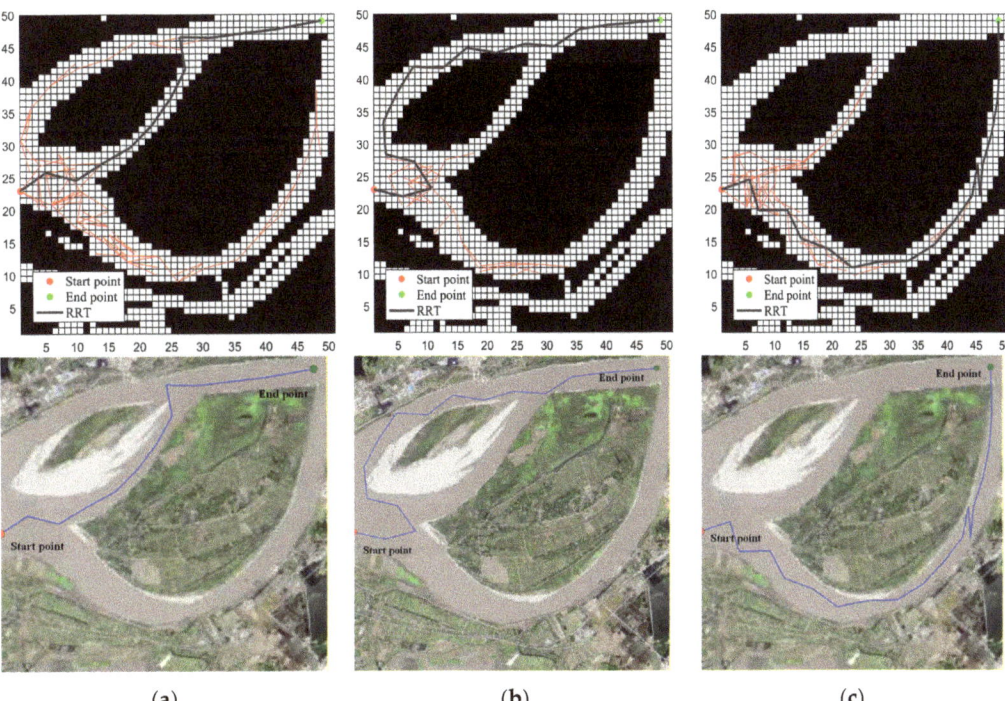

(a)   (b)   (c)

Figure 8. The results of the RRT algorithm. (**a**–**c**) are the three stochastic planning results of the RRT algorithm. The solid black line is the planned route, the solid red line records the sampling process of the algorithm, and the solid blue line shows the planning results.

Figure 9 shows the result of RRT*. RRT* needs to rewire parents to find asymptotically optimal paths. The result will be close to the optimal solution with more iterations.

Figure 10 shows the optimized trajectory of our proposed ESP. It can be seen that the optimized trajectory (the green curve in Figure 10a) meets the requirements of safety, feasibility, and smoothness.

Table 2 shows the numeric comparison in Scene 1 among ESP, A* [12], RRT [43] and RRT* [44] in terms of computation time, number of turns, and fuel consumption. It can be seen that due to the need to measure the turning cost of the path, the searching time of ESP is 0.105 s longer than that of the A* algorithm, and the algorithm's operating efficiency is reduced by 36.71%. Nevertheless, it is still 1.771 s shorter than that of the RRT algorithm, and 6.021 s shorter than that of the RRT* algorithm. The efficiency of the algorithm is improved by 4.35 times and 15.42 times, respectively. In addition, the number of turns of ESP is obviously less than that of the A* algorithm, which reduces the number of large-angle steers to 8 and improves the safety of ship navigation. The fuel consumption of ESP is 164.6008 kg less than that of the A* algorithm, 387.1543 kg less than that of the RRT algorithm, and 24.3311 kg less than that of the RRT* algorithm.

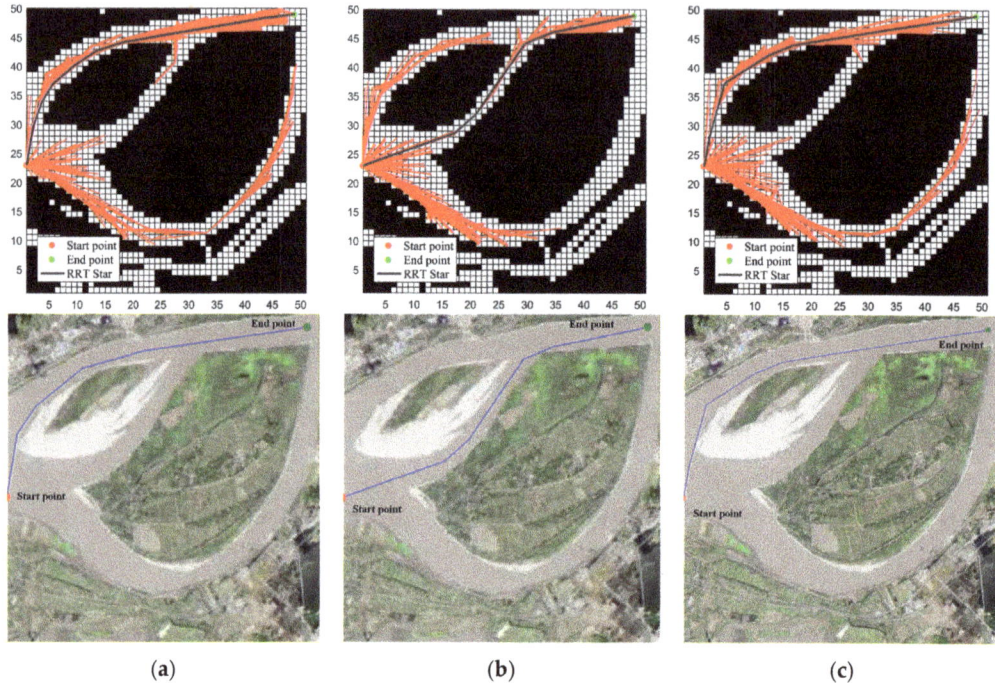

Figure 9. Three different searched paths from the RRT* algorithm. (**a**–**c**) are the three stochastic planning results of the RRT* algorithm. The solid black line is the planned route, the solid red line records the sampling process of the algorithm, and the solid blue line shows the planning results.

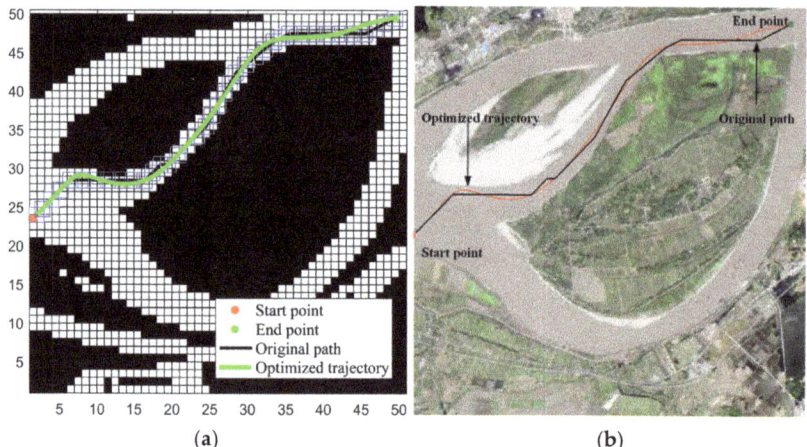

Figure 10. (**a**) The blue rectangles represent the sailing corridor. (**b**) The black line is the searched path. The optimized trajectory is shown as the red curve.

To evaluate the smoothness, Figure 11 shows the generated positions, velocities, accelerations, and jerks. All these curves are continuous and satisfy the ship's dynamic constraints.

Table 2. Performance comparison in Scene 1 with 1000 trials.

Methods	Computation Time (s)		No. of Turns	Average Fuel Consumption (kg)
A* [12]	Average	0.286	18	921.7465
	Max	0.327		
RRT [43]	Average	2.162	11	1144.3000
	Max	3.534		
RRT* [44]	Average	6.412	4	781.4768
	Max	7.282		
ESP	Average	**0.391**	8	**757.1457**
	Max	**0.532**		

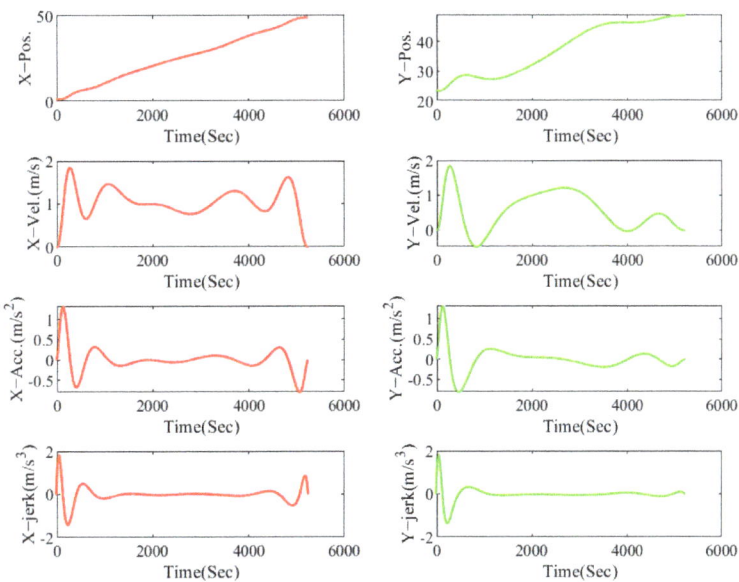

Figure 11. The optimized smooth trajectory in terms of position, velocity, acceleration, and jerk.

To highlight the effectiveness of ESP, Figure 12 compares the result of ESP with that which did not consider the ship's dynamic constraints. The blue line denotes the searched path. The green line shows that Bezier curve without considering the dynamic constraints. The red line is the optimized trajectory from ESP.

From the enlarged part, we can see that there are knots in the result without considering the dynamic constraints, making sailing control very difficult. In contrast, ESP generates a smooth and continuous trajectory with the lowest number of turns, which is safer.

Figures 13 and 14 show the collision avoidance in Scene 1 and 2. In Scene 1, ESP generates a feasible and smooth local trajectory (the orange line) that avoids the unexpected circular obstacle. In Scene 2, the original planned trajectory is very close to the shoal, increasing the risk of the ship running aground. The local replanned trajectory effectively solves the problem. In both scenarios, the goal of safely avoiding temporarily appearing static obstacles is achieved.

The computation times of the local replanning in both scenes are listed in Table 3. Over 1000 trials, the best calculation time is 48 ms, and the maximum computation time is 265 ms. The mean computation time is 192 ms in Scene 1 and 194 ms in Scene 2, ensuring real-time processing.

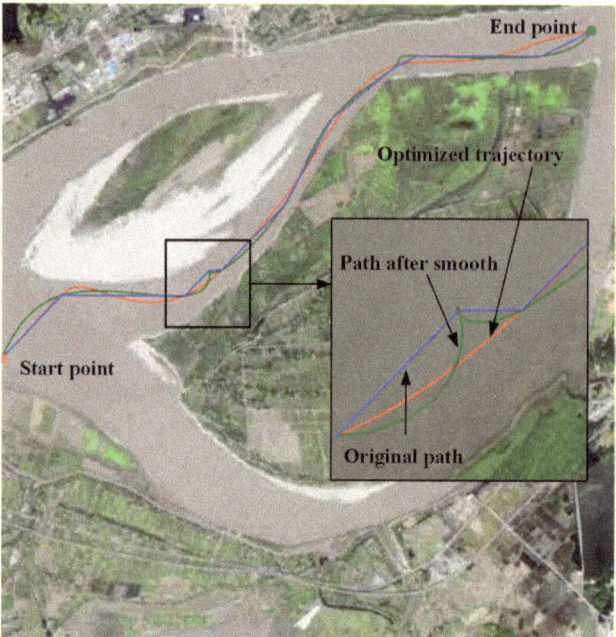

Figure 12. ESP vs. w/o considering the dynamic constraints.

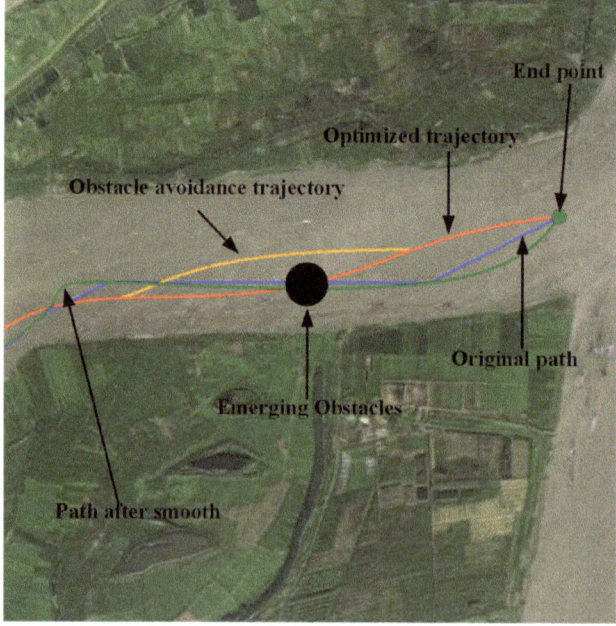

Figure 13. Avoid the unexpected circular obstacle in Scene 1.

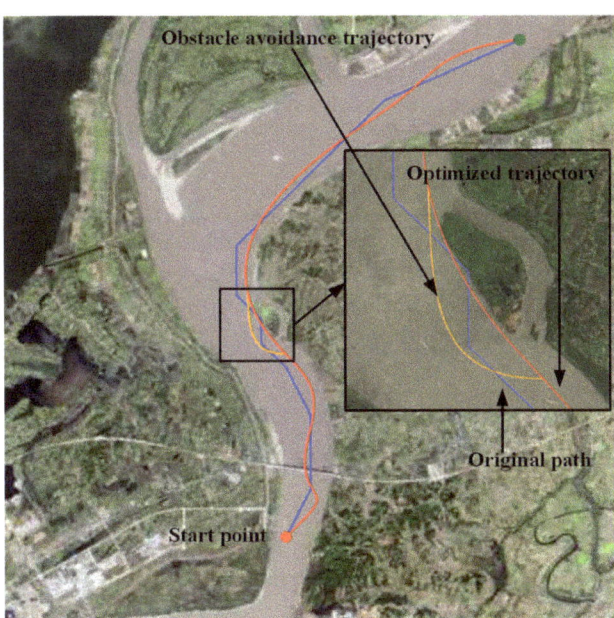

Figure 14. Avoid the shoal to reduce the risk of ship running aground in Scene 2.

Table 3. Computation time of the local replanning in both scenes with 1000 trials.

Test Scene	Max Computation Time (s)
Scene 1	0.233
Scene 2	0.265

Avoiding multiple obstacles requires multiple iterations of the above process, and the response time will be multiplied by the number of iterations.

5. Conclusions

This paper proposes a GIS-data-driven method for the efficient and safe path planning of autonomous ships in maritime transportation, which makes up for the shortcomings of existing methods that ignore the motion dynamic limitations of ships in order to achieve the shortest path, leading to sudden changes in the planned route and thus lacking practical applicability. To this end, we propose ESP, a new path planner that provides comfortable sailing while saving fuel. The key intuition of our proposal is to reduce the expensive turning in path searching. The expensiveness comes from the inertia exhibited by the huge weight of the ship. To realize the above intuitive idea, we design three components for ESP. First, we quantify the ship's turning cost based on its kinematic and dynamic model and develop a modified A* path search algorithm. Second, we formulate an optimization problem subject to dynamic ship constraints and environment constraints to produce a safe and smooth trajectory that consumes minimal fuel. Finally, we use the B-spline representation to perform real-time local replanning, enabling autonomous ships to quickly respond to unexpected risks while maintaining the previous optimization objectives. The data-driven experiments demonstrate the effectiveness of ESP. However, we currently only consider the effect of ocean currents on the dynamic ship model. In future, we will consider the influence of other environmental factors, e.g., sea wind, to build a more robust model for autonomous ships. The avoidance of dynamic obstacles and the real-time avoidance of multiple static obstacles will also be investigated on this basis.

Author Contributions: Conceptualization, X.H. and K.H.; methodology, K.H., X.H. and D.T.; software, D.T., X.H. and K.H.; validation, X.H., K.H. and D.T.; formal analysis, X.H., K.H. and D.T.; investigation, D.T., Y.Z. and Y.H.; resources, X.H., K.H., D.T. and Y.H.; data curation, X.H., K.H. and D.T.; writing—original draft preparation, X.H. and D.T.; writing—review and editing, D.T. and X.H.; visualization, D.T. and K.H.; supervision, X.H., K.H. and Y.Z.; project administration, X.H. and Y.Z.; funding acquisition, Y.Z. and Y.H. All authors have read and agreed to the published version of the manuscript.

Funding: This work was supported by the Research Project of Wuhan University of Technology Chongqing Research Institute (No. YF2021-06). This work was also supported by a grant from the National Natural Science Foundation of China (Grant No. 61801341).

Data Availability Statement: The simulation data used to support the findings of this study are available from the corresponding author upon request.

Conflicts of Interest: The authors declare no conflict of interest.

References

1. Tang, L.; Bhattacharya, S. Revisiting the Shortage of Seafarer Officers: A New Approach to Analysing Statistical Data. *WMU J. Marit. Affairs* **2021**, *20*, 483–496. [CrossRef]
2. Wilson, D.D. Evaluating Jamaica's position as a seafarer-supplying country for cruise and cargo. *Worldw. Hosp. Tour. Themes* **2022**, *14*, 124–136. [CrossRef]
3. Lee, S.-M.; Roh, M.-I.; Kim, K.-S.; Jung, H.; Park, J.J. Method for a Simultaneous Determination of the Path and the Speed for Ship Route Planning Problems. *Ocean Eng.* **2018**, *157*, 301–312. [CrossRef]
4. Liu, Y.; Bucknall, R. A survey of formation control and motion planning of multiple unmanned vehicles. *Robotica* **2018**, *36*, 1019–1047. [CrossRef]
5. Stankiewicz, P.; Kobilarov, M. A Primitive-Based Approach to Good Seamanship Path Planning for Autonomous Surface Vessels. In Proceedings of the 2021 IEEE International Conference on Robotics and Automation (ICRA), Xi'an, China, 30 May–5 June 2021; pp. 7767–7773.
6. Du, Z.; Wen, Y.; Xiao, C.; Zhang, F.; Huang, L.; Zhou, C. Motion Planning for Unmanned Surface Vehicle Based on Trajectory Unit. *Ocean Eng.* **2018**, *151*, 46–56. [CrossRef]
7. Vettor, R.; Guedes Soares, C. Development of a Ship Weather Routing System. *Ocean Eng.* **2016**, *123*, 1–14. [CrossRef]
8. Singh, Y.; Sharma, S.; Sutton, R.; Hatton, D.; Khan, A. A Constrained A* Approach towards Optimal Path Planning for an Unmanned Surface Vehicle in a Maritime Environment Containing Dynamic Obstacles and Ocean Currents. *Ocean Eng.* **2018**, *169*, 187–201. [CrossRef]
9. Niu, H.; Savvaris, A.; Tsourdos, A.; Ji, Z. Voronoi-Visibility Roadmap-Based Path Planning Algorithm for Unmanned Surface Vehicles. *J. Navig.* **2019**, *72*, 850–874. [CrossRef]
10. Jeong, M.-G.; Lee, E.-B.; Lee, M.; Jung, J.-Y. Multi-Criteria Route Planning with Risk Contour Map for Smart Navigation. *Ocean Eng.* **2019**, *172*, 72–85. [CrossRef]
11. Yang, J.-M.; Tseng, C.-M.; Tseng, P.S. Path Planning on Satellite Images for Unmanned Surface Vehicles. *Int. J. Nav. Archit. Ocean. Eng.* **2015**, *7*, 87–99. [CrossRef]
12. Song, R.; Liu, Y.; Bucknall, R. Smoothed A* Algorithm for Practical Unmanned Surface Vehicle Path Planning. *Appl. Ocean. Res.* **2019**, *83*, 9–20.
13. Zhang, L.; Mou, J.; Chen, P.; Li, M. Path Planning for Autonomous Ships: A Hybrid Approach Based on Improved APF and Modified VO Methods. *J. Mar. Sci. Eng.* **2021**, *9*, 761.
14. Liu, X.; Li, Y.; Zhang, J.; Zheng, J.; Yang, C. Self-Adaptive Dynamic Obstacle Avoidance and Path Planning for USV Under Complex Maritime Environment. *IEEE Access* **2019**, *7*, 114945–114954. [CrossRef]
15. Lazarowska, A. A Discrete Artificial Potential Field for Ship Trajectory Planning. *J. Navig.* **2020**, *73*, 233–251. [CrossRef]
16. Chen, C.; Chen, X.-Q.; Ma, F.; Zeng, X.-J.; Wang, J. A Knowledge-Free Path Planning Approach for Smart Ships Based on Reinforcement Learning. *Ocean Eng.* **2019**, *189*, 106299.
17. Wang, C.; Zhang, X.; Li, R.; Dong, P. Path Planning of Maritime Autonomous Surface Ships in Unknown Environment with Reinforcement Learning. In Proceedings of the ICCSIP 2018: International Conference on Cognitive Systems and Signal Processing, Beijing, China, 29 November–1 December 2018; Springer: Singapore, 2019; pp. 127–137.
18. Yu, L.; Qian, T.; Ye, X.; Zhou, F.; Luo, Z.; Zou, A. Research on obstacle avoidance strategy of USV based on improved grid method. In Proceedings of the 4th International Symposium on Power Electronics and Control Engineering (ISPECE 2021), Nanchang, China, 16–19 September 2021; Volume 12080, pp. 172–176.
19. Xie, L.; Xue, S.; Zhang, J.; Zhang, M.; Tian, W.; Haugen, S. A Path Planning Approach Based on Multi-Direction A* Algorithm for Ships Navigating within Wind Farm Waters. *Ocean Eng.* **2019**, *184*, 311–322. [CrossRef]
20. Wang, H.; Qi, X.; Lou, S.; Jing, J.; He, H.; Liu, W. An Efficient and Robust Improved A* Algorithm for Path Planning. *Symmetry* **2021**, *13*, 2213. [CrossRef]

21. Yu, K.; Liang, X.; Li, M.; Chen, Z.; Yao, Y.; Li, X.; Zhao, Z.; Teng, Y. USV Path Planning Method with Velocity Variation and Global Optimisation Based on AIS Service Platform. *Ocean Eng.* **2021**, *236*, 109560. [CrossRef]
22. Sang, H.; You, Y.; Sun, X.; Zhou, Y.; Liu, F. The Hybrid Path Planning Algorithm Based on Improved A* and Artificial Potential Field for Unmanned Surface Vehicle Formations. *Ocean Eng.* **2021**, *223*, 108709.
23. Han, S.; Xiao, L. An Improved Adaptive Genetic Algorithm. *SHS Web Conf.* **2022**, *140*, 01044. [CrossRef]
24. Wang, N.; Xu, H.; Li, C.; Yin, J. Hierarchical Path Planning of Unmanned Surface Vehicles: A Fuzzy Artificial Potential Field Approach. *Int. J. Fuzzy Syst.* **2021**, *23*, 1797–1808. [CrossRef]
25. Zhang, Z.; Wu, D.; Gu, J.; Li, F. A Path-Planning Strategy for Unmanned Surface Vehicles Based on an Adaptive Hybrid Dynamic Stepsize and Target Attractive Force-RRT Algorithm. *J. Mar. Sci. Eng.* **2019**, *7*, 132. [CrossRef]
26. Chen, L.; Mantegh, I.; He, T.; Xie, W. Fuzzy Kinodynamic RRT: A Dynamic Path Planning and Obstacle Avoidance Method. In Proceedings of the 2020 International Conference on Unmanned Aircraft Systems (ICUAS), Athens, Greece, 1–4 September 2020; pp. 188–195.
27. Strub, M.P.; Gammell, J.D. Adaptively Informed Trees (AIT*): Fast Asymptotically Optimal Path Planning through Adaptive Heuristics. In Proceedings of the 2020 IEEE International Conference on Robotics and Automation (ICRA), Paris, France, 31 May–31 August 2020; pp. 3191–3198.
28. Zong, C.; Han, X.; Zhang, D.; Liu, Y.; Zhao, W.; Sun, M. Research on Local Path Planning Based on Improved RRT Algorithm. *Proc. Inst. Mech. Eng. Part D J. Automob. Eng.* **2021**, *235*, 2086–2100. [CrossRef]
29. Yu, J.; Yang, M.; Zhao, Z.; Wang, X.; Bai, Y.; Wu, J.; Xu, J. Path Planning of Unmanned Surface Vessel in an Unknown Environment Based on Improved D*Lite Algorithm. *Ocean Eng.* **2022**, *266*, 112873. [CrossRef]
30. Liang, X.; Jiang, P.; Zhu, H. Path Planning for Unmanned Surface Vehicle with Dubins Curve Based on GA. In Proceedings of the 2020 Chinese Automation Congress (CAC), Shanghai, China, 6–8 November 2020; pp. 5149–5154.
31. Candeloro, M.; Lekkas, A.M.; Sørensen, A.J. A Voronoi-Diagram-Based Dynamic Path-Planning System for Underactuated Marine Vessels. *Control Eng. Pract.* **2017**, *61*, 41–54. [CrossRef]
32. Wang, N.; Zhang, Y.; Ahn, C.K.; Xu, Q. Autonomous Pilot of Unmanned Surface Vehicles: Bridging Path Planning and Tracking. *IEEE Trans. Veh. Technol.* **2022**, *71*, 2358–2374. [CrossRef]
33. Zhou, C.; Gu, S.; Wen, Y.; Du, Z.; Xiao, C.; Huang, L.; Zhu, M. Motion Planning for an Unmanned Surface Vehicle Based on Topological Position Maps. *Ocean Eng.* **2020**, *198*, 106798. [CrossRef]
34. MahmoudZadeh, S.; Abbasi, A.; Yazdani, A.; Wang, H.; Liu, Y. Uninterrupted Path Planning System for Multi-USV Sampling Mission in a Cluttered Ocean Environment. *Ocean Eng.* **2022**, *254*, 111328. [CrossRef]
35. Zheng, Y.; Zhang, X.; Shang, Z.; Guo, S.; Du, Y. A decision-making method for ship collision avoidance based on improved cultural particle swarm. *J. Adv. Transp.* **2021**, *2021*, 8898507. [CrossRef]
36. Zhou, X.; Wang, Z.; Ye, H.; Xu, C.; Gao, F. EGO-Planner: An ESDF-Free Gradient-Based Local Planner for Quadrotors. *IEEE Robot. Autom. Lett.* **2021**, *6*, 478–485. [CrossRef]
37. Zhou, M.; Jin, H. Development of a Transient Fuel Consumption Model. *Transp. Res. Part D Transp. Environ.* **2017**, *51*, 82–93. [CrossRef]
38. Wang, W.; Tu, A.; Bergholm, F. Improved Minimum Spanning Tree Based Image Segmentation with Guided Matting. *KSII Trans. Internet Inf. Syst.* **2022**, *16*, 211–230.
39. Min, B.; Zhang, X. Concise Robust Fuzzy Nonlinear Feedback Track Keeping Control for Ships Using Multi-Technique Improved LOS Guidance. *Ocean Eng.* **2021**, *224*, 108734. [CrossRef]
40. Barraquand, J.; Kavraki, L.; Latombe, J.C.; Li, T.Y.; Motwani, R.; Raghavan, P. A random sampling scheme for path planning. In Proceedings of the Robotics Research: The Seventh International Symposium, Herrsching, Germany, 21–24 October 1995; Springer: London, UK, 1996; pp. 249–264.
41. Li, Z.; Li, L.; Zhang, W.; Wu, W.; Zhu, Z. Research on Unmanned Ship Path Planning Based on RRT Algorithm. *J. Phys. Conf. Ser.* **2022**, *2281*, 012004. [CrossRef]
42. Tran, N.K.; Lam, J.S.L. Effects of Container Ship Speed on CO_2 Emission, Cargo Lead Time and Supply Chain Costs. *Res. Transp. Bus. Manag.* **2022**, *43*, 100723. [CrossRef]
43. Kaur, A.; Prasad, M.S. Path Planning of Multiple Unmanned Aerial Vehicles Based on RRT Algorithm. In *Advances in Interdisciplinary Engineering*; Kumar, M., Pandey, R.K., Kumar, V., Eds.; Springer: Singapore, 2019; pp. 725–732.
44. Zhang, X.; Chen, X. Path Planning Method for Unmanned Surface Vehicle Based on RRT* and DWA. In Proceedings of the ICMTEL 2021: International Conference on Multimedia Technology and Enhanced Learning, Virtual Event, 8–9 April 2021; Fu, W., Xu, Y., Wang, S.-H., Zhang, Y., Eds.; Springer: Cham, Switzerland, 2021; pp. 518–527.

Disclaimer/Publisher's Note: The statements, opinions and data contained in all publications are solely those of the individual author(s) and contributor(s) and not of MDPI and/or the editor(s). MDPI and/or the editor(s) disclaim responsibility for any injury to people or property resulting from any ideas, methods, instructions or products referred to in the content.

Article

Micro-Factors-Aware Scheduling of Multiple Autonomous Trucks in Open-Pit Mining via Enhanced Metaheuristics

Yong Fang and Xiaoyan Peng *

College of Mechanical and Vehicle Engineering, Hunan University, Changsha 410082, China; fangyong@hnu.edu.cn
* Correspondence: xiaoyan_p@126.com

Abstract: Traditional open-pit mineral transportation systems are typically subject to manual command, frequently leading to vehicular delays and traffic congestion. With the advancement of automation and electrification technologies, this study proposes a highly accurate scheduling method for multiple autonomous trucks in an open-pit mine. This model considers micro-level temporal and spatial factors to tackle the task of scheduling autonomous trucks within open-pit mines. The cost function of the concerned scheduling problem is a comprehensive evaluation of energy consumption, time, and output. Beyond the loading and unloading activities, the model also factors in the charging requirements of autonomous trucks in mining regions. The scheduling model integrates a Voronoi diagram search and optimal spatial path time matching, aiming to provide superior mission planning and decision-making solutions for autonomous trucks in mining regions. For an efficient solution to the scheduling problem, we propose an improved-evolution artificial bee colony (IE-ABC) algorithm. This algorithm improves the global search and re-initialization processes and conducts algorithm ablation experiments to closely examine their impact on optimization. Simulation results across various algorithms, cost function definition strategy, and encoding strategy show that our method can improve scheduling performance in energy consumption and time. Experimental results demonstrate that the proposed model and algorithm can effectively solve the scheduling decision-making problem in an unmanned open-pit mine.

Keywords: open-pit mine; autonomous truck; scheduling; artificial bee colony algorithm

Citation: Fang, Y.; Peng, X. Micro-Factors-Aware Scheduling of Multiple Autonomous Trucks in Open-Pit Mining via Enhanced Metaheuristics. *Electronics* **2023**, *12*, 3793. https://doi.org/10.3390/electronics12183793

Academic Editor: Mahmut Reyhanoglu

Received: 12 August 2023
Revised: 4 September 2023
Accepted: 5 September 2023
Published: 7 September 2023

Copyright: © 2023 by the authors. Licensee MDPI, Basel, Switzerland. This article is an open access article distributed under the terms and conditions of the Creative Commons Attribution (CC BY) license (https://creativecommons.org/licenses/by/4.0/).

1. Introduction

Open-pit mining presents benefits such as large-scale production, high resource recovery rates, and minimal environmental impact [1]. The complexity of the working environment within the open-pit mining area and the low efficiency of manual mining necessitate the introduction of autonomous mining systems [2].

In light of the rapid advancement of robotics, big data, artificial intelligence, and 5G technology, we are now observing the emergence of intelligent unmanned dump truck systems [3,4]. Given that around 50% of the gross operating costs in an open-pit mine would be spent on material transport [5], it has been an obvious trend to deploy unmanned transport tools to replace human labor [6]. Figure 1 illustrates that typical unmanned transportation tools in an open-pit mine include unmanned dump trucks, excavators, crushing stations, and charging piles. The excavator is used for mining ore, while the unmanned dump truck is designated for ore transportation. The truck moves to the location of the excavator for ore loading and then delivers it to the crushing station for unloading. Positioned within the mining area, the crushing station primarily serves to crush and pulverize the raw ore to meet the demands of further processing and utilization. When the unmanned dump truck needs recharging, it navigates to the charging pile to undergo the necessary charging process.

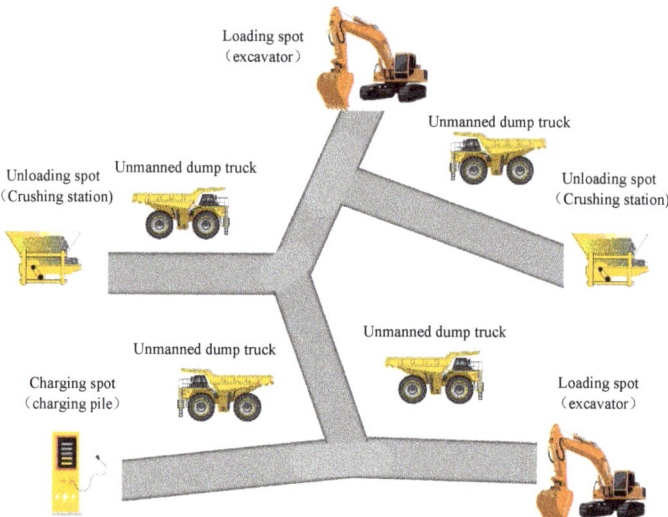

Figure 1. Schematic on an unmanned transport system in an open-pit mine.

Operating all of the aforementioned devices automatically is difficult because it requires simultaneously considering the features, principles, and capabilities of all devices when generating control commands, otherwise the automated control performance would be worse than that of human laborers [7]. In this sense, dividing the entire control scheme into multiple layers is a practical and feasible solution [8,9]. As shown in Figure 2, a scheduling module first assigns a traverse order for each of the devices; a decision-making module decides how two or more devices interact locally when their nominal trajectories are conflicting [10,11]; a planning module generates a spatio-temporal curve for each device to track [12,13]. This solution is inherently holding a decoupled strategy, i.e., the features, principles, and capabilities of all devices are no longer considered in a simultaneous way. Adopting such a decoupled strategy easily renders the loss of solution optimality, although it reduces the computational burden. Herein, the scheduling module is particularly important because a suboptimal decision made in the scheduling module would largely influence the downstream modules so that there is no chance to achieve optimality in mining operations. This analysis indicates that the scheduling module is important to guarantee the solution quality of an autonomous operating system in an open-pit mine [14]. The goal of this study is to propose a high-precision scheduling methodology with microscopic factors of each device considered, especially temporal factors. Through this, the scheduling method promises to coarsely find an ideal dispatch solution for the downstream modules efficiently without loss of optimality.

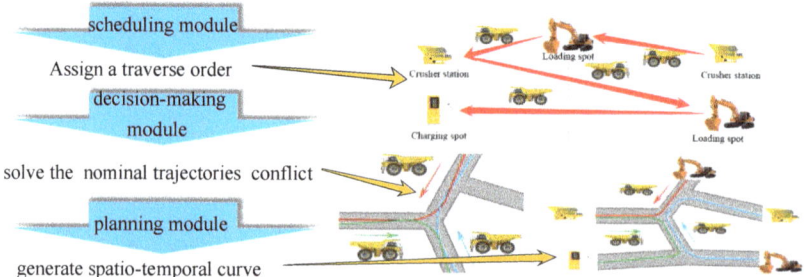

Figure 2. Overall flowchart of an unmanned transport system.

1.1. Related Work and Motivations

This subsection reviews the prevalent scheduling methods for dispatching multiple devices (especially transport vehicles) in an open-pit mine or similar scenarios.

In recent years, the scheduling problem in open-pit mines has received considerable attention. Patterson et al. [15] constructed a unique mixed-integer linear programming problem for multi-truck scheduling and used a Tabu Search algorithm to solve it, aiming to minimize the energy consumption of trucks and excavators. Zhang et al. [16], Yuan et al. [17], Wang et al. [18], Bastos et al. [19], Zhang et al. [20], and Bao et al. [21] employed similar strategies to build a model. However, the optimization objectives in these formulated problems only considered fuel consumption while ignoring factors such as consumed time and output amount. Wang et al. [22] proposed a multi-objective optimization (MOO) algorithm for truck scheduling, while Zhang et al. [23] proposed a decomposition-based constrained dominance genetic principle algorithm (DBCDP-NSGA-II) to solve the multi-objective intelligent scheduling problem for trucks in open-pit mines. Ahumada et al. [24], Chang et al. [25], and Afrapoli et al. [26] also built a multi-objective scheduling model. However, a common limitation of refs. [22–26] is that the formulated problem did not consider the refueling or charging requirements of the trucks. Zhang et al. [27] proposed a meta-heuristic search algorithm to solve a mixed-integer programming problem formulated for the concerned multi-truck scheduling scheme and demonstrated by experimentation that this approach improves the energy efficiency of the transport system in open-pit mines. Smith et al. [28] proposed a time-discretized mixed integer programming (MIP) model for the truck scheduling problem in open-pit mines, and a heuristic is used to quickly generate high-quality feasible solutions. However, the proposed model ignored the path planning between loading and unloading spots. Similarly, Zeng et al. [29], de Melo [30], and Yeganejou et al. [31] did not consider the path planning between loading and unloading spots either.

Most previous scheduling models focused on single-objective optimization, particularly energy consumption, and often overlooked the need for multiple optimization objectives. Additionally, these methods focused solely on truck-carrying activities without considering the refueling or charging requirements of the trucks. Path planning between loading and unloading spots, a crucial aspect of scheduling, has also been largely ignored in previous research. As a conclusion of this subsection, the prevalent scheduling methods do not model the concrete dynamics/kinematics and other temporal constraints of the operating devices; thus, they did not account for the actual complexity of operating an open-pit mine.

1.2. Contributions

Following the motivations introduced in the preceding subsection, we summarize the contributions of this paper and are as follows.

First, we propose a high-precision scheduling model with microscopic factors of each device considered. The model incorporates the use of a Voronoi diagram for spatial path estimation and velocity assignment. The diagram is used to enhance the scheduling accuracy so that local factors are well considered.

Second, this paper proposes a novel structure to present each of the solutions, thus enhancing the interpretability of the scheduling process and facilitating the solution generation process.

Third, a novel metaheuristic optimizer is proposed to search for a high-quality scheduling solution based on the aforementioned model and solution structure. The proposed optimizer is a special variant of the artificial bee colony (ABC) algorithm, which enhanced the ability to quickly exploit feasible solution candidates.

1.3. Organization

In the remainder of this paper, Section 2 formulates the concerned scheduling problem as a combinatorial optimization problem. Section 3 introduces the proposed optimizer to solve the formulated optimization problem together with a novel solution presentation structure. Simulation results are reported and discussed in Section 4, followed by Section 5, where conclusions are finally drawn.

2. Problem Formulation

This section formulates the concerned scheduling scheme as a combinatorial optimization problem, which is focused on how to use unmanned dump trucks and excavators to deliver mine materials efficiently in open-pit mining. The purpose of the combinatorial optimization problem is to minimize the consumed energies, maximize the mineral transport capacity, and minimize the operation time.

In this work, unmanned dump trucks and excavators are the two types of machinery operated in the concerned mining transport scheme. Given that the terrain of the open-pit mine is not even, some parts of the terrain occupied by obstacles are not drivable; thus, the uneven parts of the terrain are regarded as obstacles. Suppose that the number of deployed unmanned dump trucks is N_{truck}. This work assumes that each excavator is fixed at a loading spot, a designated location where loading operations occur. The gross number of loading spots is denoted as $N_{loading\_spots}$. Similarly, we assume that the number of unloading spots is $N_{unloading\_spots}$. Before its battery is completely depleted, an unmanned dump truck should visit a charging spot for battery recharging. The basic working states of an unmanned dump truck include moving to the loading spot, loading, moving to the unloading spot, unloading, and moving to the charging spot. Given that the excavators are fixed, operating the devices in this open-pit mine means manipulating the N_{truck} unmanned dump trucks. Thus, the concerned scheduling task is about deciding the working states and targets of each unmanned dump truck in a sequence for the next N_{step} steps.

2.1. Cost Function Formulation

The concerned scheduling scheme is inherently a combinational optimization problem, aiming to minimize overall energy consumption, maximize transport capability, and reduce vehicle idle times, for N_{step} tasks assigned to each of the N_{truck} unmanned dump trucks. The cost function of the combinational optimization problem, as described in Equation (1), is a weighted sum of three terms: each truck's transport capability, energy consumption, and the total time spent when the last truck among all the N_{truck} ones finishes its N_{step}-th task. The three terms are summed up after being multiplied by weighting coefficients w_1, w_2, and w_3:

$$\sum_{i=1}^{N_{truck}} \left(\frac{w_1}{Q_i} + w_2 \times E_i \right) + w_3 \times T, \tag{1}$$

where Q_i denotes the gross transport capability of unmanned dump truck i after finishing all of its N_{step} tasks, E_i denotes the consumed energies of unmanned dump truck i after it finishes all of its N_{step} tasks, and T denotes the earliest moment that all of the N_{truck} trucks finish their N_{step} tasks. Details behind these variables are defined as follows.

2.2. Transport Capability Composition

In Equation (2), Q_i is defined as the product of the basic loading capability of the truck i (i.e., C_i) and $N_{unloading}$, the steps among all of the N_{step} ones that involve unloading actions. Notably, $N_{unloading}$ is determined according to a specific solution.

$$Q_i = C_i \times N_{unloading}. \tag{2}$$

2.3. Energy Consumption Composition

Equation (3) shows that the energy consumption of each unmanned dump truck is composed of the energy consumption during the on-road cruising process, the energy consumption during the loading actions, and the energy consumption during the unloading actions:

$$E_i = E_{\text{cruising}} + E_{\text{loading}} + E_{\text{unloading}}. \tag{3}$$

Herein, E_{cruising} sums up the energy consumption of unmanned dump truck i during its cruising process. E_{loading} sums up the energy consumption of unmanned dump truck i during all of its loading actions. $E_{\text{unloading}}$ sums up the energy consumption of unmanned dump truck i during its unloading actions.

E_{cruising} consists of two components: energy consumed by the drive system and that of the accessory system [32]. According to the classical vehicle dynamics principle [33], the energy consumption of an unmanned dump truck is defined as

$$E_{\text{cruising}} = w_4 \times v_i^3 \times T_{\text{cruising}} + w_5 \times v_i \times T_{\text{cruising}}, \tag{4}$$

where T_{cruising} is the driving time of unmanned dump truck i to complete N_{step} tasks, which is determined as per a specific solution candidate. v_i denotes the average speed of unmanned dump truck i during cruising process. w_4 and w_5 are weighting parameters.

Equations (5) and (6) define the energy consumption during loading and unloading actions, respectively:

$$E_{\text{loading}} = w_6 \times T_{\text{loading}}, \tag{5}$$

$$E_{\text{unloading}} = w_6 \times T_{\text{unloading}}. \tag{6}$$

Herein, T_{loading} is the loading time of unmanned dump truck i during loading tasks; $T_{\text{unloading}}$ is the unloading time of unmanned dump truck i during its unloading tasks. w_6 is the coefficient that converts loading/unloading time to energy consumption.

2.4. Time Composition

In the process of completing N_{step} assigned tasks, each unmanned dump truck encounters several stages, including a driving stage, a waiting stage, a loading stage, an unloading stage, and a charging stage. Equation (7) defines the time for each truck to complete the corresponding N_{step} established tasks.

$$T_i = T_{\text{cruising}} + T_{\text{loading}} + T_{\text{unloading}} + T_{\text{waiting}} + T_{\text{charging}}, \tag{7}$$

where T_i is the total time of unmanned dump truck i to complete N_{step} tasks; $T_{\text{unloading}}$ is the waiting time of unmanned dump truck i to complete N_{step} tasks; and T_{charging} is the charging time of unmanned dump truck i to complete the scheduled N_{step} tasks.

Equation (8) presents the total time taken by N_{truck} unmanned dump trucks to complete N_{step} given tasks.

$$T \equiv \max\{T_1, T_2, T_3 \ldots T_{N_{\text{truck}}}\}. \tag{8}$$

Equation (9) indicates that the driving time of unmanned dump truck i is related to the task spots that need to be visited:

$$T_{\text{cruising}} = \sum_{j=1}^{N_{\text{step}}} TravelTime(taskspot_j, taskspot_{j+1}). \tag{9}$$

Herein, $TravelTime(a, b)$ is a function that estimates the driving time from task spot a to b. $taskspot_j$ denotes the task spot that the jth task of unmanned dump truck i is required to visit.

Equation (10) defines the loading time of unmanned dump truck i during loading tasks:

$$T_{\text{loading}} = N_{\text{step\_load}} \times T_{\text{load\_p}}, \tag{10}$$

where T_{load_p} is the loading time for unmanned dump truck i loading at each loading spot. N_{step_load} is the number of loading tasks.

Equation (11) shows the unloading time of unmanned dump truck i during unloading tasks:

$$T_{unloading} = N_{step\_unload} \times T_{unload\_p}, \quad (11)$$

where T_{unload_p} denotes the unloading time for unmanned dump truck i unloading at each unloading spot. N_{step_unload} is the number of unloading tasks.

Equation (12) indicates that waiting time includes the time waiting for loading, waiting for unloading, and waiting for charging.

$$T_{waiting} = T_{waiting\_load} + T_{waiting\_unload} + T_{waiting\_charge}, \quad (12)$$

where $T_{waiting\_load}$ denotes the time that unmanned dump truck i spends on waiting for loading during the completion of N_{step} tasks; $T_{waiting\_unload}$ denotes the time that unmanned dump truck i spends on waiting for unloading during the completion of N_{step} tasks; and $T_{waiting\_charge}$ denotes the time that unmanned dump truck i spends on waiting for charging during the completion of N_{step} tasks.

Equation (13) shows that the charging time depends on the residual capacity when the unmanned dump truck reaches the charging spot for charging:

$$T_{charging} \equiv \frac{E_{full} - E_{remain}}{q}. \quad (13)$$

Herein, E_{full} is the full electric quantity of unmanned dump truck i; E_{remain} is the residual capacity when the unmanned dump truck i reaches the charging spot; and q represents the charging efficiency.

The residual capacity of the unmanned dump truck depends on how many tasks are completed and which loading and unloading spots are visited.

3. Methodology

In the scheduling problem formulated in the previous section, the solution candidates differ from one another in their task scheduling sequences, thereby resulting in different cost function values. This section introduces how to solve the formulated problem. To that end, the first thing is to define an encoding principle so that all of the solution candidates can be presented uniformly in such a solution space. Thereafter, an efficient solver should be proposed to search for the optimal or near-optimal solution in the defined solution space. The technical details are introduced in the next few subsections.

3.1. Principle of Solution Vector Encoding

The encoding strategy for the scheduling sequence is determined by the number of loading and unloading spots in the environment. Concretely, if there are two loading spots and three unloading spots, the serial numbers for the loading spots are designated as {1, 2}. Meanwhile, the serial numbers for the unloading spots are set to {3, 4, 5}, and the serial number for the charging spot is represented as {6}. As shown in Figure 3, suppose that there are eight tasks to be scheduled for each unmanned dump truck. The scheduling sequence for the unmanned dump truck #1 might be {1, 3, 2, 4, 2, 1, 2, 5}. In such a sequence, the red part of the figure should be penalized for repeated selections of the loading spot while the green part should not be penalized. As for the unmanned dump truck #2, the red part of the figure should be penalized for repeated selections of the unloading spot while the green part should not be penalized.

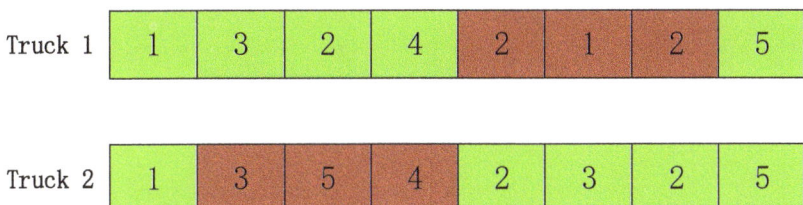

Figure 3. Schematic of solution vector encoding.

3.2. Scheduling Problem Re-Formulation

In the previous section, Equation (1) defines the cost function. However, the cost function does not adequately consider the establishment of the hard constraint. Therefore, we redefine the cost function here to ensure that the feasibility of the solution vector is not determined solely by the constraint, but also by the penalty function. This redefinition is beneficial for the subsequent utilization of the optimization algorithm, which will be discussed in detail later.

When allocating tasks to multiple unmanned dump trucks, it is vital to ensure that the optimal task scheduling sequence prevents excessive idle time for any truck. Thus, the variance of time taken by each unmanned dump truck to finish its tasks is incorporated into the cost function as a penalty term. This inclusion promotes quicker identification of the optimal task scheduling sequence, averting significant disparities in the completion times among trucks. As a result, the revised cost function is defined as follows:

$$\sum_{i=1}^{N_{truck}} \left(\frac{w_1}{Q_i} + w_2 \times E_i \right) + w_3 \times T + w_7 \times N_{punish} + w_8 \times T_{var}, \quad (14)$$

where N_{punish} is the number of repeated loading and unloading spots in the task scheduling sequence. w_7 is the weight coefficient of the penalty term for repeated spots. T_{var} represents the variance of the time taken by all unmanned dump trucks to complete the given task according to the current task scheduling sequence. w_8 is the weight coefficient of the penalty term for variance.

3.3. Improved-Evolution Artificial Bee Colony Search Procedure

The artificial bee colony (ABC) algorithm is a heuristic optimization algorithm based on bee behavior. It is used to solve complex optimization problems by simulating the foraging behavior and information exchange of bees. The local search accuracy of the conventional ABC algorithm is not satisfactory. The improved-evolution artificial bee colony (IE-ABC) algorithm used in this paper is an improved artificial bee colony algorithm. The search intensity is manipulated by adding an adaptive change multiplier to the global search equation, and the traditional re-initialization process is improved by the overall degradation strategy to obtain a better optimal solution.

Algorithm 1 uses the IE-ABC search framework to search for the optimal scheduling scheme, which includes the initialization phase (lines 1–2), the employed bee phase (lines 4–12), the calculation of the probability index to prepare for the roulette selection strategy (lines 13–20), the onlooker bee phase (lines 21–38), and the scout bee phase (lines 39–44). $trial(i)$ records the number of times an inefficient search is performed by the ith employed bee or any onlooker bee that searches around the ith employed bee. $\frac{trial(item)}{trial(item)+trial(k)}$ in the fifth line is utilized as an adaptive change multiplier to regulate the search intensity. The solution vector X represents the scheduling sequence, and Equation (14) represents the objective function GetCostFun(). P represents the probability index. During the scout bee stage, if the number of searches exceeds the limit, the position of the scout bee is reinitialized.

The pseudo-code of the IE-ABC algorithm is given as follows.

Algorithm 1. IE-ABC.

1. Set the population size PN, and maximum cycle number MCN; Set the inefficient trial time counter $trial(i) \leftarrow 1$ ($i = 1, 2, \ldots, PN/2$);
2. Randomly initialize locations of PN/2 scout bees;
3. **for** $iter = 1$ to MCN **do**
4.   **for** $item = 1$ to PN/2 **do**
5.   Generate X^*_{item} for the $item$-th employed bee to search according to
$$X^*_{item} \leftarrow X_{item} + rand(-1,1) \times (X_k - X_{item}) \times \frac{trial(item)}{trial(item)+trial(k)};$$
6. $X'_{item} \leftarrow \text{ChargeConstruct}(X^*_{item})$;
7.   **if** $\text{GetCostFun}(X'_{item}) < \text{GetCostFun}(X_{item})$ **then**
8. $X_{item} \leftarrow X^*_{item}$, and set $trial(item) \leftarrow 1$;
9.   **else**
10. $trial(item) \leftarrow trial(item) + 1$;
11.   **end if**
12.   **end for**
13.   **for** $i = 1$ to PN/2 **do**
14.   **if** $\text{GetCostFun}(X_i) \geq 0$ **then**
15. $fitness(i) \leftarrow \frac{1}{1+\text{GetCostFun}(X_i)}$;
16.   **else**
17. $fitness(i) \leftarrow 1 + |\text{GetCostFun}(X_i)|$;
18.   **end if**
19. $P(i) \leftarrow \frac{\sum_{j=1}^{i} fitness(j)}{\sum_{j=1}^{PN/2} fitness(j)}$;
20.   **end for**
21.   Set $item = 0$;
22.   Set $j = 1$;
23.   **while** $item < PN/2$ **do**
24.   **if** $P(j) > rand(0,1)$ **then**
25. $item \leftarrow item + 1$;
26.   Choose the jth employed bee to follow, and then generate,
$$Y_{item} \leftarrow X_j + rand(-1,1) \times (X_k - X_j) \times \frac{trial(j)}{trial(j)+trial(k)};$$
27. $Y'_{item} \leftarrow \text{ChargeConstruct}(Y_{item})$;
28.   **if** $\text{GetCostFun}(Y'_{item}) < \text{GetCostFun}(X_j)$ **then**
29. $X_j \leftarrow Y_{item}$, and set $trial(j) \leftarrow 1$;
30.   **else**
31. $trial(j) \leftarrow trial(j) + 1$;
32.   **end if**
33.   **end if**
34. $j \leftarrow j + 1$;
35.   **if** $j > PN/2$ **then**
36.   Set $j \leftarrow 1$;
37.   **end if**
38.   **end while**
39.   **for** $item = 1$ to PN/2 **do**;
40.   **if** $trial(item) > \text{Limit}$ **then**
41.   Re-initialize the location of the $item$-th employed bee;
42.   Set $trial(item) \leftarrow 1$;
43.   **end if**
44.   **end for**
45.   Memorize the best solution;
46. **end for**
47. Output the best solution;

During the initialization phase, the sequence for scheduling is arranged based on the loading and unloading order to streamline the search for the best solution. After establishing this sequence, the energy consumption required for the unmanned dump truck to complete its tasks is assessed. This assessment subsequently informs when the truck needs recharging.

Algorithm 2 identifies the best recharging moments, relying on the loading and unloading order of the unmanned dump truck. The algorithm starts by accepting the sequence of loading and unloading tasks as its input. In its third line, it estimates the driving time between successive task spots, as dictated by the task sequence. The next line, i.e., line 4, computes the truck's residual energy, based on the calculated driving time. Then, from lines 5 to 10, the algorithm discerns if there exists a need for recharging by considering the leftover energy. If charging is deemed necessary, it then updates the task sequence to accommodate this.

The pseudo-code for the ChargeConstruct algorithm is presented below.

Algorithm 2. ChargeConstruct

Input: X;
Output: X';
1. Set $i = 1$;
2. **while** $i < N_{step}$ **do**
3. $\quad T_{cruising} \leftarrow \text{TravelTime}(X_i, X_{i+1})$;
4. $\quad E_{remain} \leftarrow \text{Energycost}(T_{cruising})$;
5. $\quad$ **if** $E_{remain} < 0$ **then**
6. $\quad\quad X_i \leftarrow chargespot$;
7. $\quad\quad X \leftarrow newX$;
8. $\quad\quad N_{step} \leftarrow new\ N_{step}$;
9. $\quad\quad i \leftarrow 1$;
10. $\quad$ **end if**
11. **end while**
12. $X' \leftarrow X$;
13. **return**.

The energy consumption of the unmanned dump truck is related to the working time. The driving time of the unmanned dump truck between the loading spots, unloading spots, and charging piles is proportional to the path length. The planning space is modeled by using the Voronoi diagram, and the optimal path is obtained by combining the A* search algorithm [34], so as to estimate the optimal path length.

The Voronoi diagram is a fundamental geometric concept used to divide a plane based on a discrete set of spots. The diagram ensures that the distance from any spot in a given region to its corresponding spot in the discrete set is smaller than the distance to any other spot in the set.

Assuming that the set of discrete spots $D = \{d_1, d_2, d_3 \ldots d_n\}$, the mathematical expression of the Voronoi diagram is as follows:

$$\sqrt{(x-x_i)^2 + (y-y_i)^2} < \sqrt{(x-x_j)^2 + (y-y_j)^2},\ i \neq j \qquad (15)$$

where (x_i, y_i) and (x_j, y_j) represent the coordinates of any two discrete spots d_i and d_j in the set D, respectively. (x, y) represents the coordinates of any spot on the plane.

By satisfying Equation (15), the set of spots (x, y) forms the Voronoi region for the discrete spot d_i. Consequently, the plane can be divided into n polygons, where each polygon contains only one discrete spot d_i.

Furthermore, the spots lying on the edges of the Voronoi polygon satisfy specific constraints.

$$\begin{cases} \sqrt{(x-x_i)^2 + (y-y_i)^2} = \sqrt{(x-x_j)^2 + (y-y_j)^2} \\ \sqrt{(x-x_i)^2 + (y-y_i)^2} < \sqrt{(x-x_k)^2 + (y-y_k)^2} \end{cases},\ i \neq j \neq k \qquad (16)$$

where (x_i, y_i) and (x_j, y_j) represent the coordinates of adjacent discrete spots d_i and d_j, respectively. (x_k, y_k) represents the coordinates of any other discrete spot d_k.

The planning space is segmented into multiple Voronoi regions based on the placement of obstacles within the environment. Recognizing that obstacles can vary in size and shape, their boundaries are discretized using a method that employs numerous discrete spots to encapsulate these boundaries. The Delaunay triangulation algorithm constructs the Voronoi diagram from this. To enhance this representation, virtual nodes are placed near pivotal areas, including loading spots, unloading spots, and charging spots. These nodes are then integrated with the original Voronoi diagram, yielding the final representation.

The A* algorithm operates on a traversal search principle, leveraging a heuristic function. This function gauges the cost of moving from any location to the destination, steering the search towards the most viable routes. By adopting the Voronoi diagram as the model for the planning space, the A* algorithm's search is limited to traversing only the nodes within the diagram, substantially enhancing its efficiency. After identifying the best path, the algorithm can then estimate its length.

4. Simulation Results and Discussion

This section reports the simulation results, together with our in-depth discussions. Concretely, the simulation experiments will be conducted in three aspects. First, comparative experiments will be performed using various optimization algorithms. Second, comparative experiments will be carried out with different cost function definition strategies. Third, comparative experiments will be performed using different encoding strategies for the solution vector.

4.1. Simulation Setup

Simulations are implemented in a MATLAB platform and executed on an Intel(R) Core(TM) i7-7700 CPU with 16 GB RAM that runs at 8×3.6 GHz.

Critical parameters are listed in Table 1. In order to increase the diversity of unmanned dump trucks, two kinds of unmanned dump trucks are set up, which have different load capacities and loading and unloading times, respectively. Each unmanned dump truck has two average speeds during cruising.

Table 1. Parametric settings for simulations.

Parameter	Description	Setting
N_{truck}	Number of unmanned dump trucks	4
N_{step}	Number of tasks to be completed per unmanned dump truck	20
$N_{loading\_spots}$	Number of loading spot	2
$N_{unloading\_spots}$	Number of unloading spot	3
w_1, w_2, w_3	Weight coefficient in Equation (1)	$100, 2.7 \times 10^{-7}, 0.01$
w_4, w_5	Weight coefficient in Equation (4)	0.925, 430
w_6	Weight coefficient in Equation (5)	4000
C_i	The load capacity of unmanned dump truck i	1 t, 2 t
v_i	The average speed of unmanned dump truck i during cruising	10 m/s, 15 m/s
T_{load_p}	Loading time of unmanned dump truck i at loading spot	10 s, 20 s
T_{unload_p}	Unloading time of unmanned dump truck i at unloading spot	10 s, 20 s
E_{full}	Full electric quantity of unmanned dump truck i	0.25 kWh
q	Charging efficiency	3×10^4 J/s
w_7, w_8	Weight coefficient in Equation (14)	1, 0.0001

4.2. Simulation Results

The conventional ABC and the proposed IE-ABC are used to solve the proposed scheduling problem. The results are shown in Table 2. The cost function value obtained by IE-ABC is 10.31, while the cost function value obtained by ABC is 10.43. The cost function value obtained by IE-ABC is 0.67% lower than that of ABC. Additionally, the energy consumption and time required for the solution obtained by IE-ABC to complete the task are 8.93×10^6 J and 620.0 s, respectively. On the other hand, the energy consumption and time required for the solution obtained by ABC to complete the task are 9.18×10^6 J and 624.9 s, respectively. Consequently, the solution obtained by IE-ABC reduces the energy consumption and time required to complete the task by 0.23% and 0.93%, respectively, compared to ABC. The improved algorithm has achieved steady advantages in terms of both energy consumption and time. The scheduling Gantt chart of the solution obtained by IE-ABC is shown in Figure 4.

Table 2. Simulation result of different optimization algorithms.

Algorithm	Cost	Consumed Energy (J)	Time (s)
IE-ABC	10.31	8.93×10^6	620.0
ABC	10.43	9.18×10^6	624.9

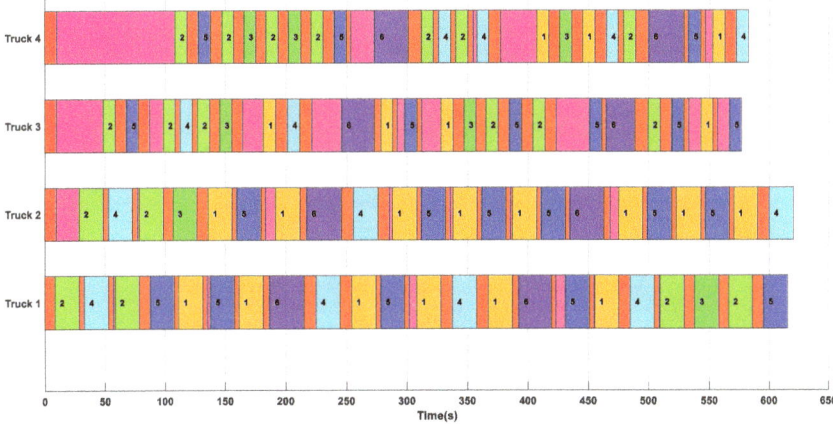

Figure 4. Vehicle scheduling results presented in a Gantt chart.

It can be seen from Figure 4 that the effectiveness of the scheduling scheme is guaranteed. The obtained scheduling sequence can arrange the loading and unloading tasks of the unmanned dump truck in an orderly manner, and the charging tasks are interspersed among them to ensure the coordination and sustainability of the tasks. As shown in Figure 4, the Gantt chart of the scheduling scheme for 20 tasks is arranged for the four unmanned dump trucks, respectively. Different colors represent the unmanned dump trucks working at different task spots, among which No. 1 and No. 2 represent the loading spots, No. 3, No. 4, and No. 5 represent the unloading spots, No. 6 represent the charging spot, red represents the unmanned dump truck driving stage, and purple represents the unmanned dump truck waiting stage.

It can be seen from the figure that the unmanned dump truck #1 first travels to the No. 2 loading spot to complete the loading task, and then travels to the No. 4 unloading spot to perform the unloading task. The unmanned dump truck #2 reaches the No. 2 loading spot later than the unmanned dump truck #1, so the unmanned dump truck #2 waits for the unmanned dump truck #1 to complete the loading task at the No. 2 loading spot, and then performs the loading task at the No. 2 loading spot.

In order to explore the influence of different factors on ABC optimization, ablation experiments are carried out. The ablation experiment results are shown in Figure 5. Compared with the conventional ABC, IE-ABC mainly changes two factors. One is to add an adaptive multiplier to the global search equation to manipulate the search intensity, and the other is to implement the traditional re-initialization process through the overall degradation strategy. In Figure 5, AE-1 only changes the first factor compared with the conventional ABC, and AE-2 only changes the second factor. It can be seen from the results in the figure that changing these two items has promoted the optimization search, and changing the second factor has a greater impact on the early optimization.

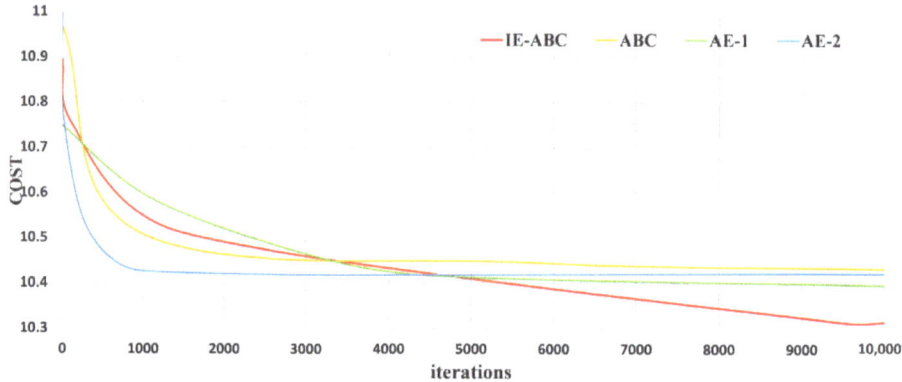

Figure 5. Ablation experimental results.

In order to explore the impact of different cost function definition strategies on solving the scheduling problem, a cost function considering energy consumption, time, and output is established. This function is compared against a cost function that disregards time and only considers energy consumption and output. The experimental results are presented in Table 3. The cost function value of the optimal solution obtained using the cost function considering energy consumption, time, and output is 10.31. In comparison, the cost function value of the optimal solution obtained ignoring time and only considering energy consumption and output is 10.63. The former is 3.1% lower than the latter. With the time-incorporated cost function, the energy consumption and time for the optimal solution to complete the task are 8.93×10^6 J and 620.0 s, respectively. Using the cost function without time, the energy consumption and time required for the optimal solution to complete the task are 9.00×10^6 J and 652.9 s, respectively. Compared to the latter, the former reduces energy consumption and task time by 0.78% and 5.04%, respectively.

Table 3. Simulation result of different cost function definition strategies.

Cost Function Definition Strategy	Cost	Consumed Energy (10^6 J)	Time (s)
Regarded-time Strategy	10.31	8.93	620.0
Disregarded-time Strategy	10.63	9.00	652.9

As shown in Figure 6, the Gantt chart illustrates the optimal scheduling scheme obtained using the cost function that disregarded time and only considered energy consumption and output. The meaning of different colors and numbers in Figure 6 is the same as that in Figure 4. Compared to the optimal scheduling scheme in Figure 4 found by comprehensively accounting for energy consumption, time, and output in the cost function, the time for each unmanned dump truck to complete its corresponding 20 tasks is increased. This indicates that incorporating time into the cost function along with energy consumption and output enabled optimization to find an improved solution.

To investigate the impact of different vector encoding strategies on the solution of the scheduling problem, we compared the solution vector encoding strategy proposed in Section 3.1 with the binary encoding strategy. In the binary encoding strategy, the allocation of tasks is represented by a binary string. For example, if there are two loading spots and three unloading spots, the scheduling sequence {01001} represents that the unmanned dump truck travels to No. 2 loading spot to load and then proceeds to No. 5 unloading spot to unload.

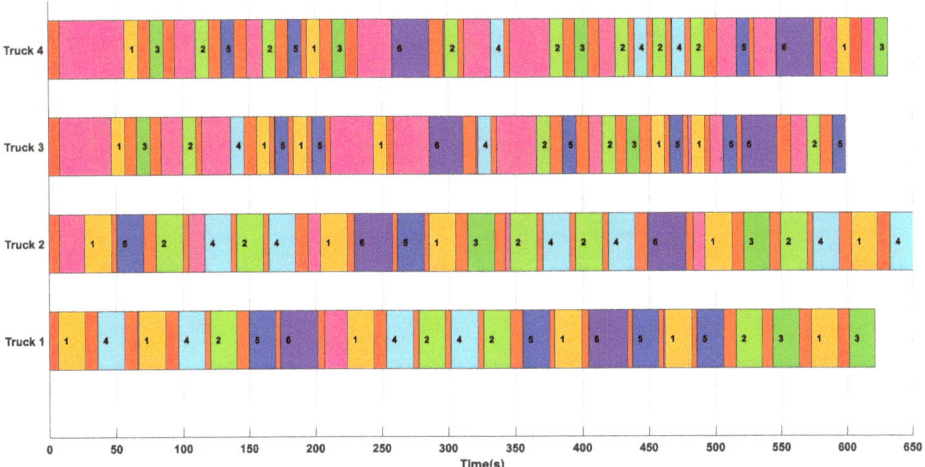

Figure 6. Scheduled Gantt chart with cost function that disregards time.

The comparative experimental results of the two encoding strategies are presented in Table 4. The cost function value of the optimal solution obtained using the proposed encoding strategy is 10.31, compared to 10.61 for the binary encoding strategy. The proposed encoding strategy's cost function value is 2.83% lower than that of the binary encoding strategy. The energy consumption and time for the proposed encoding strategy's optimal solution are 8.93×10^6 J and 620.0 s, respectively. In comparison, energy consumption and time for the binary encoding optimal solution are 9.07×10^6 J and 636.2 s, respectively. Compared to the binary encoding solution, the proposed encoding strategy's solution reduces energy consumption and task time by 1.54% and 2.55%, respectively.

Table 4. Simulation result of different encoding strategies.

Encoding Strategy	Cost	Consumed Energy (10^6 J)	Time (s)
Proposed Encoding	10.31	8.93	620.0
Binary Encoding	10.61	9.07	636.2

Figure 7 shows the Gantt chart for the optimal scheduling scheme obtained through binary encoding. The meaning of different colors and numbers in Figure 7 is the same as that in Figure 4. Compared to the Gantt chart in Figure 4 using the proposed encoding strategy, a larger time gap existed between the earliest finishing unmanned dump truck #3 and the latest finishing unmanned dump truck #2 for their respective 20 assigned tasks. Additionally, the total time is longer with the binary encoding strategy. This indicates that the encoding strategy used in this study enabled obtaining an improved scheduling scheme.

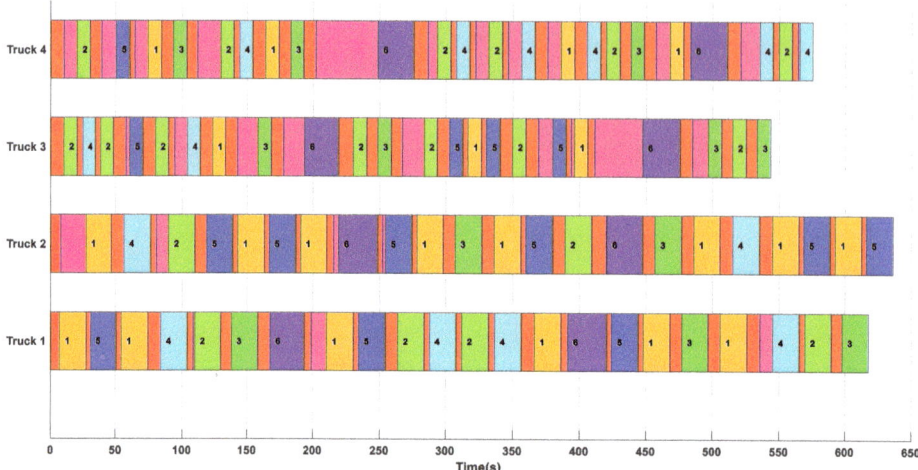

Figure 7. The scheduling Gantt chart with binary encoding strategy.

4.3. Discussion

The developed scheduling algorithm is suitable for the unmanned dump truck scheduling problem in a complex open-pit mine environment. First, for the complex terrain and road network in the open-pit mine area, the algorithm can select the best path according to the location and target of the unmanned dump truck to minimize energy consumption and time cost. Second, the algorithm can intelligently allocate the driving route for the unmanned dump truck according to the energy situation of the unmanned dump truck, so as to avoid the occurrence of energy exhaustion. In addition, the algorithm can reasonably arrange the driving sequence and work allocation of unmanned dump trucks according to the location and task requirements between unmanned dump trucks, so as to maximize the overall production efficiency. However, it is important to note that the algorithm may face certain technical limitations in the open-pit mine environment. Factors such as terrain changes and weather conditions can impact the algorithm's performance.

5. Conclusions

This article has proposed a high-precision multi-vehicle collaborative scheduling proposition model considering micro-space-time factors to solve the unmanned dump truck scheduling problem in open-pit mines.

The optimization objective comprehensively considers energy consumption, time, and output. In addition to the loading and unloading activities, the unmanned dump truck also considers charging demand in the scheduling model. The model incorporates Voronoi graph search and optimal time matching of spatial paths. It aims to provide a better task decision planning solution for unmanned dump trucks in mining areas. To effectively solve the scheduling problem, an improved artificial bee colony algorithm is proposed. The original algorithm is enhanced in the global search process and the re-initialization process. An ablation experiment is conducted to explore the impact of these improvements on the optimization process. The ablation experiment result shows that changing these two items has promoted the optimization search. In addition, comparative simulation experiments are conducted using different algorithms, cost function definition strategies, and encoding strategies. Comparative simulations indicate the proposal can reduce energy consumption and time. Compared to the models utilizing ABC algorithms, cost function strategy definition without considering time, and binary encoding strategy, the proposed model and method achieved reductions of 0.67%, 3.1%, and 2.83% in the comprehensive cost functions of energy consumption, time, and output, respectively. Moreover, simulation results demonstrate that the proposed model and method offer an effective solution for scheduling decisions in mining areas.

There are numerous factors that impact the overall cost of the unmanned dump truck scheduling problem. In the future, it is worth exploring the optimization of speed and scheduling arrangements in the event of unmanned dump truck failures. These areas present promising avenues for further research.

Author Contributions: Conceptualization, Y.F. and X.P.; methodology, X.P.; validation, Y.F. and X.P. All authors have read and agreed to the published version of the manuscript.

Funding: This work was supported by the National Natural Science Foundation of China under Grant 52275105.

Institutional Review Board Statement: Not applicable.

Informed Consent Statement: Not applicable.

Data Availability Statement: The simulation data used to support the findings of this study are available from the corresponding author upon request.

Conflicts of Interest: The authors declare no conflict of interest.

References

1. Tang, J.; Zhao, B.; Yang, C.; Zhou, C.; Chen, S.; Ni, X.; Ai, Y. An Architecture and Key Technologies of Autonomous Truck Dispatching System in Open-pit Mines. In Proceedings of the 2022 International Conference on Cyber-Physical Social Intelligence (ICCSI), Nanjing, China, 18–21 November 2022; pp. 122–127.
2. Zhang, B.; Li, D.; Ba, T.; Jiang, D.; Qiu, X.; Wang, T. Obstacle Detection and Tracking of Unmanned Mine Car Based on 3D Lidar. In Proceedings of the 2022 International Conference on Cyber-Physical Social Intelligence (ICCSI), Nanjing, China, 18–21 November 2022; pp. 222–228.
3. Ma, Y.; Wang, Z.; Yang, H.; Yang, L. Artificial intelligence applications in the development of autonomous vehicles: A survey. *IEEE/CAA J. Autom. Sin.* **2020**, *7*, 315–329. [CrossRef]
4. Chen, L.; Hu, X.; Wang, G.; Cao, D.; Li, L.; Wang, F.Y. Parallel Mining Operating Systems: From Digital Twins to Mining Intelligence. In Proceedings of the 2021 IEEE 1st International Conference on Digital Twins and Parallel Intelligence (DTPI), Beijing, China, 15 July–15 August 2021; pp. 469–473.
5. Moradi Afrapoli, A.; Askari-Nasab, H. Mining Fleet Management Systems: A Review of Models and Algorithms. *Int. J. Min. Reclam. Environ.* **2019**, *33*, 42–60. [CrossRef]
6. Alarie, S.; Gamache, M. Overview of Solution Strategies Used in Truck Dispatching Systems for Open Pit Mines. *Int. J. Surf. Min. Reclam. Environ.* **2002**, *16*, 59–76. [CrossRef]
7. Li, B.; Zhang, Y.; Shao, Z.; Jia, N. Simultaneous Versus Joint Computing: A Case Study of Multi-vehicle Parking Motion Planning. *J. Comput. Sci.* **2017**, *20*, 30–40. [CrossRef]
8. Ge, S.; Wang, F.Y.; Yang, J.; Ding, Z.; Wang, X.; Li, Y.; Teng, S.; Liu, Z.; Ai, Y.; Chen, L. Making standards for smart mining operations: Intelligent vehicles for autonomous mining transportation. *IEEE Trans. Intell. Veh.* **2022**, *7*, 413–416. [CrossRef]
9. Gleixner, A. Solving Large-Scale Open Pit Mining Production Scheduling Problems by Integer Programming. Ph.D. Thesis, Zuse Institute Berlin (ZIB), Berlin, Germany, 2009.
10. Li, B.; Yin, Z.; Ouyang, Y.; Zhang, Y.; Zhong, X.; Tang, S. Online Trajectory Replanning for Sudden Environmental Changes During Automated Parking: A Parallel Stitching Method. *IEEE Trans. Intell. Veh.* **2022**, *7*, 748–757. [CrossRef]
11. Tian, F.; Zhou, R.; Li, Z.; Li, L.; Gao, Y.; Cao, D.; Chen, L. Trajectory Planning for Autonomous Mining Trucks Considering Terrain Constraints. *IEEE Trans. Intell. Veh.* **2021**, *6*, 772–786. [CrossRef]
12. Li, B.; Ouyang, Y.; Li, X.; Cao, D.; Zhang, T.; Wang, Y. Mixed-integer and Conditional Trajectory Planning for an Autonomous Mining Truck in Loading/Dumping Scenarios: A Global Optimization Approach. *IEEE Trans. Intell. Veh.* **2022**, *8*, 1512–1522. [CrossRef]
13. Li, B.; Ouyang, Y.; Li, L.; Zhang, Y. Autonomous Driving on Curvy Roads Without Reliance on Frenet Frame: A Cartesian-Based Trajectory Planning Method. *IEEE Trans. Intell. Transp. Syst.* **2022**, *23*, 15729–15741. [CrossRef]
14. Chen, L.; Xie, J.; Zhang, X.; Deng, J.; Ge, S.; Wang, F.Y. Mining 5.0: Concept and Framework for Intelligent Mining Systems in CPSS. *IEEE Trans. Intell. Veh.* **2023**, *8*, 3533–3536. [CrossRef]
15. Patterson, S.R.; Kozan, E.; Hyland, P. Energy Efficient Scheduling of Open-pit Coal Mine Trucks. *Eur. J. Oper. Res.* **2017**, *262*, 759–770. [CrossRef]
16. Zhang, L.; Shan, W.; Zhou, B.; Yu, B. A Dynamic Dispatching Problem for Autonomous Mine Trucks in Open-pit Mines Considering Endogenous Congestion. *Transp. Res. Part C Emerg. Technol.* **2023**, *150*, 104080. [CrossRef]

17. Yuan, W.; Li, D.; Jiang, D.; Jia, Y.; Liu, Z.; Bian, W. Research on Real-Time Truck Dispatching Model in Open-pit Mine Based on Improved Genetic Algorithm. In Proceedings of the 2022 International Conference on Cyber-Physical Social Intelligence (ICCSI), Nanjing, China, 18–21 November 2022; pp. 234–239.
18. Wang, Z.; Wang, J.; Zhao, M.; Guo, Q.; Zeng, X.; Xin, F.; Zhou, H. Truck Dispatching Optimization Model and Algorithm Based on 0-1 Decision Variables. *Math. Probl. Eng.* **2022**, *2022*, 5887672. [CrossRef]
19. Bastos, G.S.; Souza, L.E.; Ramos, F.T.; Ribeiro, C.H. A Single-dependent Agent Approach for Stochastic Time-dependent Truck Dispatching in Open-pit Mining. In Proceedings of the 2011 14th International IEEE Conference on Intelligent Transportation Systems (ITSC), Washington, DC, USA, 5–7 October 2011; pp. 1057–1062.
20. Zhang, X.; Chen, L.; Ai, Y.; Tian, B.; Cao, D.; Li, L. Scheduling of Autonomous Mining Trucks: Allocation Model Based Tabu Search Algorithm Development. In Proceedings of the 2021 IEEE International Intelligent Transportation Systems Conference (ITSC), Indianapolis, IN, USA, 19–22 September 2021; pp. 982–989.
21. Bao, H.; Zhang, R. Study on Optimization of Coal Truck Flow in Open-pit Mine. *Adv. Civil Eng.* **2020**, *2020*, 8848140. [CrossRef]
22. Wang, Y.; Liu, W.; Wang, C.; Fadzil, F.; Lauria, S.; Liu, X. A Novel Multi-Objective Optimization Approach with Flexible Operation Planning Strategy for Truck Scheduling. *IJNDI* **2023**, *2*, 100002. [CrossRef]
23. Zhang, S.; Lu, C.; Jiang, S.; Shan, L.; Xiong, N.N. An Unmanned Intelligent Transportation Scheduling System for Open-Pit Mine Vehicles Based on 5G and Big Data. *IEEE Access* **2020**, *8*, 135524–135539. [CrossRef]
24. Ahumada, G.I.; Herzog, O. Application of Multiagent System and Tabu Search for Truck Dispatching in Open-pit Mines. *ICAART* **2021**, *1*, 160–170.
25. Chang, Y.; Ren, H.; Wang, S. Modelling and Optimizing an Open-pit Truck Scheduling Problem. *Discrete Dyn. Nat. Soc.* **2015**, *2015*, 745378. [CrossRef]
26. Afrapoli, A.M.; Tabesh, M.; Askari-Nasab, H. A Multiple Objective Transportation Problem Approach to Dynamic Truck Dispatching in Surface Mines. *Eur. J. Oper. Res.* **2019**, *276*, 331–342. [CrossRef]
27. Zhang, X.; Guo, A.; Ai, Y.; Tian, B.; Chen, L. Real-Time Scheduling of Autonomous Mining Trucks via Flow Allocation-Accelerated Tabu Search. *IEEE Trans. Intell. Veh.* **2022**, *7*, 466–479. [CrossRef]
28. Smith, A.; Linderoth, J.; Luedtke, J. Optimization-based Dispatching Policies for Open-pit Mining. *Optim. Eng.* **2021**, *22*, 1347–1387. [CrossRef]
29. Zeng, W.; Baafi, E.Y.; Fan, H. A Simulation Model to Study Truck-allocation Options. *J. S. Afr. Inst. Min. Metall.* **2022**, *122*, 741–750. [CrossRef] [PubMed]
30. de Melo, W.B. Optimization of truck allocation in open pit mines using differential evolution algorithm. *Int. J. Innov. Res.* **2021**, *9*, 338–350. [CrossRef]
31. Yeganejou, M.; Badiozamani, M.; Moradi-Afrapoli, A.; Askari-Nasab, H. Integration of Simulation and Dispatch Modelling to Predict Fleet Productivity: An Open-pit Mining Case. *Min. Technol.* **2022**, *131*, 67–79.
32. Li, N.; Zhang, J.; Zhang, S.; Hou, X.; Liu, Y. The Influence of Accessory Energy Consumption on Evaluation Method of Braking Energy Recovery Contribution Rate. *Energy Convers. Manag.* **2018**, *166*, 545–555. [CrossRef]
33. Liu, T.; Zou, Y.; Liu, D. Energy Management for Battery Electric Vehicle with Automated Mechanical Transmission. *Int. J. Veh. Des.* **2016**, *70*, 98–112. [CrossRef]
34. Li, B.; Tang, S.; Zhang, Y.; Zhong, X. Occlusion-Aware Path Planning to Promote Infrared Positioning Accuracy for Autonomous Driving in a Warehouse. *Electronics* **2021**, *10*, 3093. [CrossRef]

Disclaimer/Publisher's Note: The statements, opinions and data contained in all publications are solely those of the individual author(s) and contributor(s) and not of MDPI and/or the editor(s). MDPI and/or the editor(s) disclaim responsibility for any injury to people or property resulting from any ideas, methods, instructions or products referred to in the content.

Article

Joint Dispatching and Cooperative Trajectory Planning for Multiple Autonomous Forklifts in a Warehouse: A Search-and-Learning-Based Approach

Tantan Zhang [1,*], Hu Li [1], Yong Fang [1], Man Luo [2] and Kai Cao [2]

1. College of Mechanical and Vehicle Engineering, Hunan University, Changsha 410082, China; lihu@hnu.edu.cn (H.L.); fangyong@hnu.edu.cn (Y.F.)
2. Dongfeng USharing Technology Co., Ltd., Wuhan 430056, China; tc-luoman@dfmc.com.cn (M.L.); caok@dfmc.com.cn (K.C.)
* Correspondence: zhangtantan@hnu.edu.cn

Abstract: Dispatching and cooperative trajectory planning for multiple autonomous forklifts in a warehouse is a widely applied research topic. The conventional methods in this domain regard dispatching and planning as isolated procedures, which render the overall motion quality of the forklift team imperfect. The dispatching and planning problems should be considered simultaneously to achieve optimal cooperative trajectories. However, this approach renders a large-scale nonconvex problem, which is extremely difficult to solve in real time. A joint dispatching and planning method is proposed to balance solution quality and speed. The proposed method is characterized by its fast runtime, light computational burden, and high solution quality. In particular, the candidate goals of each forklift are enumerated. Each candidate dispatch solution is measured after concrete trajectories are generated via an improved hybrid A* search algorithm, which is incorporated with an artificial neural network to improve the cost evaluation process. The proposed joint dispatching and planning method is computationally cheap, kinematically feasible, avoids collisions with obstacles/forklifts, and finds the global optimum quickly. The presented motion planning strategy demonstrates that the integration of a neural network with the dispatching approach leads to a warehouse filling/emptying mission completion time that is 2% shorter than the most efficient strategy lacking machine-learning integration. Notably, the mission completion times across these strategies vary by approximately 15%.

Keywords: autonomous forklift; cooperative trajectory planning; joint dispatching and planning; Hybrid A* search algorithm; artificial neural network

Citation: Zhang, T.; Li, H.; Fang, Y.; Luo, M.; Cao, K. Joint Dispatching and Cooperative Trajectory Planning for Multiple Autonomous Forklifts in a Warehouse: A Search-and-Learning-Based Approach. *Electronics* **2023**, *12*, 3820. https://doi.org/10.3390/electronics12183820

Academic Editor: Felipe Jiménez

Received: 12 July 2023
Revised: 1 September 2023
Accepted: 7 September 2023
Published: 9 September 2023

Copyright: © 2023 by the authors. Licensee MDPI, Basel, Switzerland. This article is an open access article distributed under the terms and conditions of the Creative Commons Attribution (CC BY) license (https:// creativecommons.org/licenses/by/ 4.0/).

1. Introduction

The increasing demands in the logistics industry all over the world have driven researchers and engineers to focus on developing intelligent transportation systems aimed at enhancing logistics efficiency [1,2]. One prominent application in this domain is unmanned warehouse systems [3]. As a typical component in an unmanned warehouse, an autonomous forklift transports parcels more efficiently than one driven manually because the former does not induce objective mistakes such as fatigue, anxiety, impatience, or anger [4]. Multiple autonomous forklifts should work together when the delivery burden is heavy [5]. Deploying multiple autonomous forklifts enhances delivery efficiency if the inter-forklift cooperation potential can be maximized. The typical modules that influence delivery efficiency include delivery task dispatch [6], cooperative trajectory planning [7,8], and control [9]. This paper is focused on the dispatching and cooperative trajectory planning schemes.

1.1. Related Work

1.1.1. Dispatching Methods for Multiple Forklifts

A complete multi-forklift delivery planning system consists of two functions: delivery task dispatch and cooperative trajectory planning. Before cooperative trajectory planning, the goal point of each forklift is assigned in the dispatching phase [10]. Weidinger et al. [6] proposed a metaheuristic-based method in which assignment candidates are pruned a priori to facilitate the solution process [11]. A similar idea was proposed by Zhang et al. [12] to dispatch multiple automated guided vehicles (AGVs) in a matrix manufacturing workshop. However, both methods run slowly; thus, they cannot meet the real-time computation demand in a warehouse [6,12]. Lin et al. built a multi-AGV dispatching system via network structure together with simplex decision variables; in this system, an evolutionary algorithm minimizes the completion time of all AGVs in a formulated network optimization problem. However, References [6,12,13] shared a common limitation of assuming a uniform speed for AGVs. Moreover, the inter-vehicle collision avoidance problem is reduced to an oversimplified constraint, ensuring solely nonoverlapping time intervals per stopover.

Furthermore, recent studies have focused on the task integration of one AGV instead of considering how multiple AGVs cooperatively operate within the confined area. Bao et al. [14] proposed a heuristic method based on an auction strategy for a multi-AGV task dispatch scheme considering complex factors (such as pod repositioning). The concerned dispatch scheme is inherently an optimization problem with complex cost terms and constraints facilitated by the proposed auction strategy. Lee et al. [15] proposed a two-stage dispatching method. In particular, the first stage deals coarsely with the delivery efficiency and delivery flow balance by solving a bi-objective optimization problem. The result indicates how the parcels to be picked can be clustered. At stage 2, vehicles are dispatched to complete the clustered missions. Dividing the original scheme into two stages largely reduces the number of dispatch candidates without losing the optimum. Machine-learning-based dispatching methods have also been proposed [16]. The formulated reward functions efficiently simplify the dispatch scheme, particularly when complex factors are considered [17,18]. However, few vehicle kinematics is considered in [14,15], and the dispatching phase is fully separated from the trajectory planning strategy. This approach results in difficulties in maximizing overall delivery efficiency.

1.1.2. Cooperative Trajectory Planning Methods for Multiple Forklifts

Cooperative trajectory planning follows the aforementioned dispatching phase. The prevalent cooperative trajectory planners are based on model predictive control [19], which is highlighted by its fast feature while strictly satisfying safety-related constraints. The artificial potential field method is similarly widely applied in trajectory planning, but it may encounter difficulty finding paths through narrow passages [20].

Ma et al. [21] converted constrained time-varying nonlinear programming problems to general unconstrained optimization problems by properly designing a penalty function. Thereafter, a particle swarm optimization method was employed to plan the motion of multiple robots sequentially in a double warehouse with two elevators. However, the optimization phase, along with other methods based on optimization, can significantly increase the computational burden [22–24], which can be reduced by forming model-based paths because warehouses are generally structured.

Yang et al. [1] proposed a strategy in which a time-varying dynamic evaluation function is formed based on a network congestion diffusion model to quantify the degree of road congestion. Hereafter, an improved A* search algorithm and a time window algorithm were combined as a hierarchical planning method to search the idle path and avoid collisions. A path planning framework was designed by Zhou et al. [25] to simultaneously reduce the cost of operation and the path for AGVs in airport parcel loading scenarios. An ant colony optimization method was used to optimize the parcel pickup sequencing by ignoring other moving vehicles, while Dijkstra's algorithm was employed to determine the shortest route of each AGV. Zacharia et al. developed a joint routing and motion planning method for AGVs that addresses uncertainties in demands and travel times. Their approach combines a scheduler for updating destination resources during navigation and integrates a fuzzy-based genetic algorithm with A* search to handle capacity and distance variations [26]. Nonetheless, the vehicle kinematics considered in [1,25,26] remain oversimplified for industrial use even in the trajectory planning phase.

In other studies, the paths of vehicles were assumed to be predefined, and only the longitude trajectories were investigated. Kneissl et al. [27] formulated a method in which potential collision zones are continuously detected. Moreover, the right-of-way is granted to the first arriving vehicle while all the other vehicles involved stop and wait. Dresner et al. proposed confronting the analogous problem of conflict zones with a reservation-based system; in this system, vehicles request and receive time slots from the intersection while they pass [28]. Similarly, a discrete-event logic, which is comparable with a conventional right-handed bidirectional traffic system, was designed by Guney et al. [29] to handle the priorities of the AGVs in a warehouse dynamically. Thus, the need for computationally demanding heuristic searches is eliminated to ease strategy implementation in real-life industrial applications. Furthermore, Digani et al. [30] proposed an obstacle-free path generation method to deal with local deviations from the predetermined path. In the proposed method, new paths are generated via polar spline curves. However, the aisles in a warehouse cannot be fully exploited when certain traffic laws in [27–30] are strictly enforced.

1.1.3. Joint Dispatching and Planning Methods for Multiple Forklifts

Most of the existing dispatching studies, e.g., [6,10–18], cannot accurately evaluate the candidate choices, possibly preventing the downstream planning module from achieving global optimality [1,21–30]. Thus, combining the dispatching and planning phases is naturally considered. The multi-agent path finding (MAPF) problem in its classical form is an effective approach for simplifying complex warehouse scenarios and facilitating cooperative solutions for dispatching and planning. In the MAPF problem, time is discretized into steps, allowing vehicles to either move or wait during each step [31]. Consequently, it becomes challenging to plan trajectories for vehicles with varying velocities or based on specific kinematic constraints. To address this limitation, researchers have explored extensions of the MAPF problem to accommodate such complexities. Among those extensions, Zhang et al. [32] designed a joint strategy to deal with an automatic valet parking system, in which a travel-distance-related reward function combined with a deep reinforcement learning technique was used to allocate the target parking spaces. The parking lot was segmented into local regions, and a rule-based right-of-way assignment strategy was applied to solve collisions and deadlocks. A simplified trajectory planning algorithm based on the car-following model [33] served as a tool to solve the trajectories of multiple AGVs when no potential collision was involved. A similar strategy was proposed by Lee et al. [34] for a supply-chain-connected warehouse. In their work, a cloud-based semiautomatic warehouse management system assigns tasks to mobile robots to optimize resource allocation. A robot control system executes an improved A* search algorithm to generate the path of each AGV. Then, potential collisions, named stay-on, head-on, and cross-conflict, are identified and solved by following certain priority-based rules. Redispatchment of the AGV with low priority is triggered as the conflict cannot be prevented by those basic rules.

With regard to joint strategies, the studies above [32,34] can deal with large-scale AGV-based scenarios. However, these studies were concerned with vehicles possessing the simple kinematics of unicycle (differential-drive) robots [35] and generally focused on the construction of maneuverable systems, ignoring the overall optimal solutions concerning warehouse operations. Furthermore, during the trajectory planning phase, they initially planned only the paths and ignored other moving obstacles. Such considerations substantially reduce the risk of collisions and simplify the evaluation of the traveling difficulties pertaining to one potential task relative to the corresponding traveling distance. In other words, it remains unknown the specific trajectory pertaining to each AGV as the jointed strategy is finished, and this trajectory is dependent on local scenes when conflict zones are involved. Thus, the results become unreliable when such strategies are applied to a warehouse of automated forklifts with complex kinematics. As a conclusion of this subsection, it deserves to develop a joint dispatching and planning method to balance the forks' motion quality and reaction speed in an unmanned warehouse.

1.2. Motivations

This study aims to substantially improve the efficiency of cooperative operations among multiple autonomous forklifts by seamlessly integrating the dispatching and cooperative trajectory planning phases. Our primary objective is to address the limitations of existing dispatching methods, which often overlook low-level forklift kinematic capability. To overcome this challenge, we opt for the implementation of a graph search process in this phase. Moreover, to ensure a robust solution that avoids getting trapped in local optima, we chose to incorporate a machine-learning-based technique. In the trajectory planning phase, we recognize that optimization-based methods are computationally expensive. As such, our secondary objective is to develop an alternative search algorithm that employs a model-based approach. This algorithm is designed to be both velocity-aware and sequentially solvable, striking a balance between accuracy and computational efficiency.

1.3. Contributions

The core contribution of this paper is the proposal of a joint framework, which is promising to reduce the computational burden because all formulations involved are explicitly expressed. Concretely, the dispatching stage can enhance the multi-vehicle task solution quality because it considers the future trajectory pertaining to each forklift. Moreover, the kinematically feasible and safe trajectory of each forklift can be quickly generated through our proposed method at the trajectory planning stage, due to the removal of optimization-based methods.

1.4. Organization

In the rest of this paper, Section 2 formulates the in-warehouse delivery problem. Section 3 provides the score-based dispatching technique, in which ANN is applied to avoid deadlocks in evaluating the cost of each candidate dispatching option. Section 4 introduces the trajectory planning method, namely a model-based velocity-aware hybrid A* search algorithm. Section 5 integrates the two aforementioned methods to develop a joint dispatching and cooperative trajectory planning framework, followed by Section 6, where comparative simulation results are present. Conclusions are drawn in Section 7, finally.

2. Problem Statement

Forklifts are used to deliver goods between fixed picking stations and predetermined shelf areas during delivery tasks in warehouses. The passages are generally designed to be narrow, and they merely allow turning maneuvers with a minimum radius and the passing of only two vehicles. Hence, conflicts arise when multiple forklifts cooperatively operate within a single warehouse.

Within one subtask during the filling of one warehouse, there are two stages: first, the initial pose and the final one should be assigned to one forklift as the dispatching stage; second, the trajectory planning stage generates a trajectory by avoiding collisions with any static or moving objects and satisfying the vehicle kinematics.

2.1. Warehouse Layout

A typical small warehouse layout is schematized in Figure 1. In this warehouse, six separated shelf clusters, denoted as s_d ($d = 1, 2, \cdots, 6$), are placed and provide areas to store goods. Continuous lines indicate shelf walls, through which forklifts cannot move. Two firewalls, represented by rows of grey squares, are present between shelf cluster s_1 and

s_3 as well as cluster s_2 and s_4. Meanwhile, four picking stations, marked with slender solid rectangles and p_r ($r = 1, 2, 3, 4$), are located in both extremities of the vertical and wide passage in Figure 1. Four forklifts can enter the passage of each row of the shelf clusters, as long as neither stored stacks nor other vehicles block the route.

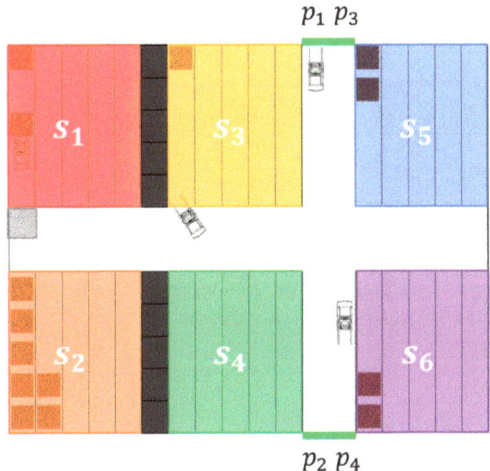

Figure 1. Schematic of a warehouse layout.

As presented in Figure 2, each shelf cluster has the capacity to accommodate varying numbers of goods stacks with strategically positioned notches in the arrangement designed to suit forklift kinematics during turns. This aspect will be elaborated on in Section 2.2. Each stack is marked with a number in Figure 2 to represent the order in which a shelf cluster can be filled. The shelf filling state and the vehicle state can be effectively expressed by noting the covered grids when the warehouse is divided by the squares outlined by grey dashed lines in Figure 2. In addition, when one forklift is unloading goods within a shelf cluster, its fork side should point to the stack position (cf. upper left forklift schematic and within shelf cluster s_1 in Figure 1). Similarly, when one forklift is picking goods at a picking station, the fork side should point to the station position (cf. upper forklift schematic and picking station p_1 in Figure 1).

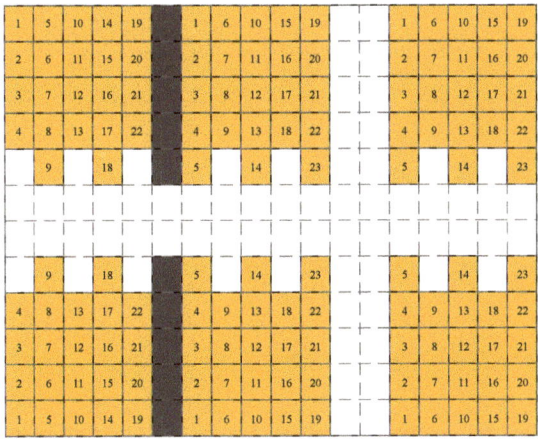

Figure 2. Filling sequence pertaining to each shelf cluster.

2.2. Kinematics of a Forklift Vehicle

As reported in Figure 3, a forklift can be described as a front-steering vehicle if the fork part of the vehicle is treated as the rear side. The corresponding kinematic formulas write:

$$\begin{cases} \frac{dx(t)}{dt} = v(t) \cdot \cos \theta(t) \\ \frac{dy(t)}{dt} = v(t) \cdot \sin \theta(t) \\ \frac{dv(t)}{dt} = a(t) \\ \frac{d\theta(t)}{dt} = \frac{v(t) \cdot \tan \phi(t)}{l} \\ \frac{d\phi(t)}{dt} = w(t) \end{cases} \quad (1)$$

where t is time; P, located at coordinate (x, y), indicates the mid-point of the rear wheel axis; and θ, v, a, ϕ, and ω respectively denote the orientation angle, linear velocity pertaining to point P, acceleration, steering angle of the front wheels, and steering rate. Furthermore, l stands for the wheelbase length, m denotes the rear overhang length, n refers to the front overhang length, and $2b$ is the car width. Given that the initial values as well as $\omega(t)$ and $a(t)$ are provided, the state variables can be calculated through integration over the dynamic process.

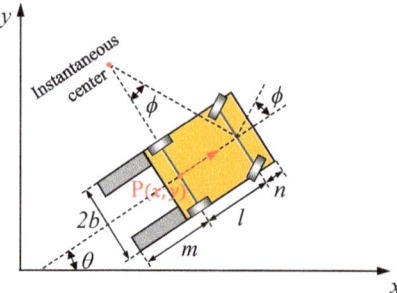

Figure 3. Parametric notations related to vehicle size and kinematics.

Meanwhile, a few boundaries are imposed on the state profiles over the entire simulation period throughout all dynamic maneuvers:

$$\begin{cases} |a(t)| \leq a_{max} \\ |v(t)| \leq v_{max} \\ |\phi(t)| \leq \phi_{max} \\ |\omega(t)| \leq \omega_{max} \end{cases} \quad (2)$$

where a_{max}, v_{max}, ϕ_{max}, and ω_{max} respectively indicate the upper limits of the corresponding variables.

3. ANN-Combined Score-Based Dispatching Approach

Filling a warehouse in an orderly manner requires several forklifts to perform multiple deliveries and return subtasks. In the current work, subtasks are assigned to different vehicles sequentially. With regard to such subtasks, selecting which vehicle will be used to plan the new trajectory and determining which goal coordinates the forklift is going to should comprise the fundamental initialization. Therefore, a dispatching system is necessary to solve these problems.

The dispatching approach (cf. Algorithm 1) utilizes the planned trajectories **T** of all vehicles, along with map information map, and vector **F** representing filled stacks for different clusters, as its inputs. The core of this approach lies within a while loop, wherein the potential subtask undergoes iterative updates (with a preset maximum iteration number $iter_{dispatch}$) until it reaches an optimal state, as defined by the proposed method.

Outside of the loop, the function rank() sorts all vehicles based on **t**, arranging them in ascending order from first to last. This sorting process generates a ranking vector **R** consisting of four vehicle indices. If $fail = 1$, indicating a failure in the trajectory search between $P_i(x_i, y_i)$ at instant t_i and $P_f(x_f, y_f)$, the corresponding vehicle is repositioned at the end in **R** and flagged as having been selected as $n_{current}$. In the loop, the algorithm is divided into two parts. The first one (lines 4 to 13) concerns the selection of the current investigated vehicle with an index of $n_{current}$, and the second part (lines 14 to 30), featuring a scoring system combined with the results of an ANN method, determines the goal coordinate for the current subtask. Notably, Algorithm 1 is applied when the ANN correction system, which is elaborated on in Section 3.2, is enforced.

Algorithm 1: ANN combined score-based dispatching algorithm

$\left[n_{current}, P_i, P_f, t_i\right] \leftarrow \text{Dispatch}\left(\mathbf{T}, \mathbf{F}, fail, P_i, P_f, map\right)$

1. Initialize $\alpha \leftarrow 0$;
2. $\mathbf{R} \leftarrow \text{rank}\left(\mathbf{T}, P_i, P_f, fail\right)$;
3. **while** $iter < iter_{dispatch}$, **do**
4.   **if** $\alpha = 0$, **then**
5. $[n_{current}, t_i] \leftarrow \text{SelectInitialState}(\mathbf{R})$;
6.   **else**
7.   **if** $\text{CheckSelection}(\mathbf{R}) > 0$, **then**
8. $[n_{current}, t_i] \leftarrow \text{SelectAlteredState}(\mathbf{R})$;
9.   **else**
10. $[n_{current}, t_i] \leftarrow \text{SelectBackupState}(\mathbf{R})$;
11.   **end if**
12.   **end if**
13. $P_i = \text{SetInitialPose}(\mathbf{T}, n_{current}, map)$;
14.   **if** $\text{CheckDeliverTask}(P_i)$ is **true**, **then**
15. $S_d = \text{PreAstarDeliver}(P_i, \mathbf{F}, map)$;
16.   **if** $\max(S_d) > 0$, **then**
17. $P_f = \text{SetFinalPose}(S_d, map)$;
18.   **return**;
19.   **else**
20. $\alpha \leftarrow 1$;
21.   **end if**
22.   **else**
23. $S_r = \text{PreAstarReturn}(P_i, map)$;
24.   **if** $\max(S_r) > 0$, **then**
25. $P_f = \text{SetFinalPose}(S_r, map)$;
26.   **return**;
27.   **else**
28. $\alpha \leftarrow 1$;
29.   **end if**
30.   **end if**
31. **end while**
32. **return**;

3.1. Vehicle Selection and Initial Pose of a New Subtask

The function SelectInitialState() selects the vehicle, ranking the first one as $n_{current}$, and sets t_i, which is the initial instant of the trajectory to be planned, as the ending instant of the last trajectory pertaining to $n_{current}$. Failure may occur in the determination of the new trajectory for the fork $n_{current}$ in the new subtask because other forklifts may block the only corresponding route for a considerable time. Under such circumstances, a flag variable α is set to 1, and all vehicles in **R** are checked to see if they have been selected as $n_{current}$ once for the current subtask by CheckSelection(). The function SelectAlteredState()

is then utilized. In this function, the first motion-finished vehicle is discarded, and the other forklifts are subsequently selected in turn as $n_{current}$ on the basis of the rankings in **R** until the trajectory can be formed. Furthermore, no trajectory can be successfully planned for all forklifts at certain moments. In this case, SelectBackupState() is applied, in which the vehicle that ranks last in **R** is selected, and the t_i of the new trajectory is postponed for a fixed time length of Δt_i relative to the end of the last subtask for vehicle $n_{current}$. The vehicle $n_{current}$ final stopping pose $P_i(x_i, y_i, \theta_i)$ is set as the initial pose of the new subtask by function SetInitialPose().

3.2. Scoring System

A scoring system is applied to decide the goal coordinate of the new subtask. First, the function CheckDeliverTask(P_i) is initially employed to ascertain whether the planned trajectory involves heading to clusters or returning. The functions PreAstarDeliver() and PreAstarReturn() are then applied separately depending on whether the goal is a rack cluster or a picking station. In both functions, the grid networks, outlined by light colors in Figure 2, indicate the nodes used to define the location of a vehicle and stacks. The resolution of such nodes is purposely reduced with the aim of lowering computation costs. Given that the nodes are defined, a time dimension involved preliminary A* search algorithm, whose expansion manner is presented in Figure 4, is used to generate preliminary trajectories that link the starting pose $P_i(x_i, y_i)$ at instant t_i to each potential target coordinate. Five patterns in total for this search algorithm are applied to vaguely indicate the possible maneuvers a vehicle could perform. In particular, manner 5 in Figure 4 expands only in the time dimension, representing the stopping condition of a virtual forklift. The time consumed derived from this algorithm for a virtual vehicle represented by one node is then applied as a parameter to evaluate the difficulty grades of reaching different goal poses. Other vehicles and walls are treated as obstacles during the search. Notably, the orientation angles θ for the initial and final poses are not required to be determined in such a system. Thus, θ is not considered a dimension in this search for the sake of calculation simplification.

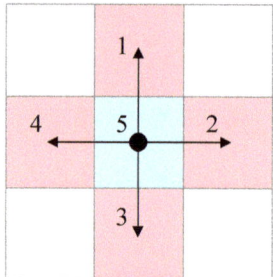

Figure 4. Expansion manner of the preliminary hybrid A* search.

In the function PreAstarDeliver(), the function $S_{d,0}$ is applied to evaluate the scores pertaining to different target stack locations s_d (cf. Figure 1) when a subtask with one stack as the goal is considered. It writes:

$$\begin{cases} S_{d,0} = C_1 G_d + J_{d,0} \\ J_{d,0} = \begin{cases} H + C_2 I_d + C_3 & \text{preliminary trajectory is planned} \\ C_4 & \text{preliminary trajectory planning is failed within } iter_{pre} \end{cases} \\ H = -\Delta t_{cover} \\ I_d = \begin{cases} t_{find} - t_i + C_5 & s_d \in \{s_1, s_2\} \\ t_{find} - t_i & s_d \in \{s_3, s_4, s_5, s_6\} \end{cases} \end{cases} \quad (3)$$

where G_d denotes the number of stacks to be filled/emptied in the target shelf cluster s_d in order to balance the warehouse filling/emptying mission in different clusters. $J_{d,0}$ stands for the approximate difficulty to reach different goal coordinates in shelf cluster s_d. C_4 is a negative constant applied when the goal in practice is not reachable. H indicates time length Δt_{cover}. Δt_{cover} counts the time units during which other vehicles occupy the shelf entrance node (cf. the grey grid for the upper left forklift in s_1 in Figure 1) of the goal pose within a fixed time length Δt_{entry} after the virtual vehicle entering the rack passage. The reason Δt_{cover} is added as a parameter is that, in practice, the availability of the above-mentioned node is critical during the planning of the final trajectory that considers kinematics. I_d is a function of the time length $\left(t_{find} - t_i\right)$ consumed to arrive at the target in this preliminary search. Notably, this function is built to normalize the results for different shelf clusters because shelf clusters s_1 and s_2 are far from the picking stations, and delivering goods to these locations consumes much time. Furthermore, a negative constant C_4 is assigned to $J_{d,0}$ when the target cannot be reached through the search within a predefined maximum iteration number $iter_{pre}$. Meanwhile, $C_1, C_2, \ldots, C_5$ are calibration parameters. Among these, a substantial weighting coefficient, C_1, is allocated to regulate the stacks filled in each shelf cluster; aiming for balance, C_2, C_3, C_4, and C_5 are designed to quantitatively assess scores with respect to time considerations. If $S_{d,0} \leq 0$ is derived within $iter_{pre}$, the vehicle kinematics-considered trajectory is difficult to find. Thus, the flag variable α is set to 1 in the initialization cycle, where ANN is not enforced, and $n_{current}$ should be reassigned.

Similarly, in the function PreAstarReturn(), only the time consumed with respect to different picking stations p_r (cf. Figure 1) in the A* search is used in the evaluation of S_r, which is expressed as:

$$S_r = \begin{cases} I_r & \text{preliminary trajectory is planned} \\ 0 & \text{preliminary trajectory planning is failed within } iter_{pre} \end{cases} \quad (4)$$

When ANN is not enforced, the potential targets are initially scored solely by means of Equations (3) and (4). The greater the functions to be evaluated are, the greater the likelihood of subsequent trajectory planning is and the faster the entire warehouse can be filled. In this case, arg max $S_{d,0}$ and arg max S_r are selected as the goal poses for delivery and return subtasks, respectively.

3.3. ANN Correction Method

The function PreAstarDeliver() is employed to refine goal score evaluations through a multilayer perceptron (MLP) network, which is elaborated upon as follows.

Figure 5 presents a typical MLP network of ANN with one hidden layer. Mathematically, with the trajectory planning states as known variables, the MLP network of the type reported in Figure 5 can be expanded step by step as follows:

$$\hat{y}(w, W) = F(\sum_{j=1}^{m} W_j h_j(w) + W_0) = F(\sum_{j=1}^{m} W_j f_j(\sum_{i=1}^{n} w_{ji} z_i + w_{j0}) + W_0) \quad (5)$$

where w_{ji} and W_j denote the weights assigned to the connection of the neurons. W_o and w_{j0} are linked to the bias, whose values are simply the constant 1.

The ANN correction in the current study is designed for the scoring system for the goal pose determination of delivery subtasks. In the initialization phase of the ANN correction system, excluding the filling balance parameter of G_d in Equation (3), $\{J_{1,0}, J_{2,0}, \cdots, J_{6,0}\}$ respectively denote the base values of the output elements in $J = \{J_{1,est}, J_{2,est}, \cdots, J_{6,est}\}$ (cf. $\hat{y}$ in Figure 5) for six MLP networks. Among the six elements, the one whose corresponding pose is selected as the target for vehicle kinematics-considered trajectory planning is further fixed based on the corresponding trajectory variables. The values of the other elements remain as unchanged as the results in Equation (3). Suppose that s_d is the shelf cluster investigated within a subtask. Given that the base value $S_{d,0}$ is derived with Equation (3), if s_d that corresponds to arg max $S_{d,0}$ is selected as the goal to determine the trajectory, then the trajectory with the goal of s_d will be planned, and the corresponding element in **J** will be derived with Equation (6) as follows:

$$J_{d,est} = \begin{cases} J_{d,0} - C_6(t_f - t_i - \bar{t} + C_7) & \text{trajectory is planned, and } d \in \{1,2\} \\ J_{d,0} - C_6(t_f - t_i - \bar{t} + C_8) & \text{trajectory is planned, and } d \in \{3,4,5,6\} \\ C_5 & \text{trajectory planning is failed within } iter_{pre} \\ J_{d,0} & \text{trajectory is not planned} \end{cases} \quad (6)$$

where t_f is the ending instant of the trajectory, $\bar{t}$ stands for the average value of the time length pertaining to all previously derived trajectories, C_6 is a calibrated constant intended to balance $J_{d,0}$, and the latter solely accounts for time consumed to reach a goal without factoring in the distance covered. C_7 and C_8 are respectively used to normalize the difference in distances corresponding to various shelf clusters and the expected moving time period of each subtask.

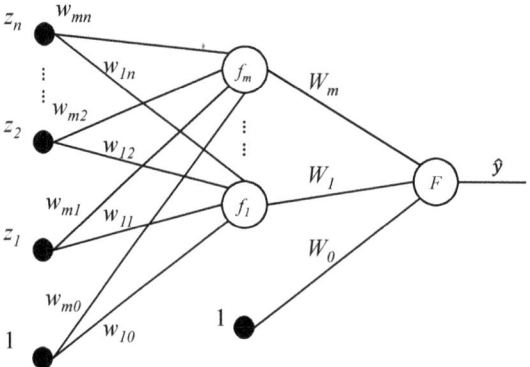

Figure 5. Schematic of an MLP network.

Similar to the parameter G_d in Equation (3), the inputs (cf. $z_1, z_2, \cdots, z_n$ in Figure 5) of the ANN system are the number of stacks filled within each shelf cluster. A vector **U** with six elements $\{u_1, u_2, \cdots, u_6\}$, which correspond to six shelf clusters s_d, is used. Suppose that $u_d \in \mathbf{U}$ is considered, it can be formulated as

$$u_d = \begin{cases} G_d & \text{delivery subtask is linked} \\ -G_d & \text{return subtask or subtasks of both types are linked} \\ 0 & \text{no subtask is linked} \end{cases} \quad (7)$$

The integers other than 0 can represent the filling states of the shelf clusters, which are associated with the currently moving forklifts. The vehicle motion states are also observed. This input of the MLP network vaguely provides information associated with the possible area the vehicles may be located in, given that each shelf cluster should be filled by following certain orders. Furthermore, at least two shelf clusters are not connected to any subtask because only four forklifts in the warehouse are employed. Evidently, these shelf clusters have no impact on the trajectory planning, and this situation is in line with the circumstances, where the investigated shelf group is filled. Thus, 0 is assigned to u_d in this case.

As the number of hidden neurons is set to 12 according to an empirical technique [36], the MLP system used to score shelf cluster s_d can be expressed in the form of Equation (5) as

$$\hat{J}_{d,est}(w_d, W_d) = F(\sum_{j=1}^{12} W_{d,j} h_j(w_d) + W_0) = F(\sum_{j=1}^{12} W_{d,j} f_j(\sum_{i=1}^{6} w_{d,ji} u_{d,i} + w_{d,j0}) + W_0) \quad (8)$$

The Levenberg-Marquardt method is then used to train the MLP and the values assigned to the elements of $\mathbf{W}$ and $\mathbf{w}$.

In the following warehouse filling cycles and the final dispatching system, the scoring system described in Equation (3) is replaced by the expression below to determine the optimal goal pose for vehicle $n_{current}$. Equation (9) is the final scoring equation of the discussed dispatching system, wherein C_9 and C_{10} act as the calibration constants. These constants are determined via a trial-and-error approach to yield results. The MLP contributes without excessively disrupting performance concerning warehouse filling/emptying time.

$$\begin{cases} S_d = C_1 G_d + J_d \\ J_d = C_9 J_{d,0} + C_{10} \hat{J}_{d,est}(\mathbf{w}_d, \mathbf{W}_d) \end{cases} \quad (9)$$

After the initialization cycle of warehouse filling (referring to Figure 6), the training process can continue until C_y cycles are finished. In the following cycles, the MLP has already been established depending on the data pertaining to the previous cycles, and the saved values of $\mathbf{W}$ and $\mathbf{w}$ are used to estimate the values $\hat{J}_{d,est}()$ via Equation (8). Therefore, the base values of the MLP outputs $J_{d,0}$ are replaced by J_d in Equation (9) when shelf cluster s_d is considered. The expression of the corresponding element in $\mathbf{J}$ for the following cycles writes:

$$J_{d,est} = \begin{cases} J_d - C_6(t_f - t_i - \bar{t} + C_7) & \text{trajectory is planned, and } d \in \{1,2\} \\ J_d - C_6(t_f - t_i - \bar{t} + C_8) & \text{trajectory is planned, and } d \in \{3,4,5,6\} \\ C_5 & \text{trajectory planning is failed within } iter_{pre} \\ J_d & \text{trajectory is not planned} \end{cases} \quad (10)$$

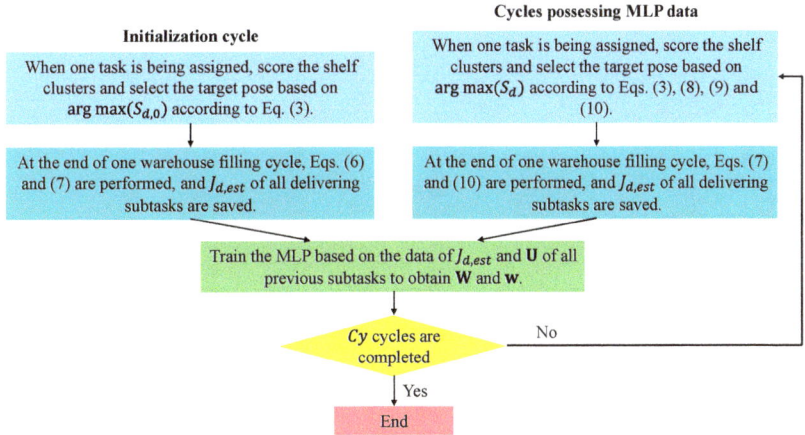

Figure 6. Flowchart of MLP training.

3.4. Final Pose Selection

The final pose of each subtask is determined using the SetFinalPose() function. For subtasks with the purpose of delivery, the destination coordinate $P_f(x_f, y_f)$ is chosen based on the arg max S_d criterion; for return subtasks, the coordinate $P_f(x_f, y_f)$ corresponding to arg max S_r is selected. Finally, the corresponding θ information should be added to complete the destination pose.

4. Improved Hybrid A* Search Algorithm

Hybrid A* algorithm [37] is an extension of the conventional 2D A* search algorithm due to its consideration of kinematics during node expansions over time. Different from the original hybrid A* search, the velocity variation is considered, and the node expansion manner is determined based on action purposes in the algorithm (cf. Algorithm 2) applied in this study. Furthermore, this improved heuristic method directly determines trajectory details throughout a subtask without an optimization stage.

Algorithm 2 is used to expand a parent node (P_p, t_p) in one manner N_c through the improved hybrid A* search with the target of P_f. Focusing on the motion of one vehicle, $List_{open}$ is used to store the data pertaining to all nodes (P_{open}, t_{open}), which has been explored and can be further expanded. The information contains the corresponding parent nodes, expansion manner (cf. Section 4.1), time expansion data (cf. Section 4.2), and costs (cf. Section 4.4). On the contrary, the node (P_{closed}, t_{closed}) cannot be expanded anymore and is stored in $List_{closed}$. Furthermore, the function DetectCollision() is discussed in Section 4.3. In addition, the function AddNode() is used to add a node with its affiliating data into $List_{open}$ or $List_{closed}$.

Algorithm 2: Improved hybrid A* search algorithm

$[\sigma, List_{open}] \leftarrow$ SearchAStar $\left(\mathbf{T}, N_c, P_p, P_f, t_p, List_{open}, List_{closed}, map\right)$

1. $\left[t_{high,c}, t_{mid,c}, t_{zero,c}\right] \leftarrow$ FixMovingTime $(List_{open}, P_p, t_p, N_c)$;
2. $List_{open} \leftarrow$ SetCost $(List_{open}, P_p, t_p, N_c)$;
3. $\gamma \leftarrow$ DetectCollision $(\mathbf{T}, N_c, P_p, t_p, List_{open}, map)$;
4. **if** $(P_c, t_{zero,c}) \in List_{closed}$, **then**
5.   return;
6. **end if**
7. **if** $(P_c, t_{zero,c}) \in List_{open}$ **and** $\gamma = 0$, **then**
8.   **if** $f_c < f_{pre}$, **then**
9. $List_{open} \leftarrow$ ReplaceNode $(List_{open}, P_p, t_p)$;
10.   **end if**
11. **else**
12.   **if** $\gamma = 0$, **then**
13. $List_{open} \leftarrow$ AddNode $(List_{open}, P_c, t_{zero,c})$;
14.   **if** $P_c = P_f$, **then**
15. $\sigma \leftarrow 1$;
16.   return;
17.   **end if**
18.   **else if** $\gamma = 1$, **then**
19. $List_{closed} \leftarrow$ AddNode $(List_{closed}, P_c, t_{zero,c})$;
20.   return;
21.   **end if**
22. **end if**
23. return;

4.1. Node Expansion Method

This section elaborates on the various possible expansion manners denoted as N_{all}, with each individual possibility represented by N_c. The drivable area is initially mapped with the above-discussed grid networks (cf. Figure 2). The dimensions of the grid should be skillfully coupled with the size of the vehicle, with the aim of reducing occupied cells during a certain action and enhancing the utility rate of the space. The case demonstrated in Figure 7 is a well-designed example. In this case, one vehicle covers two grid cells. Thus, on a 2D space domain, the orientation angle θ involved in forklift movements can be easily described. In addition, through one maneuver, the vehicle body occupies a small number of grids. This can reduce the possibility of interference with other forklifts during motion planning.

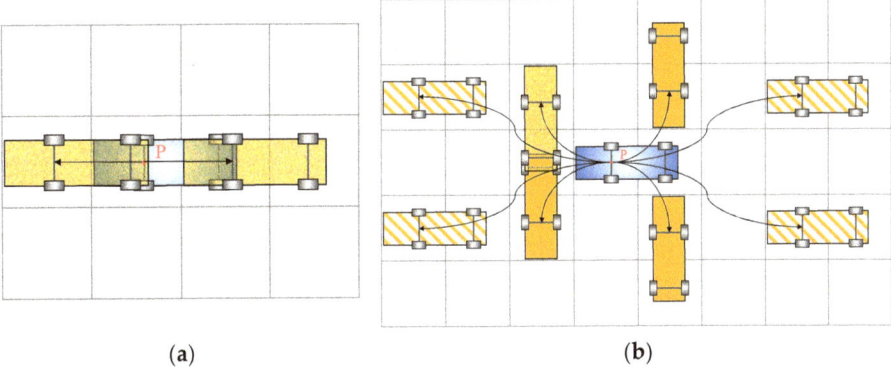

(a)   (b)

Figure 7. Node expansion manners of improved hybrid A* algorithm: (**a**) Stopping and going straight; (**b**) lane changing and turning.

As reported in Figure 7a,b, the nodes expanding manners N_{all} in the space domain can be divided into eleven patterns, which are summarized as four categories covering all possible maneuvers a forklift may intend to conduct. These include stopping, going straight (cf. semitransparent solid rectangles in Figure 7a), lane changing (cf. rectangles with diagonal stripes in Figure 7b), and turning (cf. solid rectangles in Figure 7b).

As far as the expanding manners are concerned, when v remains 0, one vehicle stops. As v is other than 0 with steering angle $\phi = 0$, one fork can go or reverse straight. Figure 8 demonstrates one modeled forklift path depicted by consecutive outlines, when the vehicle goes upward and turns to the left from the right lower side to the center left. Through this maneuver, ϕ is varied to gain an identical final location relative to grids as the initial state, despite the change of $\pi/2$ in θ. Under such circumstances, the consecutive motions can be easily established. Meanwhile, the modeled maneuver of reverse turning to the left can also be noted in Figure 8, when the initial and final body outlines are exchanged.

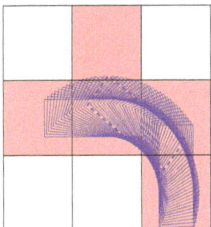

Figure 8. Path of turning.

Similar to turning actions, a typical lane-changing maneuver is also modeled and outlined in Figure 9. The forklift goes forward and changes to the left lane, which is shown on the top row in Figure 9. The other possible node-expanding paths of turning and lane changing are modeled by mirroring or rotating the examples reported in Figs. 8 and 9.

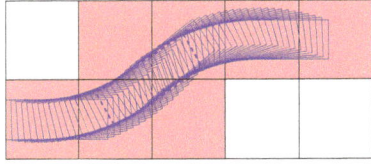

Figure 9. Path of lane changing.

The smoothness of the modeled paths can be enhanced by adopting advanced techniques, and the identical heuristic trajectory search rule of this work can also be applied to the newly modeled paths.

4.2. Velocity Planner

The velocity planner determines the node expansion rule in the time domain and the potential node arrival instants. When one maneuver starts or finishes, the vehicle speed solely falls into one of three determined constants (v_{high}, v_{mid} and v_{zero}), which correspond to high speed, mid speed, and zero speed. Figure 10 reports all possible velocity selections linked to the start and end of the maneuvers. v_{high} can only be selected with the expansion pattern of going straight to fulfill the kinematic constraints given by Equations (1) and (2). The stop maneuver is an expansion of finite value solely on the time domain, it follows that only v_{zero} can be applied to this action.

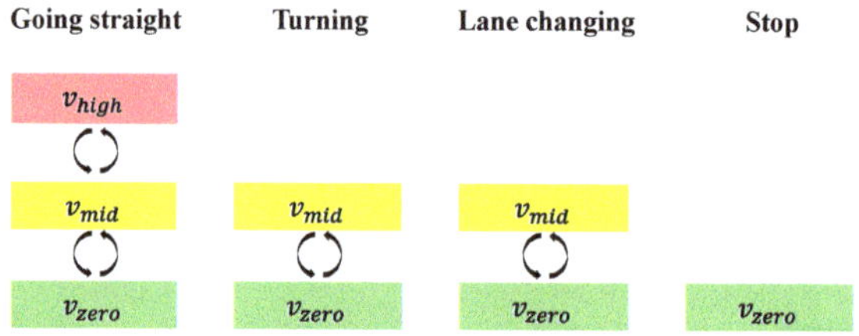

Figure 10. Velocity selections when one maneuver starts or ends.

The velocity level over one maneuver can only be linked to itself or an adjacent one, although all maneuvers can be arbitrarily linked along a trajectory. In other words, during one maneuver, the vehicle speed can maintain or alter among those three velocities, but the direct change between v_{zero} and v_{high} is illegal. For instance, if the vehicle finishes a series of going straight actions with speed v_{zero}, its ending velocity of the last second action should then be v_{mid} or v_{zero}. Likewise, if turning follows going straight with an original velocity v_{high}, the speed of the last going straight maneuver should reduce to v_{mid} to gain an initial velocity v_{mid} for turning.

Following the velocity selection rules, all possible speed variations through one maneuver are shown in Figure 11. It is noteworthy that the link of $v_{high} \to$ Going straight $\to v_{zero}$ and the opposite link are infeasible. Thus, 16 connection choices in total are applicable. Furthermore, velocity is merely an intermediate state used to determine a time dimension expansion, although it is a vital parameter for motion planning. The detailed velocity time history during one maneuver can be modeled by applying varied techniques to obtain the required initial and finishing speeds. In the current work, deceleration or acceleration along one action between the same velocity levels is modeled with identical time lengths, with the aim of simple calculation. The number of time expansion selections (cf. T_e with $e \in \{1, 2 \cdots, 12\}$ in Table 1) thus reduces to 12.

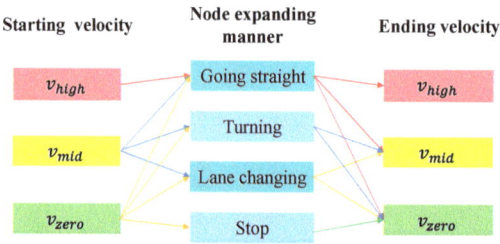

Figure 11. Velocity variations through one maneuver.

Table 1. Node expansion selections in time domain.

Abbreviation	Description
T_1	$v_{zero} \to$ Going straight $\to v_{zero}$
T_2	$v_{zero} \to$ Going straight $\to v_{mid}$ or $v_{mid} \to$ Going straight $\to v_{zero}$
T_3	$v_{mid} \to$ Going straight $\to v_{mid}$
T_4	$v_{mid} \to$ Going straight $\to v_{high}$ or $v_{high} \to$ Going straight $\to v_{mid}$
T_5	$v_{high} \to$ Going straight $\to v_{high}$
T_6	$v_{zero} \to$ Turning $\to v_{zero}$
T_7	$v_{zero} \to$ Turning $\to v_{mid}$ and $v_{mid} \to$ Turning $\to v_{zero}$
T_8	$v_{mid} \to$ Turning $\to v_{mid}$
T_9	$v_{zero} \to$ Lane changing $\to v_{zero}$
T_{10}	$v_{zero} \to$ Lane changing $\to v_{mid}$ or $v_{mid} \to$ Lane changing $\to v_{zero}$
T_{11}	$v_{mid} \to$ Lane changing $\to v_{mid}$
T_{12}	$v_{zero} \to$ Stop $\to v_{zero}$

The function FixMovingTime() determines time expansion data based on Table 2, which is elaborated as follows. When planning a trajectory, the maximum feasible velocities are consistently selected for all maneuvers with zero speed supplied to the start and end of the trajectory. With regard to one node expansion, three values of $t_{high,p}$, $t_{mid,p}$ and $t_{zero,p}$ are initially saved in the parent node as the possible start instant of the maneuver, and these respectively correspond to ending velocities of v_{high}, v_{mid}, and v_{zero}, if they exist. In the cases where v_{high} is not reachable, the value stored in $t_{high,p}$ will be the minimum time length during the maneuver, with ending velocity v_{mid}. Consequently, $t_{high,p} = t_{mid,p}$. Similar result of $t_{mid,p} = t_{zero,p}$ is obtained when the maximum realizable ending velocity is v_{zero}. As the parent node is initialized, values of $t_{high,c}$, $t_{mid,c}$, and $t_{zero,c}$ are derived and saved for the child node with the identical manner for the parent node. Furthermore, the starting velocity of a new expansion may be imposed as v_{zero}, combined with the node expansion type of the previous maneuver. A stopping flag μ is then set as 1. Such a case happens when an expansion of stop occurs or moving direction of a vehicle is reversed. In the rest of the working conditions, the flag μ remains 0. Notably, a node is of four dimensions, x, y, θ, and t; among them, t is an index used in the graph search, and more than one value stored in one index could complicate the problem. Consequently, only the v_{zero}-related time instant t_{zero} is stored as the node index.

Table 2. Model-based approach for saving time consumption data.

Current Maneuver	$\mu=0$	$\mu=1$
Stop	-	$t_{zero,c} \leftarrow t_{zero,p} + T_{12}$ $t_{mid,c} \leftarrow t_{zero,p} + T_{12}$ $t_{high,c} \leftarrow t_{zero,p} + T_{12}$

Going straight	if $t_{high,p} = t_{mid,p}$, then
	$t_{zero,c} \leftarrow t_{mid,p} + T_2$
	$t_{mid,c} \leftarrow t_{mid,p} + T_3$
	$t_{high,c} \leftarrow t_{mid,p} + T_4$
	else
	$t_{zero,c} \leftarrow t_{mid,p} + T_2$
	$t_{mid,c} \leftarrow t_{high,p} + T_4$
	$t_{high,c} \leftarrow t_{high,p} + T_5$
		$t_{zero,c} \leftarrow t_{zero,p} + T_1$
		$t_{mid,c} \leftarrow t_{zero,p} + T_2$
		$t_{high,c} \leftarrow t_{zero,p} + T_2$

Turning	$t_{zero,c} \leftarrow t_{mid,p} + T_7$ $t_{mid,c} \leftarrow t_{mid,p} + T_8$ $t_{high,c} \leftarrow t_{mid,p} + T_8$	$t_{zero,c} \leftarrow t_{zero,p} + T_6$ $t_{mid,c} \leftarrow t_{zero,p} + T_7$ $t_{high,c} \leftarrow t_{zero,p} + T_7$
Lane changing	$t_{zero,c} \leftarrow t_{mid,p} + T_{10}$ $t_{mid,c} \leftarrow t_{mid,p} + T_{11}$ $t_{high,c} \leftarrow t_{mid,p} + T_{11}$	$t_{zero,c} \leftarrow t_{zero,p} + T_9$ $t_{mid,c} \leftarrow t_{zero,p} + T_{10}$ $t_{high,c} \leftarrow t_{zero,p} + T_{10}$

Meanwhile, each node stores three ending time instants. Thus, one algorithm should be applied to select the exact ending instant of each node. This algorithm first chooses the minimum time t_{high} of each maneuver over the path. The deceleration phase is then imposed on the last two nodes of the trajectory, and the corresponding t_{mid} and t_{zero} are in turn assigned to these two nodes as the maneuver-finishing time instants. As the above methods are implemented from the start to the end of the initialized trajectory, a few modifications are imposed to fix the node time indices t in certain circumstances. For instance, when $\mu = 1$, the maximum achievable ending time t_p for the former maneuver node is determined by selecting the value of $t_{mid,p}$. In this case, the ending velocity of this maneuver turns to be v_{mid}. Meanwhile, in the same combination of maneuvers, if a previous maneuver exists and $t_{pp} > t_{mid,pp}$, $t_{mid,pp}$ is assigned to t_{pp}, as the prior maneuver cannot achieve a higher velocity.

4.3. Collision Detection Strategy

The pseudocode of the collision detection function DetectCollision() is recorded in Algorithm 3. A variable γ is used as a flag to indicate the validity of the node expansion as well as the type of collisions that may have occurred. When the current expansion N_c is valid, γ is set as 0. $\gamma = 1$ means that at least the ending pose of the expansion risks colliding, and $\gamma = 2$ signifies that the invalidity is only found in the link between the starting and final locations. This algorithm consists of two parts, which separately refer to collisions that occurred with ending and intermediate pose of the vehicle. As regards both circumstances, collisions should be avoided throughout $t \in \left[t_{mid,p}, t_{zero,c}\right]$.

Algorithm 3: Collision detection algorithm

$\gamma \leftarrow \text{DetectCollision}(\mathbf{T}, N_c, P_p, t_p, \text{List}_{open}, map)$

1. Initialize $\gamma \leftarrow 0$;
2. $P_c \leftarrow \text{FindEndingPose}(P_p, N_c)$;
3. $P_{in} \leftarrow \text{FindIntermediatePose}(P_p, N_c)$;
4. **if** CheckStaticCollision(P_c, map) is **true**, **then**
5. $\gamma \leftarrow 1$;
6.   **return**;
7. **end if**
8. $\left[t_{zero,c}, t_{mid,p}\right] \leftarrow \text{FixParkingTime}(P_p, t_p, \text{List}_{open}, N_c)$;
9. $[\text{List}_{cur,1}] \leftarrow \text{FindFinalGrids}(P_p, N_c)$;
10. $[\text{List}_{other}] \leftarrow \text{FindObstaclesGrids}\left(\mathbf{T}, t_{zero,c}, t_{mid,p}\right)$;
11. **if** CheckDynamicCollision($\text{List}_{cur,1}, \text{List}_{other}$) is **true**, **then**
12. $\gamma \leftarrow 1$;
13.   **return**;
14. **end if**
15. **if** CheckStaticCollision(P_{in}, map) is **true**, **then**
16. $\gamma \leftarrow 2$;
17.   **return**;
18. **end if**
19. $[\text{List}_{cur,2}] \leftarrow \text{FindIntermediateGrids}(P_p, N_c)$;
20. **if** CheckDynamicCollision($\text{List}_{cur,2}, \text{List}_{other}$) is **true**, **then**
21. $\gamma \leftarrow 2$;
22.   **return**;
23. **end if**
24. **return**;

Static obstacles correspond to the walls of the shelves and warehouse, and the related collision detection method is expressed on lines 4 to 7 in Algorithm 3. FindEndingPose() is used to obtain the current pose $P_c(x_c, y_c, \theta_c)$ based on the parent pose $P_p(x_p, y_p, \theta_p)$ and the current expansion manner N_c. The function CheckStaticCollision() is applied to check if collisions with static obstacles exist. This check is first judged by identifying the validity of the final pose. When it comes to maneuvers of turning and lane changing, an additional intermediate pose (cf. thick dashed rectangular in Figures 8 and 9), obtained through FindIntermediatePose(), should be further examined (cf. lines 15 to 18 in Algorithm 3). Focusing on the collision avoidance constraints formulation between the point Q_j ($j = 1, \cdots, N_{obs}$ and N_{obs} denotes the obstacle point number) and the current investigated vehicle featuring vertexes A, B, C and D. A collision forms when Q_j enters the rectangle $ABCD$. The restriction that Q_j is located outside of the rectangle $ABCD$ can be formulated by applying a triangle-area-based criterion [38],

$$S_{\triangle P_i AB} + S_{\triangle P_i BC} + S_{\triangle P_i CD} + S_{\triangle P_i DA} > S_{\square ABCD} \tag{11}$$

where $S_\triangle$ indicates the triangle area, and $S_\square$ refers to the rectangle area. Applying Equation (11) to every node expansion with respect to every obstacle point, the static obstacle collision judgement is yielded.

Dynamic obstacles in the current study are only forklifts, whose motions have been saved in **T**, and the function FixParkingTime() is applied to find $t_{mid,p}$ in $List_{open}$ and $t_{zero,c}$ by calling Table 2. As reported in Figures 8 and 9, the highlighted grids indicate the approximate area occupied during the movements of turning and lane changing. Similarly, the covered grids pertaining to other maneuvers in the same category can be determined by mirroring or rotating the highlighted grids (cf. Figures 8 and 9). The functions FindFinalGrids() and FindIntermediateGrids() are respectively used to record the covered grids of the final pose P_c and the intermediate pose P_{in} for the current node expansion patterns.

Without loss of generality, let us focus on the collision avoidance constraint formulation between the vehicle $n_{current}$ and the vehicle n_k ($n_{current}, n_k = 1, \cdots, 4$ and $n_{current} \neq n_k$). The function FindObstaclesGrids() is first utilized to search for the covered grids by the vehicles n_k, whose trajectories are not being planned, with respect to time. It is possible that no actions of the forklift n_k have been determined during the period when the current maneuver of vehicle $n_{current}$ could occupy. Under such circumstances, the grids that the vehicle n_k finally parks are treated as the covered ones.

Finally, CheckDynamicCollision() is used to detect if $List_{cur,1}$ or $List_{cur,2}$ is going to simultaneously cover the grids already stored in $List_{oth}$ over the valid time. Notably, during the collision detection, the current action is virtually regarded as the final maneuver with v_{zero} set as the finishing velocity with $t_{zero,c}$ selected, thus an equal or longer time length of this maneuver in actual operation is considered. This treatment can enhance the safety performance to certain extents.

Notably, referring to Algorithm 3, the types of failure through an expansion have been noticed when one expansion has failed. This provides a tool to distinguish if one child node should be closed or skipped because that failure can be simply due to the intermediate trajectory of the action is interfered, while the corresponding child node could be valid in other situations.

4.4. Trajectory Cost Function

The function SetCost() is explained in this section. With regard to one node expansion, the cost function f is the sum of two parts, as reported in Equation (12),

$$f = g + C_{11} h \tag{12}$$

in which g stands for the cumulative cost from the initial pose to the current pose and h indicates the estimated cost from the child node of the current expansion to the target node of the trajectory being defined. C_{11} is a calibrated weighting aimed at achieving a balance between the computational resources used for searching and the resultant trajectory's quality. In total, it is expected that one forklift complete a delivery or return subtasks subjected to the minimum travelling time of $\left(t_f - t_i\right)$. Meanwhile, times of turning, lane changing, and speed inverse maneuvers should be minimized in order to reduce unnecessary movements, which may decrease the traffic capacity of the passages.

As far as the function of trajectory to the goal is concerned, Equation (13) is applied. We used the Manhattan distance plus another function of θ_c. Both the distance and the angle are evaluated in times of certain characteristic dimensions.

$$h = |x_g - x_c|/\Delta x + |y_g - y_c|/\Delta y + |\theta_g - \theta_c|/(\pi/2) \tag{13}$$

where $\Delta x = \Delta y$, indicating the grid side length; x_g, y_g, and θ_g correspond to the goal space coordinates; x_c, y_c, and θ_c refer to the coordinates of the current expansion child node.

The passed trajectory cost function that characterizes the time consumed from the starting pose of the trajectory to the current one writes:

$$\begin{cases} g = t_{zero,c} - t_i + p \\ p = p_{turn} \cdot m_{turn} + p_{lane} \cdot m_{lane} + p_{inv} \cdot m_{inv} \end{cases} \tag{14}$$

where p_{turn}, p_{lane}, and p_{inv} respectively indicate the predefined penalties pertaining to single time of turning, lane changing, and speed inverse. m_{turn}, m_{lane}, and m_{inv} denote the cumulative number of corresponding maneuvers from the initial pose to the current child node along the trajectory being defined.

In addition, if one node originally stored in $List_{open}$ is reached a second time with a reduced value of f_c during a node expansion compared to the previous saved f_{pre} for the same nodes, the function ReplaceNode() is used to switch the parent node to the parent node of current expansion, with the aim of reducing calculation time length and optimizing the trajectory being planned.

5. Joint Dispatching and Cooperative Trajectory Planning Framework

The trajectories of forklift vehicles in the warehouse are sequentially determined in the complete cooperative operative algorithm (cf. Algorithm 4). Generally, the dispatching technique is first applied to select the current vehicle index as well as the goal coordinate in the space domain for the current trajectory planning subtask. Subsequently, an improved hybrid A* search algorithm is used to determine all details pertaining to the newly planned trajectory **T**. The above methods are repeated until all stack locations in the warehouse are filled, as the warehouse filling state **F** is being updated.

The function CheckFilling() is used to derive the number of unfilled stacks. The function FindInitialPose() is then applied to find the parent node (P_p, t_p) for the next expansion with minimum cost f found in $List_{open}$ and to simultaneously remove this node from $List_{open}$.

A maximum number of iterations $iter_{search}$ in the improved hybrid A* algorithm is induced to break the endless iterations that may derive a trajectory involving an unacceptable waiting period. The improved hybrid A* search algorithm is finished by satisfying any one of the three criteria, which are, respectively, the goal coordinate reached as a new child node, the iteration number exceeding $iter_{search}$, and no node saved in $List_{open}$. Among them, only the first criterion indicates the trajectory of the current subtask is successfully planned, and a corresponding flag σ is thus set as 1 to declare this success.

The function GenerateTraj() is employed to generate the trajectory with the newly planned trajectory by backtracking from P_f to P_i with $t_{f,zero}$ as the ending instants. During the backtracking, the maximum realizable velocities of the intermediate nodes are selected based on the manner depicted in Figure 11 and Table 2.

Finally, the function UpdateFillingState() updates **F** as the stack located at $P_f\left(x_f, y_f, \theta_f\right)$ has been filled.

Algorithm 4: Cooperative operation algorithm

$[\mathbf{T}, \mathbf{F}] \leftarrow \text{OperateCooperative}(\mathbf{T}, \mathbf{F}, iter_{search}, map)$

1. **while** $\text{CheckFilling}(\mathbf{F}) > 0$
2. $\left[n_{current}, P_i, P_f, t_i\right] \leftarrow \text{Dispatch}\left(\mathbf{T}, \mathbf{F}, fail, P_i, P_f, map\right);$
3. $\sigma \leftarrow 0;$
4. $fail \leftarrow 0;$
5. $t_{high,p} \leftarrow t_i, t_{mid,p} \leftarrow t_i, t_{zero,p} \leftarrow t_i;$
6. $List_{open} \leftarrow (P_i, t_i), List_{closed} \leftarrow \varnothing;$
7.   **while** $List_{open} \neq \varnothing$ **or** $iter \leq iter_{search}$ **or** $\sigma \neq 1$, **do**
8. $[List_{open}, P_p, t_p] \leftarrow \text{FindInitialPose}(List_{open});$
9. $List_{closed} \leftarrow \text{AddNode}(List_{closed}, P_p, t_p);$
10.   **for each** $N_c \in N_{all}$, **do**
11. $[\sigma, List_{open}] \leftarrow \text{SearchAStar}\left(\mathbf{T}, N_c, P_p, P_f, t_p, List_{open}, List_{closed}, map\right);$
12.   **end for**
13.   **end while**
14.   **if** $\sigma = 1$, **then**
15. $\mathbf{T} \leftarrow \text{GenerateTraj}\left(P_i, P_f, t_i\right);$
16. $\mathbf{F} \leftarrow \text{UpdateFillingState}(\mathbf{T});$
17.   **else**
18. $fail \leftarrow 1;$
19.   **end if**
20. **end while**
21. **return;**

6. Numerical Experiments

Simulations of filling and emptying missions were conducted on the MATLAB 2021b platform, utilizing the parametric settings reported in Table 3. Each warehouse filling or emptying mission began with fully empty or filled initial conditions, respectively. The motion planning for each forklift in the simulation environment solely takes into account obstacles such as warehouse walls and other forklifts.

Table 3. Parametric settings regarding model and approach.

Parameter	Description	Setting
n	Forklift front overhang length	0.3 m
m	Forklift rear overhang length	1 m
l	Forklift wheelbase	1.5 m
$2b$	Forklift width	1 m
$[lb_x, ub_x]$	Horizontal boundaries of map	$[-18, 18]$ m
$[lb_x, ub_y]$	Vertical boundaries of map	$[-12, 12]$ m
$resol_{xy}$	Node resolution for search algorithms	2 m
$iter_{pre}$	Maximum iteration in the time dimension involved A* search	500
$iter_{dispatch}$	Maximum iteration of redispatching	10
$iter_{search}$	Maximum iteration in the improved A* search	5000
$\{C_1, C_2, \cdots, C_{11}\}$	Calibration parameters	$\{6, 1.5, 80, -40, 6, 0.5, -4, 2, 0.5, 0.5, 3\}$
$\{T_1, T_2, \cdots, T_{12}\}$	Modeled time lengths of maneuvers	$\{4, 2, 1.25, 0.75, 0.5, 8, 5, 3, 12, 8, 5, 1\}$ s

Table 3. Cont.

Parameter	Description	Setting
p_{turn}	Penalty for turning maneuver	4
p_{lane}	Penalty for lane changing maneuver	6
p_{inv}	Penalty for speed inverse maneuver	6
Δt_i	Time postponed when one trajectory planning is failed	10 s
Δt_{cover}	Time in Equation (3)	20 s
Δt_{entry}	Time length after the virtual vehicle entering the rack passage to evaluate Δt_{cover}	20 s
$\{t_{claim}, t_{unload}\}$	Time period for picking goods and unloading goods	{5,5} s

The simulation results are divided into two parts. The first part focuses on assessing the consistency of the motion planning strategy by presenting three short-time trajectories for multiple forklifts. In the second part, various dispatching strategies were benchmarked using different scores specifically designed for the current warehouse scenario. This evaluation was conducted to determine the performance and efficiency of different dispatching strategies in the given warehouse scenario.

6.1. On the Performance of the Trajectory Planning Technique

Figure 12 reports how forklift 1 returns to a picking station from shelf cluster s_1, where it has just unloaded goods. Before the start of the scenario, forklifts 2 and 4 are in the same state of return, and forklift 3 is unloading in shelf cluster s_4. When the scene in Figure 12 starts, forklift 1 accelerates and decelerates to reverse to the entrance of shelf cluster s_1. At the meantime, forklift 4 turns and heads to the picking station p_4, leaving the crossing of the wide passages empty. With the purple square showing the newly filled stack, forklift 3 leaves shelf cluster s_4 and enters s_3 to finish turning round. When it is inside of s_3, forklift 1 goes through the passage between s_3 and s_4. Subsequently, forklift 1 reverses and turns to the picking station p_2, and forklift 3 then enters the wide passages to return to the assigned picking station, which is p_3. Finally, forklift 2 is the first vehicle that arrives at the crossing area of wide passages with the goods picked. During the scenario, all trajectories are directly determined by means of the approach elaborated on in Section 4. The least priority is dynamically offered to the newly departed vehicle. For instance, in this scenario, forklift 3 initially has the lowest priority. Thus, it should judge if there is enough time for the vehicle to turn around. Therefore, if the time is limited, vehicle 3 would wait at the entrance of s_4, until the passing of forklift 1.

Figure 13 shows how the last stack is filled. Because there is nowhere to be filled hereafter, forklift 3 is parked in one picking station, and forklift 1 heads to picking station s_4 and stops. Initially, forklift 4 has just finished one unloading process, then it enters shelf cluster s_2 to turn over. After forklift 4 leaves s_2, forklift 2 enters the shelf cluster to complete the last delivery subtask. Meanwhile, forklift 4 accelerates, decelerates, and turns to picking station s_3 for final parking.

Figure 14 shows a scenario where the trajectory decision for forklift 3 encounters a failure. In this case, forklift 1 reaches shelf s_4 to unload goods, while forklift 3 completes its unloading process earlier than forklift 1. Consequently, forklift 3 should be assigned a higher priority to plan the trajectory based on function rank(). However, at this moment, forklift 1 remains stationed at the entrance of shelf s_4, obstructing the passage for forklift 3. As a result, a decision failure occurs, leading to an update in the trajectory planning priority stored in **R**. Only after the completion of the trajectory planning for forklift 1 can the trajectory planning for forklift 3 proceed accordingly.

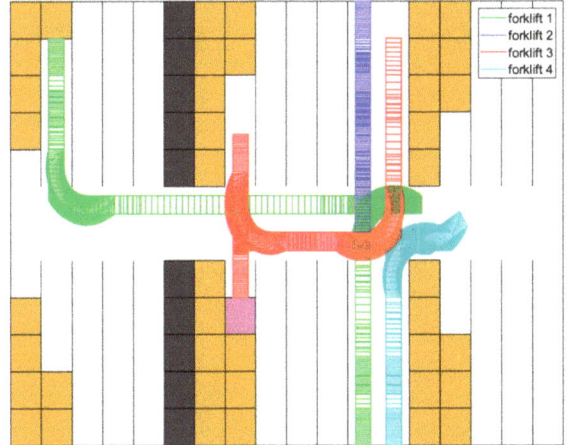

Figure 12. Planned trajectories between $t = 687.25$ s and $t = 715.25$ s.

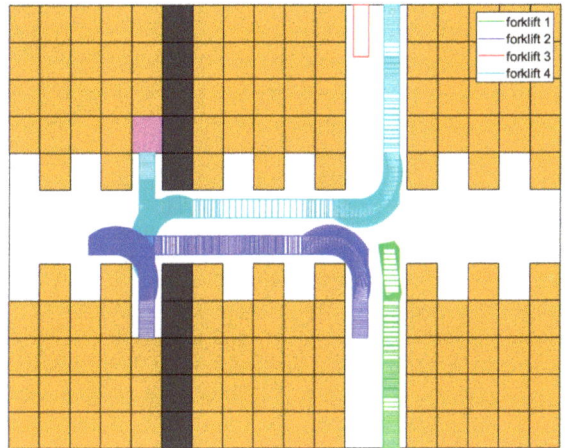

Figure 13. Planned trajectories between $t = 2327.25$ s and $t = 2354.50$ s.

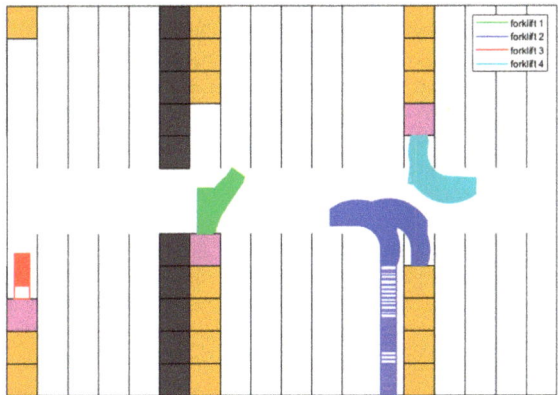

Figure 14. Planned trajectories between $t = 284.75$ s and $t = 303.75$ s.

6.2. On the Performance of Dispatching Strategies

Table 4 benchmarks five distinct dispatching strategies for both filling and emptying tasks. It is worth noting that these dispatching methods are applicable only to specific coupling scenarios between grids and vehicle maneuvers, as they incorporate a preliminary A* search during the dispatching process.

Table 4. Benchmark of the different dispatching strategies.

Strategy Name	Filling		Emptying	
	Decision Failure Times	End Time (s)	Decision Failure Times	End Time (s)
ANN combined strategy	38	2388.25	27	2597.25
Comprehensive strategy	38	2441.00	21	2658.25
Greedy strategy	135	2768.75	75	3419.00
Traffic jam removing strategy	120	2709.00	51	3117.75
Balance strategy	55	2557.75	35	2660.50

The approaches differ in the assignment methods pertaining to delivering subtasks. While the decision failure times in Table 4 indicate the number of times $n_{current}$ or t_i is changed without a solid trajectory decided throughout the entire warehouse filling mission, the end time is the finishing timing of the same task. Among the strategies, ANN combined determines the goal through $\arg\max(S_d)$ (cf. S_d in Equation (9)) when the coefficients of the MLP are determined. Comprehensive strategy applies $S_{d,0}$ in Equation (3), and the target is determined based on $\arg\max(S_{d,0})$. The other three strategies only concern a few parameters in Equation (3). Greedy, traffic jam removal, and balance strategies select the goal pose through $\arg\max(I_d)$, $\arg\max(J_{d,0})$, and $\arg\max(G_d)$, respectively.

As reported in Table 4, the end time of the emptying task is approximately 10% longer than that for filling tasks. However, the occurrence of decision failure times is generally lower. This indicates that during emptying tasks, instances of a forklift obstructing routes and blocking other vehicles for extended periods are rare. Yet, with the current configuration of $iter_{pre}$ and $iter_{search}$, all vehicles might experience longer wait times on average during each subtask.

The ANN combined strategy is the optimal method in terms of end time for both filling and emptying tasks, requiring approximately 2% less time than the comprehensive strategy. However, it does exhibit comparable or even slightly greater decision failure times when compared to the comprehensive strategy. In general, a reduction in the first parameter can reduce the computation burden of the hardware by performing fewer useless calculations in searching a trajectory. The second parameter directly shows the efficiency of the cooperative operation of multiple forklifts. Notably, the ANN approach applied to train the dispatching system only concerns the efficiency of planning a trajectory for the current vehicle. Therefore, it is possible that some stacks in only one shelf cluster are left to be filled by selecting the best solution multiple times. In this case, multiple forklifts have to cooperatively fill one shelf cluster, and the waiting period of each vehicle must increase. Therefore, both the decision failure times and the end time may sharply increase.

In Figures 15–18, the sequence in which the warehouse is filled with different methods is reported. The colored blocks plotted in the schematic warehouse stand for the stacks filled during different periods (red for the first quarter of time, blue for the second, green for the third, and cyan for the last). It is observed that the filling of different shelf clusters for the ANN combined strategy (cf. Figure 15) in different quarters of the period is balanced to a certain extent. There are, respectively, 35, 37, 34, and 30 stacks filled during each quarter. Particularly in the last quarter of the period, similar numbers of stacks filled are distributed within all six shelf clusters, and this provides a possibility for a feasible and fast solution to filling the warehouse.

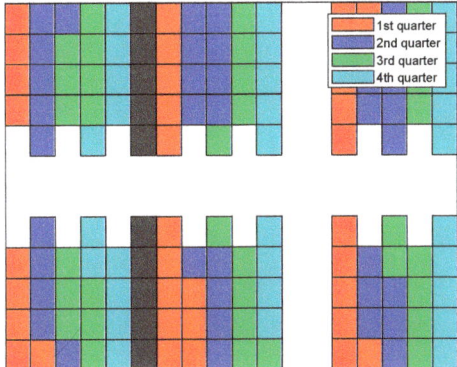

Figure 15. Filling sequence of ANN combined strategy.

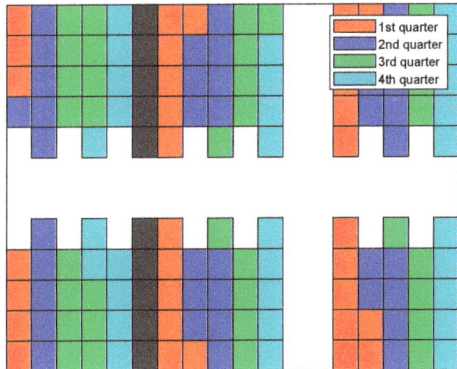

Figure 16. Filling sequence of comprehensive strategy.

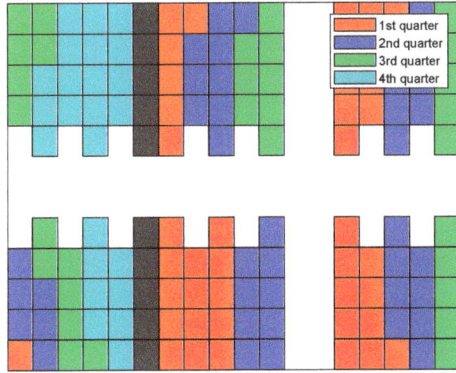

Figure 17. Filling sequence of traffic jam removing strategy.

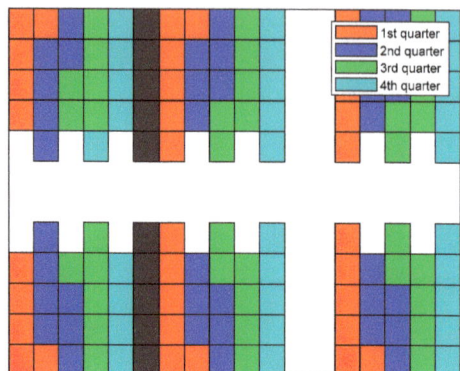

Figure 18. Filling sequence of balance strategy.

By excluding the MLP of ANN, a comprehensive strategy is also capable of assigning subtasks to each forklift explicitly. Although the performance is slightly worse than the ANN combined one, it does not require the data pertaining to previous cycles. Figure 16 demonstrates the filling sequence for this approach. Similar to that of the ANN combined approach, the filling state of the shelf clusters in the warehouse is relatively balanced throughout the period. A few seconds are added in terms of the end time for this strategy without the fine adjustment of the MLP.

Figure 17 exhibits the filling sequence of the traffic-jam-removing strategy. The difference between this method and the greedy strategy is that the previous one simultaneously considers H in Equation (3). One stack in shelf cluster s_2 is filled in the first quarter of the period, and both the decision failure times and the end time pertaining to the traffic jam removal approach reduce compared to the greedy strategy. Thus, a simple additional function of H can slightly further characterize the difficulty of the trajectory planning. Meanwhile, because the filling sequences of these two strategies are similar, the figure pertaining to the greedy method is omitted. As shown in Figure 17, it basically evenly fills shelf clusters s_4, s_5, and s_6 in the first quarter of the period. The majority of stacks in s_1 and around a third of stacks in s_2 are filled in the last quarter. Considering four vehicles cooperatively delivering goods, violations between trajectories evidently arise, and the times of failure in trajectory decisions and the waiting time can increase markedly.

The filling sequence of the balance strategy is plotted in Figure 18. The filling of different shelf clusters is nearly perfectly balanced. Evidently, this strategy can better arrange the trajectories to fill the last few stacks of the warehouse compared to any other approach. Therefore, the number of stacks filled during the last quarter of the warehouse filling cycle is comparable to that during other periods. In particular, when the 5th, 14th, and 23rd stacks of shelf clusters s_3 and s_4 (cf. Figure 2) are being filled, the paths between the shelf clusters (s_1 and s_2) in the left side of the warehouse (cf. Figure 1) and picking stations are cut. Meanwhile, the first delivery vehicle reaching these certain stack locations between shelf clusters s_3 and s_4 can simultaneously block the routes to certain shelf clusters. This phenomenon also appears in shelf clusters s_5 and s_6 when the identical stack indexes are considered. Those are the unique conflicts that arose to cause the failure decisions and the lengthening of the end time for the balance strategy.

7. Conclusions

This study presents a joint dispatching and cooperative trajectory planning framework. In this framework, the dispatching method applies a time dimension involved A* search, which is used to score different goal poses during a delivery subtask. ANN is simultaneously implemented to evaluate the difficulties for a forklift to arrive at a goods unloading pose from a picking station. In addition, the stacks to be filled pertaining to different shelf clusters are used as the third parameter to balance the shelf filling states, with the aim of avoiding deadlocks at the final stage of the mission.

As far as the trajectory planning approach within the framework is concerned, a model-based improved A* search algorithm is used to sequentially determine the trajectory of each vehicle without further optimization-based techniques. The node expansion manners are determined on the basis of the purpose of a maneuver, and the speed pertaining to the start and end of a maneuver is divided into stages.

Different dispatching strategies are benchmarked. The ANN combined strategy shows the best performance in warehouse filling efficiency, but the decision failure times of this method are comparable to those of the comprehensive strategy. Meanwhile, it is observed that the balance of the filling state pertaining to shelf clusters is as important as the evaluation of the goal-reaching difficulties. Furthermore, the trajectories of multiple vehicles in a short time are presented. It is shown that although the priority of a vehicle during one subtask is predefined, the vehicles are capable of cooperatively and continuously finding feasible trajectories.

In the future, our joint strategy will be tested in a downsized unmanned warehouse setup. Additionally, we aim to implement a zonal shutdown feature for enhanced safety in case pedestrians enter the area. We also plan to integrate a forklift failure detection mechanism to enable uninterrupted warehouse operations even in the presence of a few malfunctioning forklifts.

Author Contributions: Conceptualization, T.Z., M.L. and K.C.; methodology, H.L. and Y.F.; validation, T.Z., Y.F. and K.C. All authors have read and agreed to the published version of the manuscript.

Funding: This research was funded by Fundamental Research Funds for the Central Universities: 531118010692; National Natural Science Foundation of China: 62103139; Fundamental Research Funds for the Central Universities: 531118010509; Natural Science Foundation of Hunan Province, China: 2021JJ40114.

Conflicts of Interest: The authors declare no conflict of interest.

References

1. Bozek, P.; Karavaev, Y.L.; Ardentov, A.A.; Yefremov, K.S. Neural network control of a wheeled mobile robot based on optimal trajectories. *Int. J. Adv. Robot. Syst.* **2020**, *17*, 2. [CrossRef]
2. Blatnický, M.; Dižo, J.; Sága, M.; Gerlici, J.; Kuba, E. Design of a Mechanical Part of an Automated Platform for Oblique Manipulation. *Appl. Sci.* **2020**, *10*, 8467. [CrossRef]
3. Yang, Q.; Lian, Y.; Xie, W. Hierarchical planning for multiple AGVs in warehouse based on global vision. *Simul. Model. Pract. Theory* **2020**, *104*, 102124. [CrossRef]
4. Li, G.; Wang, X.; Yang, J.; Wang, B. A new method to design fuzzy controller for unmanned autonomous forklift. In Proceedings of the 2015 IEEE International Conference on Robotics and Biomimetics (ROBIO), Zhuhai, China, 6–9 December 2015; pp. 946–951. [CrossRef]
5. Zhong, Z.; Wan, M.; Pei, Z.; Qin, X. Spatial and temporal optimization for smart warehouses with fast turnover. *Comput. Oper. Res.* **2021**, *125*, 105091. [CrossRef]
6. Weidinger, F.; Boysen, N.; Briskorn, D. Storage assignment with rack moving mobile robots in KIVA warehouses. *Transp. Sci.* **2018**, *52*, 1297–1588. [CrossRef]
7. Yu, W.; Wang, S.; Wang, R.; Tan, M. Generation of temporal-spatial Bezier curve for simultaneous arrival of multiple unmanned vehicles. *Inf. Sci.* **2017**, *418–419*, 34–45. [CrossRef]
8. Ouyang, Y.; Li, B.; Zhang, Y.; Acarman, T.; Guo, Y.; Zhang, T. Fast and optimal trajectory planning for multiple vehicles in a nonconvex and cluttered environment: Benchmarks, methodology, and experiments. In Proceedings of the 2022 IEEE International Conference on Robotics and Automation (ICRA), Philadelphia, PA, USA, 23–27 May 2022; pp. 10746–10752.
9. Chen, B.M. On the trends of autonomous unmanned systems research. *Engineering* **2022**, *12*, 20–23. [CrossRef]
10. Miyamoto, T.; Inoue, K. Local and random searches for dispatch and conflict-free routing problem of capacitated AGV systems. *Comput. Ind. Eng.* **2016**, *91*, 1–9. [CrossRef]
11. Stutzle, T.; Ruiz, R. Iterated Local Search. In *Handbook of Heuristics*; Marti, R., Pardalos, P., Resende, M., Eds.; Springer: Cham, Switzerland, 2018. [CrossRef]
12. Zhang, X.; Sang, H.; Li, J.; Han, Y.; Duan, P. An effective multi-AGVs dispatching method applied to matrix manufacturing workshop. *Comput. Ind. Eng.* **2022**, *163*, 107791. [CrossRef]
13. Lin, L.; Shinn, S.W.; Gen, M.; Hwang, H. Network model and effective evolutionary approach for AGV dispatching in manufacturing system. *J. Intell. Manuf.* **2006**, *17*, 465–477. [CrossRef]

14. Bao, Y.; Jiao, G.; Huang, M. Cooperative optimization of pod repositioning and AGV task allocation in Robotic Mobile Fulfillment Systems. In Proceedings of the 2021 33rd Chinese Control and Decision Conference (CCDC), Kunming, China, 22–24 May 2021; pp. 2597–2601. [CrossRef]
15. Lee, I.G.; Chung, S.H.; Yoon, S.W. Two-stage storage assignment to minimize travel time and congestion for warehouse order picking operations. *Comput. Ind. Eng.* **2020**, *139*, 106129. [CrossRef]
16. Fragapane, G.; de Koster, R.; Sgarbossa, F.; Strandhagen, J.O. Planning and control of autonomous mobile robots for intralogistics: Literature review and research agenda. *Eur. J. Oper. Res.* **2021**, *294*, 405–426. [CrossRef]
17. Hu, H.; Jia, X.; He, Q.; Fu, S.; Liu, K. Deep reinforcement learning based AGVs real-time scheduling with mixed rule for flexible shop floor in industry 4.0. *Comput. Ind. Eng.* **2020**, *149*, 106749. [CrossRef]
18. Zhou, T.; Tang, D.; Zhu, H.; Zhang, Z. Multi-agent reinforcement learning for online scheduling in smart factories. *Robot. Comput.-Integr. Manuf.* **2021**, *72*, 102202. [CrossRef]
19. Lian, Y.; Yang, Q.; Liu, Y.; Xie, W. A spatio-temporal constrained hierarchical scheduling strategy for multiple warehouse mobile robots under industrial cyber-physical system. *Adv. Eng. Inform.* **2022**, *52*, 101572. [CrossRef]
20. Liu, Y.; Du, Y.; Dou, S.; Peng, L.; Su, X. Research summary of intelligent optimization algorithm for warehouse AGV path planning. In *Lecture Notes in Operations Research*; Shi, X., Bohács, G., Ma, Y., Gong, D., Shang, X., Eds.; Springer: Singapore, 2022. [CrossRef]
21. Ma, Y.; Wang, H.; Xie, Y.; Guo, M. Path planning for multiple mobile robots under double-warehouse. *Inf. Sci.* **2014**, *278*, 357–379. [CrossRef]
22. Xidias, E.K. A decision algorithm for motion planning of car-like robots in dynamic environments. *Cybern. Syst.* **2021**, *52*, 1909844. [CrossRef]
23. Li, B.; Acarman, T.; Zhang, Y.; Ouyang, Y.; Yaman, C.; Kong, Q.; Zhong, X.; Peng, X. Optimization-based trajectory planning for autonomous parking with irregularly placed obstacles: A lightweight iterative framework. *IEEE Trans. Intell. Transp. Syst.* **2022**, *23*, 11970–11981. [CrossRef]
24. Li, B.; Zhang, Y.; Feng, Y.; Zhang, Y.; Ge, Y.; Shao, Z. Balancing computation speed and quality: A decentralized motion planning method for cooperative lane changes of connected and automated vehicles. *IEEE Trans. Intell. Veh.* **2018**, *3*, 340–350. [CrossRef]
25. Zhou, Y.; Huang, N. Airport AGV path optimization model based on ant colony algorithm to optimize Dijkstra algorithm in urban systems. *Sustain. Comput. Inform. Syst.* **2022**, *35*, 100716. [CrossRef]
26. Zacharia, P.T.; Xidias, E.K. AGV routing and motion planning in a flexible manufacturing system using a fuzzy-based genetic algorithm. *Int. J. Adv. Manuf. Technol.* **2020**, *109*, 1801–1813. [CrossRef]
27. Kneissl, M.; Madhusudhanan, A.K.; Molin, A.; Esen, H.; Hirche, S. A multi-vehicle control framework with application to automated valet parking. *IEEE Trans. Intell. Transp. Syst.* **2021**, *22*, 5697–5707. [CrossRef]
28. Dresner, K.; Stone, P. Multiagent traffic management: A reservation-based intersection control mechanism. In Proceedings of the Third International Joint Conference on Autonomous Agents and Multiagent Systems, New York, NY, USA, 23 July 2004; pp. 530–537.
29. Guney, M.A.; Raptis, I.A. Dynamic prioritized motion coordination of multi-AGV systems. *Robot. Auton. Syst.* **2021**, *139*, 103534. [CrossRef]
30. Digani, V.; Caramaschi, F.; Sabattini, L.; Secchi, C.; Fantuzzi, C. Obstacle avoidance for industrial AGVs. In Proceedings of the 2014 IEEE 10th International Conference on Intelligent Computer Communication and Processing (ICCP), Cluj-Napoca, Romania, 4–6 September 2014; pp. 227–232. [CrossRef]
31. Stern, R. Multi-Agent Path Finding—An Overview. In *Artificial Intelligence. Lecture Notes in Computer Science*; Osipov, G., Panov, A., Yakovlev, K., Eds.; Springer: Cham, Switzerland, 2019; Volume 11866. [CrossRef]
32. Zhang, J.; Li, Z.; Li, L.; Li, Y.; Dong, H. A bi-level cooperative operation approach for AGV based automated valet parking. *Transp. Res. Part C* **2021**, *128*, 103140. [CrossRef]
33. Meng, Y.; Li, L.; Wang, F.; Li, K.; Li, Z. Analysis of cooperative driving strategies for nonsignalized intersections. *IEEE Trans. Veh. Technol.* **2018**, *67*, 2900–2911. [CrossRef]
34. Lee, C.K.M.; Lin, B.; Ng, K.K.H.; Lv, Y.; Tai, W.C. Smart robotic mobile fulfillment system with dynamic conflict-free strategies considering cyber-physical integration. *Adv. Eng. Inform.* **2019**, *42*, 100998. [CrossRef]
35. Rochel, P.; Rios, H.; Mera, M.; Dzul, A. Trajectory tracking for uncertain Unicycle Mobile Robots: A Super-Twisting approach. *Control Eng. Pract.* **2022**, *122*, 105078. [CrossRef]
36. Zhang, T. An estimation method of the fuel mass injected in large injections in Common-Rail diesel engines based on system identification using artificial neural network. *Fuel* **2022**, *310*, 122404. [CrossRef]
37. Dolgov, D.; Thrun, S.; Montemerlo, M.; Diebel, J. Path planning for autonomous vehicles in unknown semi-structured environment. *Int. J. Robot. Res.* **2010**, *29*, 485–501. [CrossRef]
38. Li, B.; Shao, Z. A unified motion planning method for parking an autonomous vehicle in the presence of irregularly placed obstacles. *Knowl. Based Syst.* **2015**, *86*, 11–20. [CrossRef]

Disclaimer/Publisher's Note: The statements, opinions and data contained in all publications are solely those of the individual author(s) and contributor(s) and not of MDPI and/or the editor(s). MDPI and/or the editor(s) disclaim responsibility for any injury to people or property resulting from any ideas, methods, instructions or products referred to in the content.

Article

Tube-Based Event-Triggered Path Tracking for AUV against Disturbances and Parametric Uncertainties

Yuheng Chen [1] and Yougang Bian [1,2,*]

[1] College of Mechanical and Vehicle Engineering, Hunan University, Changsha 410082, China; chenyuheng@hnu.edu.cn
[2] Wuxi Intelligent Control Research Institute (WICRI), Hunan University, Wuxi 214072, China
* Correspondence: byg19@hnu.edu.cn

Abstract: In order to enhance the performance of disturbance rejection in AUV's path tracking, this paper proposes a novel tube-based event-triggered path-tracking strategy. The proposed tracking strategy consists of a speed control law and an event-triggered tube model predictive control (tube MPC) scheme. Firstly, the speed control law using linear model predictive control (LMPC) technology is obtained to converge the nominal path-tracking deviation. Secondly, the event-triggered tube MPC scheme is used to calculate the optimal control input, which can enhance the performance of disturbance rejection. Considering the nonlinear hydrodynamic characteristics of AUV, a linear matrix inequality (LMI) is formulated to obtain tight constraints on the AUV and the feedback matrix. Moreover, to enhance real-time performance, tight constraints and the feedback matrix are all calculated offline. An event-triggering mechanism is used. When the surge speed change command does not exceed the upper bound, adaptive tight constraints are obtained. Finally, numerical simulation results show that the proposed tube-based event-triggered path-tracking strategy can enhance the performance of disturbance rejection and ensure good real-time performance.

Keywords: autonomous underwater vehicle; tube model predictive control; path tracking

1. Introduction

Autonomous underwater vehicles (AUVs) have been widely used in marine scientific research, underwater resource exploration, underwater oil and gas pipeline and structure overhaul, seabed hydrothermal research, and military fields [1,2]. When AUVs perform underwater tasks, they usually need to complete path-tracking tasks [3].

The 6-DOF motion of AUV in three-dimensional underwater space is coupled and nonlinear, and the parameters of the model are often difficult to obtain precisely. In model-based control methods, the control performance will suffer from parametric uncertainties [4]. Moreover, external disturbances caused by ocean currents will also degrade the control performance [5,6]. Therefore, it is a challenge to enhance the robustness against external disturbances and parametric uncertainties in model-based control methods [7]. Until now, researchers have applied strategies for improving the robustness of model-based control methods such as the model predictive control (MPC) technique [8,9] and sliding mode control (SMC) technology [10] in the path-tracking control of AUVs. Note that MPC can easily handle the physical constraints of the AUV when formulating the optimal control problem. It is also well-known that MPC technology can provide some assistance for the disturbance rejection [11]. In other words, the MPC technology itself is robust against disturbance. Therefore, MPC is widely used in the path-tracking control of AUVs [12,13].

Zhang proposed a 3D path-tracking control method for AUVs using a linear model predictive control (LMPC) [13]. The LMPC controller is used to calculate the speed control law. Then, the control inputs of the AUV were directly calculated based on the dynamics model, where the physical constraints on the control input failed to be considered. In [14,15], the speed control law was generated by the kinematics LMPC, and the control inputs were

generated by the dynamic LMPC. These physical constraints on the control input can be considered when formulating the optimal control problem. Compared with [13], the method in [14,15] can also enhance the robust performance against disturbances, by the robustness of the nominal MPC technology itself. However, there is no direct disturbance rejection strategy, such as disturbance estimation [16,17] or robust MPC technology [18]. The robustness of the nominal MPC technology itself is limited. These disturbance rejection strategies can significantly improve the robustness performance, compared with the nominal MPC technology. Therefore, a direct disturbance rejection strategy can be introduced in the nominal MPC technology to improve the tracking control performance.

The extended Kalman filter technology is used to estimate external current disturbances [17]. Based on the 12-dimensional kinematic model and kinetic model, a NMPC controller is proposed to calculate the optimal control law using these results of disturbance estimation. However, the disturbance estimation will bring extra dimensions, which may lead to poor real-time performance. To overcome the challenge, disturbance estimation is only based on the 5-dimensional kinematic model using MPC, which can save online optimization computing time [10]. Note that control inputs are calculated using adaptive sliding mode control technology, which is sensitive to the noise in the actual control system. The control performance may suffer from the chattering problem in the practical application [19].

Tube MPC, as a disturbance rejection strategy, was first proposed by Blanchini [20]. Compared with the disturbance estimation, the robustness improvement is achieved by its own relatively stable mechanism. Suffering from external disturbance and parametric uncertainties, there is a model mismatch between the nominal model and the actual model. A robust positively invariant (RPI) set is proposed to measure the boundedness of the mismatch [21]. In the tube MPC scheme, the tight constraint is calculated by tightening the constraints of the actual system by an RPI set. The control law of the tube MPC scheme consists of a nominal optimal control law and a feedback control law. The nominal control law is obtained by solving a receding horizon optimal control problem with a tight constraint. The feedback control law is used to address the deviation of the nominal and actual states due to the model mismatch. The traditional tube MPC scheme [21,22] is proposed for AUV's path tracking [18]. Note that the RPI set is obtained based on the assumed disturbance upper bound. Hence, the corresponding tight constraints may become too conservative to degrade the path-tracking performance. Based on the coupled 6-dimensional AUV model, both the RPI set and the terminal feasible set are easy to have no solution. Moreover, online calculating tight constraints of the nominal model brings too much computing time, which will also lead to poor real-time performance.

Since the inherent robustness of the nominal MPC to address the model mismatch is limited, the tube MPC has the potential to improve robustness against model mismatches. However, the control performance suffers from poor real-time performance and no solution for the RPI set. Our motivation is to apply the tube MPC to enhance the robustness of AUV's path tracking, with these issues addressed. This study proposes a tube-based event-triggered path-tracking strategy, which consists of a kinematics LMPC controller and a tube MPC controller. To converge the nominal path-tracking deviation, the kinematics LMPC controller is used to calculate the optimal speed control law. The tube MPC controller is used to compute the control input of the AUV to track the speed control law. Compared with the tube MPC technology used in [18], to avoid no solution to the RPI, the coupled kinetic model is decoupled into three Lipschitz nonlinear models [23]: a surge speed control model, a heading control model, and a depth control model. With the corresponding Lipschitz constant obtained, nonlinear properties of these models when formulating a linear matrix inequality (LMI) are used to calculate the RPI set and the feedback matrix. The terminal feasible set is obtained based on linear differential inclusion (LDI) technology. In order to achieve good real-time performance, constraints on the nominal model and the feedback matrix are all calculated offline. Note that the hydrodynamic force of the AUV is related to the surge speed. The mismatch may depend on the surge speed change command.

These offline calculated invariant constraints are too conservative to achieve better control performance. Then, an event-triggering mechanism is used. When the surge speed change command does not exceed the upper bound, two decision variables are introduced to formulate a flexible tube. Then, adaptive constraints on the nominal model are obtained to address the mismatch. When the surge speed change command exceeds the upper bound, the offline tight constraints will be used. The main contributions of this work are as follows:

1. A tube-based event-triggered path-tracking strategy, which consists of a LMPC controller and a tube MPC controller, is proposed to enhance the robustness against disturbances and parametric uncertainties. The LMPC controller is used to calculate the speed control law to converge the path-tracking deviation, and the tube MPC controller is used to track the speed control law.
2. In the tube MPC controller, with nonlinear characteristics of AUV hydrodynamic force considered, tight constraints in the nominal control law and the feedback matrix in the feedback control law are obtained by formulating two LMIs. To achieve real-time performance, these linear matrix inequalities are all calculated offline.
3. To overcome control performance degradation brought by conservative tight constraints calculated offline, an event-triggering mechanism is used to dynamically adjust these constraints in the nominal control law according to the surge speed change command. Compared with conservative tight constraints, better path tracking can be achieved, and the real-time performance is also satisfied.

The remainder of this paper is organized as follows. In Section 2, preliminaries are given. In Section 3, the AUV's motion model and the path-tracking problem are given. In Section 4, the detail design of the tube-based event-triggered path-tracking strategy is given. In Section 5, the numerical simulation analysis is shown.

2. Preliminaries

The actual nonlinear continuous-time dynamics is described as a Lipschitz nonlinear system [23]:

$$\dot{x} = f(x, u, \omega) = Ax + Bu + g(x) + B_\omega \omega \tag{1}$$

with $x \in R^{n \times 1}$ and $u \in R^{m \times 1}$. $\omega \in \mathbb{w} = \left\{ \omega \in R^{n \times 1} : \|\omega\|_\infty < c_\omega \right\}$ denotes the bounded external disturbance. Positive constant c_ω is the disturbance upper bound. System (1) is also subject to state and control input constraints:

$$x \in \mathcal{X} \subset R^{n \times 1}, u \in \mathcal{U} \subset R^{m \times 1} \tag{2}$$

where $\mathcal{U}$ is a compact set and $\mathcal{X}$ is bounded. Here $g(x)$ is a Lipschitz nonlinear function with a Lipschitz constant $L > 0$ such that:

$$\|g(x_1) - g(x_2)\| \leq L\|x_1 - x_2\|, \forall x_1, x_2 \in \mathcal{X} \tag{3}$$

The overline format of a variable denotes its nominal value, e.g., $\bar{x}$ denotes the nominal value of x. The continuous-time nominal model is given by:

$$\dot{\bar{x}} = f(\bar{x}, \bar{u}, 0) \tag{4}$$

and the corresponding discrete-time system models are given by:

$$x_{t+1} = f_d(x_t, u_t, \omega_t) \tag{5}$$

$$\bar{x}_{t+1} = f_d(\bar{x}_t, \bar{u}_t, 0) \tag{6}$$

Define $\mathbb{K}_{N_1:N_2} := \{N_1, N_1+1, \cdots, N_2-1, N_2\}$. The nominal cost function of predicted state sequence, $\bar{x}_{k|t}$, $k \in \mathbb{K}_{0:N_T}$, and control input sequence, $\bar{u}_{k|t}$, $k \in \mathbb{K}_{0:N_T-1}$, is given as:

$$J = \sum_{k=0}^{N_T-1} l(\bar{x}_{k|t}, \bar{u}_{k|t}) + V_f(\bar{x}_{N_T|t}) \tag{7}$$

where N_T is the predictive horizon. l is the positive definite stage cost and V_f is the terminal cost:

$$l(\bar{x}_{k|t}, \bar{u}_{k|t}) = \left\|\bar{x}_{k|t}\right\|_{Q_T}^2 + \left\|\bar{u}_{k|t}\right\|_{R_T}^2, V_f(\bar{x}_{N_T|t}) = \left\|\bar{x}_{N_T|t}\right\|_{P_T}^2 \tag{8}$$

The state deviation between the actual system and the nominal actual is denoted by $z = x - \bar{x}$. The deviation system is given as:

$$\dot{z} = \dot{x} - \dot{\bar{x}} = f(x, u, \omega) - f(\bar{x}, \bar{u}, 0) \tag{9}$$

In a tube MPC controller, the control law consists of a nominal MPC control law $\bar{u}$ and a state feedback control law $\kappa(\bar{x}, x)$:

$$u := \bar{u} + \kappa(\bar{x}, x) \tag{10}$$

where $\bar{u}$ is obtained by solving an optimal control problem, and $\kappa(\bar{x}, x)$ is used to converge the state deviation z.

Definition 1. (Robust positively invariant (RPI) set): A set $\Omega \subset \mathcal{X}$ is the RPI set of deviation system (9), if there exists a feedback control law $\kappa(\bar{x}, x) \in \mathcal{U}$, such that for all $z_{t_0} \in \Omega$ and $\omega \in \mathbb{w}$, it holds that $z_t \in \Omega$ for all $t \geq t_0$.

Then the constraints of nominal system (6) are given with an RPI set Ω as:

$$\bar{x} \in \bar{\mathcal{X}} := \mathcal{X} \ominus \Omega, \bar{u} \in \bar{\mathcal{U}} := \{\bar{u} | \bar{u} + \kappa(\bar{x}, x) \in \mathcal{U}\} \tag{11}$$

where $\bar{\mathcal{X}}$ and $\bar{\mathcal{U}}$ are tight constraint sets, which can be expressed as:

$$(\bar{x}, \bar{u}) \in M := \left\{ (\bar{x}, \bar{u}) \in R^{(n+m) \times 1} \middle| h_j(\bar{x}, \bar{u}) \leq 0, j = 1, 2, \cdots, p \right\} \tag{12}$$

Considering linear constraints, these constraints can also be expressed as a polytope:

$$M = \left\{ \begin{bmatrix} \bar{x} \\ \bar{u} \end{bmatrix} \in R^{(n+m) \times 1} : c_j \bar{x} + d_j \bar{u} \leq 1, j = 1, 2, \cdots, p \right\} \tag{13}$$

Considering tight constraints (11) and nominal system dynamics (6), the following optimal control problem is formulated to calculate the nominal MPC control law:

$$\min_{\bar{u}_{k|t}, k \in \mathbb{K}_{0:N_T-1}} J \tag{14}$$

$$s.t. \quad \begin{aligned} &\bar{x}_{0|t} = \bar{x}_0, \bar{u}_{0|t} = \bar{u}_0 \\ &\bar{x}_{k+1|t} = f_d(\bar{x}_{k|t}, \bar{u}_{k|t}, 0) \\ &h_j(\bar{x}_{k|t}, \bar{u}_{k|t}) \leq 0, j = 1, 2, \cdots, p \\ &\bar{x}_{N_T|t} \in X_f \end{aligned}$$

where X_f is the terminal feasible set. The optimal control input sequence $\bar{u}_{k|t}^*$, $k \in \mathbb{K}_{0:N_T-1}$, is the solution to optimal control problem (14), and the nominal optimal MPC control law $\bar{u}$ is obtained by:

$$\bar{u} = \bar{u}_{0|t}^* \tag{15}$$

3. AUV Motion Model and Problem Formulation

In this section, the kinematics model and kinetic model of the AUV are given, where both external disturbances and parametric uncertainty are considered in the kinetic model. In the proposed tube-based event-triggered path-tracking strategy, described in Section 4, based on the kinematics model, a speed control law is designed to converge the nominal path-tracking deviation. Then, based on the kinetic model, the control input of the AUV is calculated to track the speed control law. Correspondingly, two problems treated in this study are formalized.

3.1. AUV Motion Model

The global coordinate and the local coordinate frame are defined, and the coordinate transformation relationship is shown in Figure 1. Here $E - \xi \eta \zeta$ denotes the global coordinate system, and $O - xyz$ denotes the local coordinate system [24].

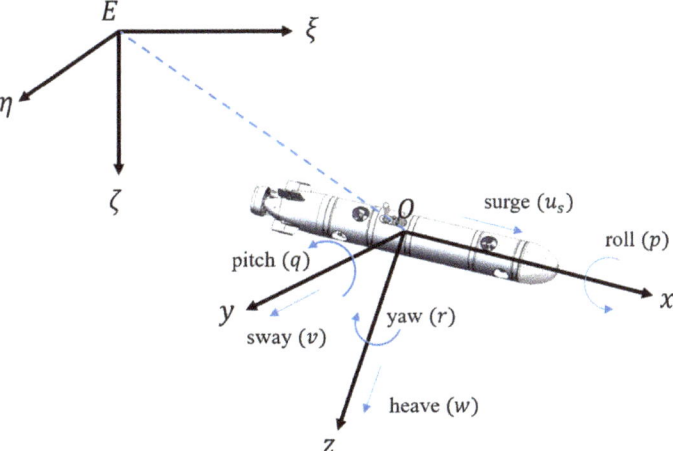

Figure 1. AUV coordinate system [24].

Note that the roll motion is self-stable, and the roll motion attitude is also small, meaning that the roll angle ϕ and the roll speed p can all be regarded as 0. Therefore, the roll motion is not considered in this paper. The speed vector is denoted by $v = (u_s, v, w, q, r)^T$ in the motion coordinate $O - xyz$, where u_s, v, w, q, and r are respectively surge speed, sway speed, heave speed, pitch speed, and yaw speed. The position and attitude angle (pitch and yaw angles) vector is denoted by $\eta = (x, y, z, \theta, \psi)^T$ in global coordinate system $E - \xi \eta \zeta$. The kinematics model is given as:

$$\eta^T = J_{v\eta} v^T \quad (16)$$

where $J_{v\eta}$ is a transformation matrix from $O - xyz$ to $E - \xi \eta \zeta$:

$$J_{v\eta} = \begin{bmatrix} \cos\psi\cos\theta & -\sin\psi\cos\theta + \cos\psi\sin\theta\sin\psi & \cos\psi\sin\theta & 0 & 0 \\ \sin\psi & \cos\psi & -\cos\psi\sin\psi + \sin\theta\sin\psi & 0 & 0 \\ -\sin\theta & 0 & \cos\psi & 0 & 0 \\ 0 & 0 & 0 & 1 & 0 \\ 0 & 0 & 0 & 0 & \frac{1}{\cos\theta} \end{bmatrix} \quad (17)$$

The kinetic model is given as:

$$M\dot{v} = (C(v) + D(v) + \Delta F_{CD})v + F + G(\tau) + \tau_d \quad (18)$$

where $\mathcal{M} \in R^{5\times5}$ is the inertial matrix, $C(\nu) \in R^{5\times5}$ is the Coriolis and centripetal matrix, $D(\nu) \in R^{5\times5}$ is the hydrodynamic damping matrix, and $F \in R^{5\times1}$ is the hydrostatic force and moment. $\tau = (F_x, \delta_r, \delta_s)^T$ is the AUV's control vector, where F_x is the stern thruster force, δ_r is the vertical plane deflection, and δ_s is the translational plane deflection. $G(\tau): R^{3\times1} \to R^{5\times1}$ is the active control force in the AUV's motion coordinate system $O - xyz$. $\tau_d \in R^{5\times1}$ is the external disturbance. $\Delta F_{CD} \in R^{5\times5}$ represents the disturbance brought by parametric uncertainties [10].

3.2. Model Decoupling

Note that the degrees of freedom of the AUV are coupled in nonlinear model (18). In order to simplify the design of the controller, the 5 DOF nonlinear dynamic model (18) of the AUV is decoupled for surge speed control, heading control, and depth control. Considering that the AUV always maintains a constant surge speed for path tracking, the nominal surge speed $\overline{u}_s$ in the heading nominal control model and depth nominal control model above is set as a constant. Then, these decoupled models can all be described as a Lipschitz nonlinear system. The hydrodynamic coefficients in these models are given in our previous research [24].

1. Surge speed nominal control model:

$$\dot{\overline{x}}_u = A_u \overline{x}_u + B_u \overline{U}_u + g_x(\overline{x}_u) \tag{19}$$

where the state is denoted by $\overline{x}_u = \overline{u}_s$. The control input is denoted by the nominal stern thruster force $\overline{U}_u = \overline{F}_x$. $A_u = x_u/(m - X_{\dot{u}})$, $B_u = 1/(m - X_{\dot{u}})$, and $g_x(\overline{x}_u) = X_{uu}\overline{u}_s|\overline{u}_s|$. m is the mass of the AUV. $X_{\dot{u}}$ is the added mass. X_{uu} is hydrodynamic damping coefficient.

2. Heading nominal control model:

$$\dot{\overline{x}}_y = A_y \overline{x}_y + B_y \overline{U}_y + g_y(\overline{x}_y) \tag{20}$$

where the state is denoted by $\overline{x}_y = (\overline{v}, \overline{r})^T$. The control input is denoted by the nominal vertical plane deflection: $\overline{U}_y = \overline{\delta}_r$.

$$A_y = \begin{bmatrix} m - Y_{\dot{v}} & -Y_{\dot{r}} \\ -N_{\dot{v}} & I_{zz} - N_{\dot{r}} \end{bmatrix}^{-1} \begin{bmatrix} Y_{uv}\overline{u}_s & (m + Y_{ur})\overline{u}_s \\ N_{uv}\overline{u}_s & N_{ur}\overline{u}_s \end{bmatrix},$$

$$B_y = \begin{bmatrix} m - Y_{\dot{v}} & -Y_{\dot{r}} \\ -N_{\dot{v}} & I_{zz} - N_{\dot{r}} \end{bmatrix}^{-1} \begin{bmatrix} Y_{uu\delta_s} \\ N_{uu\delta_s} \end{bmatrix},$$

$$g_y(\overline{x}_y) = \begin{bmatrix} m - Y_{\dot{v}} & -Y_{\dot{r}} \\ -N_{\dot{v}} & I_{zz} - N_{\dot{r}} \end{bmatrix}^{-1} \begin{bmatrix} Y_{vv}\overline{v}|\overline{v}| + Y_{rr}\overline{r}|\overline{r}| \\ N_{vv}\overline{v}|\overline{v}| + N_{rr}\overline{r}|\overline{r}| \end{bmatrix} \tag{21}$$

where I_{zz} is the rotational inertia. $Y_{\dot{v}}$, $N_{\dot{r}}$, $Y_{\dot{r}}$ and $N_{\dot{v}}$ are the added mass. Y_{uv}, Y_{ur}, N_{uv}, N_{ur}, $Y_{uu\delta_s}$ and $N_{uu\delta_s}$ are hydrodynamic coefficients. Y_{vv}, Y_{rr}, N_{vv} and N_{rr} are hydrodynamic damping coefficients.

3. Depth nominal control model:

$$\dot{\overline{x}}_z = A_z \overline{x}_z + B_z \overline{U}_z + g_z(\overline{x}_z) \tag{22}$$

where the state is denoted by $\overline{x}_z = (\overline{w}, \overline{q})^T$. The control input is denoted by the nominal translational plane deflection: $\overline{U}_z = \overline{\delta}_s$.

$$A_z = \begin{bmatrix} m - Z_{\dot{w}} & -Z_{\dot{q}} \\ -M_{\dot{w}} & I_{yy} - M_{\dot{q}} \end{bmatrix}^{-1} \begin{bmatrix} Z_{uw}\overline{u}_s & (-m + Z_{uq})\overline{u}_s \\ M_{uw}\overline{u}_s & M_{uq}\overline{u}_s \end{bmatrix},$$

$$B_z = \begin{bmatrix} m - Z_{\dot{w}} & -Z_{\dot{q}} \\ -M_{\dot{w}} & I_{yy} - M_{\dot{q}} \end{bmatrix}^{-1} \begin{bmatrix} Z_{uu\delta_s} \\ M_{uu\delta_s} \end{bmatrix},$$

$$g_z(\bar{x}_z) = \begin{bmatrix} m - Z_{\dot{w}} & -Z_{\dot{q}} \\ -M_{\dot{w}} & I_{yy} - M_{\dot{q}} \end{bmatrix}^{-1} \begin{bmatrix} Z_{ww}\overline{w}|\overline{w}| + Z_{qq}\overline{q}|\overline{q}| \\ M_{ww}\overline{q}|\overline{q}| + M_{qq}\overline{q}|\overline{q}| \end{bmatrix} \quad (23)$$

where I_{yy} is the rotational inertia. $Z_{\dot{w}}$, $M_{\dot{w}}$, $Z_{\dot{q}}$ and $M_{\dot{q}}$ are the added mass. Z_{uw}, Z_{uq}, M_{uw}, M_{uq}, $Z_{uu\delta_s}$ and $M_{uu\delta_s}$ are hydrodynamic coefficients. Z_{ww}, Z_{qq}, M_{ww} and M_{qq} are hydrodynamic damping coefficients.

3.3. Problem Statement

Problem 1. *Given a vector $\eta_r \in R^{5\times 1}$ that stands for the reference position and attitude angle and a vector $v_r \in R^{5\times 1}$ that stands for the reference speed, the nominal path-tracking deviation vector is denoted by $\bar{e}_\eta := \bar{\eta} - \eta_r$ and the surge speed deviation vector is denoted by $\bar{e}_{u_s} := \bar{u}_s - u_{sr}$. The speed control law $v_d := (u_{sd}, v_d, w_d, q_d, r_d)^T = \kappa_\eta(\bar{e}_\eta, \bar{e}_v)$ needs to be obtained to converge the nominal path-tracking deviation: $\lim_{t\to\infty} \bar{e}_\eta(t) = 0$.*

Problem 2. *The speed control law deviation vector is denoted by $e_{v_d} = \bar{v} - v_d$ and the actual path-tracking deviation is denoted by $e_\eta := \eta - \eta_r = (e_x, e_y, e_z, e_\theta, e_\psi)^T$. Based on decoupling models (19), (20), and (22), AUV's control vector $\tau = (F_x, \delta_r, \delta_s)^T$ needs to be obtained to respond to the speed control law: $\lim_{t\to\infty} e_{v_d}(t) = 0$. Finally, the actual path-tracking deviation e_η can be converged: $\lim_{t\to\infty} e_\eta(t) = 0$.*

4. Methodology

In order to address the external ocean current disturbance and parametric uncertainties of the Coriolis and centripetal matrix and the hydrodynamic damping matrix, a tube-based event-triggered path-tracking strategy consisting of a LMPC controller and a tube MPC controller is developed. The scheme of the proposed path-tracking strategy is shown in Figure 2.

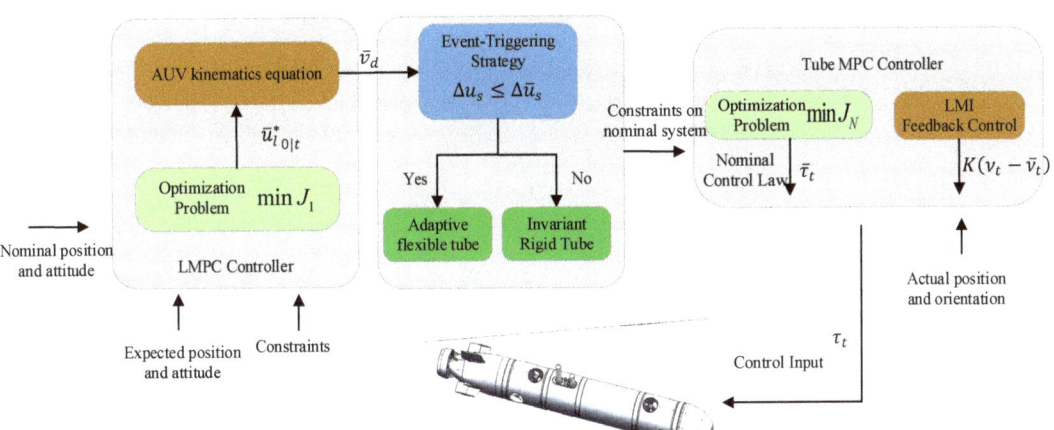

Figure 2. Scheme of the proposed path-tracking strategy.

Based on the kinematics model, the LMPC controller is used to address Problem 1, whose inputs are the reference waypoint and real-time nominal states of the AUV, and outputs are the speed control law.

Based on decoupled models (19), (20), and (22), according to the speed control law and real-time states of the AUV, the tube MPC controller is used to compute optimal control inputs of the AUV. The nominal control law is obtained by solving a constrained optimal control problem. According to these decoupled models, the surge speed will have a great effect on the AUV's state, and the mismatch of these decoupled models may depend on the surge speed. These offline calculated invariable constraints on the nominal system may be too conservative. According to the change in the surge speed command, an event-triggering strategy is used to formulate an adaptive flexible tube to deal with the mismatch.

4.1. LMPC Controller

First, nominal kinematics model (1) is discretized as:

$$\bar{\eta}_{t+1} = \bar{\eta}_t + T J_{v\eta} \bar{v}_t \quad (24)$$

where T is the sampling time.

Limited by the AUV's kinetics characteristics, sharp changes in speed v are not allowed. Then, the increment of speed v is used as the control input:

$$\bar{u}_{lt} = \Delta \bar{v}_t = \bar{v}_t - \bar{v}_{t-1} \quad (25)$$

To minimize the path-tracking deviation and avoid sharp changes in speed, the objective function is designed as follows:

$$J_{LMPC} = \sum_{k=0}^{N_l-1} \left(\left\| \bar{e}_{\eta_{k|t}} \right\|_{Q_\eta}^2 + \left\| \bar{e}_{u_s k|t} \right\|_{Q_{us}}^2 + \left\| \bar{u}_{l k|t} \right\|_{R_v}^2 \right) + \left\| \bar{e}_{\eta_{N_l|t}} \right\|_{P_\eta}^2 \\ + \left\| \bar{e}_{u_s N_l|t} \right\|_{P_v}^2 \quad (26)$$

where N_l is the predictive horizon in the LMPC controller.

The constraints of the control input and speed are given as:

$$\bar{u}_l \in \mathfrak{u}_1 = \{ \bar{u}_l : \bar{u}_{l min} \leq \bar{u}_l \leq \bar{u}_{l max} \} \quad (27)$$

$$\bar{v} \in V = \{ \bar{v} : \bar{v}_{min} \leq \bar{v} \leq \bar{v}_{max} \} \quad (28)$$

where $\bar{u}_{l max}$ and $\bar{u}_{l min}$ are the control input's upper bound and low bound, which satisfy $\bar{u}_{l min} = -\bar{u}_{l max}$, and $\bar{v}_{max}$ and $\bar{v}_{min}$ are the speed's upper bound and low bound.

Then an optimal control problem is designed to calculate the nominal speed control law $\bar{v}_{d_t}$:

$$\min_{u_{l k|t}, k \in \mathbb{K}_{0:N_l-1}} J_{LMPC}$$

$$s.t. \begin{array}{l} \bar{\eta}_{0|t} = \bar{\eta}_t, \bar{u}_{0|t} = \bar{u}_t \\ \bar{\eta}_{k+1|t} = \bar{\eta}_{k|t} + T J \bar{v}_{k|t} \\ \bar{v}_{k|t} = \bar{v}_t + \sum_{j=0}^{k} \bar{u}_{l j|t} \\ \bar{u}_{l k|t} \in \mathfrak{u}_1 \\ \bar{v}_{k|t}, \bar{v}_{N_l|t} \in V \end{array} \quad (29)$$

where $\bar{u}^*_{l k|t}$ is the solution to the optimal control problem. Then the speed control law v_{d_t} is obtained as:

$$v_{d_t} = \bar{u}^*_{l 0|t} + \bar{v}_t \quad (30)$$

where $\bar{u}^*_{l 0|k}$ is the increment of speed in the present moment.

4.2. Tube MPC Controller

Like the tube MPC scheme given in (10)–(14), the nominal control law is used to track the speed control law, which is obtained by solving an optimal control problem. The nonlinear hydrodynamic characteristics are considered in the state transition constraint, $\bar{x}_{k+1|t} = f_d(\bar{x}_{k|t}, \bar{u}_{k|t}, 0)$, which are decoupled in (19), (20), and (22). Moreover, the terminal feasible set X_f and the RPI set Ω also need to be obtained when formulating the optimal control problem. The state feedback control law, which considers the nonlinear characteristics of the AUV, is used to converge the deviation of the actual state x and the nominal state $\bar{x}$. In a Lipschitz nonlinear system, the Lipschitz constant L can be used to describe the nonlinear characteristics. Following [23], with the Lipschitz constant L set, a feedback matrix used to calculate the stated feedback control law and the (RPI) set Ω can be obtained by formulating an LMI. Following [25], another optimal control problem is formulated to obtain the terminal feasible set X_f considering the linear differential inclusion characteristics of the AUV. To ensure real-time performance, they are calculated offline. A brief derivation of the LMI and the optimal control problem is given as follows.

Assumption 1. *In these decoupling models, there always exists a corresponding constant L to satisfy the condition of the Lipschitz nonlinear function (3).*

Lemma 1 [23]. *For a Lipschitz nonlinear system (1), there exists a positive definite matrix $X \in R^{n \times n}$, a matrix $Y \in R^{m \times n}$, and scalars $\lambda_0 > \lambda > 0$ and $\mu > 0$ such that:*

$$\begin{bmatrix} (AX+BY)^T + AX + BY + \lambda X & B_\omega \\ B_\omega^T & -\mu I_n \end{bmatrix} \leq 0, L \leq \frac{(\lambda - \lambda_0)\alpha_{min}(P_R)}{2\|P_R\|} \tag{31}$$

with $P_R = X^{-1}$, and the feedback matrix $K = X^{-1}Y$.

With parameters λ_0 and μ set, LMI (31) is solved to obtain the matrices X, Y and the parameter λ. Matrices X and Y are used to calculate the feedback matrix K. With disturbance upper bound c_ω set, together with λ and X, the RPI set Ω and the feedback control law $\kappa(\bar{x}, x)$ can be obtained:

$$\begin{aligned} \Omega &= \{x \in R^n | x^T P_R x \leq \tfrac{\mu c_\omega}{\lambda}\} \\ \kappa(\bar{x}, x) &= K(x - \bar{x}) \end{aligned} \tag{32}$$

where the RPI set Ω and the feedback matrix K are all invariant. With the RPI set Ω calculated, the constraint on the nominal state is obtained, which is equivalent to constraints on the nominal control input using the Minkowski Operation [26]. Then, constraints of nominal system (11) can be obtained.

With the RPI set Ω obtained, the constraints of nominal system (6) are invariant, which are treated as an invariant rigid tube. The LDI of nominal system (6) is defined:

$$\Theta(M) = \left\{ F_\Theta(i) := [\mathcal{A}(i), \mathcal{B}(i)] = \left[\frac{\partial \bar{f}}{\partial \bar{x}}, \frac{\partial \bar{f}}{\partial \bar{u}}\right], \begin{bmatrix}\bar{x}\\ \bar{u}\end{bmatrix} \in M, i \in [k+\mathcal{N}, \infty) \right\} \tag{33}$$

The minimum convex polytope is denoted by $\text{Co}\Theta(M)$:

$$\text{Co}\Theta(M) = \left\{ F_\Theta(i) \in R^{n \times (n+m)} : F_\Theta(i) = \sum_{j=1}^{\mathcal{N}} \beta_j F_{\Theta j} = \sum_{j=1}^{\mathcal{N}} \beta_i [\mathcal{A}_j, \mathcal{B}_j] \right.$$

$$\left. \beta_i \geq 0, \sum_{j=1}^{\mathcal{N}} \beta_i = 1, i \in [k+\mathcal{N}, \infty) \right\} \tag{34}$$

where $F_{\Theta j}$ is the extreme matrix of the minimum convex polytope $\text{Co}\Theta(M)$, and $\mathcal{N}$ is the number of the extreme matrix.

The terminal feasible set $X_f \subset R^n := \{x \in R^n | x^T P_T x \leq \gamma\}$ is obtained as follows:

Lemma 2 [25]. Suppose the LDI of nominal system (6) is given by (33), and the constraints of nominal system (6) are obtained by (11) and (31). There exist matrices $0 < W_1 \in R^{n \times n}$ and $W_2 \in R^{m \times n}$ such that:

$$\begin{bmatrix} -F_{\Theta_j}W^T - WF_{\Theta_j}^T & W_1 Q^{\frac{1}{2}} & W_2^T \\ \begin{bmatrix} \left(Q^{\frac{1}{2}}\right)^T W_1 \\ W_2 \end{bmatrix} & \begin{bmatrix} I_n & 0 \\ 0 & R^{-1} \end{bmatrix} \end{bmatrix} \geq 0, j = 1, 2, \cdots, \mathcal{N} \tag{35}$$

and

$$\begin{bmatrix} 1/\gamma & c_j W_1 + d_j W_2 \\ (c_j W_1 + d_j W_2)^T & W_1 \end{bmatrix} \geq 0, \; j = 1, 2, \cdots, p \tag{36}$$

are satisfied with $W = \begin{bmatrix} W_1 & W_2^T \end{bmatrix}$. The terminal weighting matrix P_T is set as $P_T = W_1^{-1}$. $Q \in R^{n \times n}$ and $R \in R^{m \times m}$ are positive definite diagonal matrices.

Note that the determinant, $\det(\gamma W_1)$, represents the volume of the terminal region X_f, and a too-small terminal region will easily lead to the infeasibility of nominal optimal control problem (14). To enlarge the terminal region, another optimal control problem is formulated as:

$$\min_{\gamma, W_1, W_2} \log \det(\gamma W_1)^{-1} \tag{37}$$

$$\text{s.t.} \begin{matrix} \text{constraints (35), (36)} \\ \gamma > 0, W_1 > 0 \end{matrix}$$

According to speed control law (30), surge speed step signal Δu_s can be obtained. When the surge speed step signal exceeds the upper bound, i.e., $\Delta u_s > \Delta \bar{u}_s$, the constraints of nominal system (6) will be used. When the surge speed step signal Δu_s does not exceed the upper bound, i.e., $\Delta u_s \leq \Delta \bar{u}_s$, an adaptive flexible tube, treated in the form of inequalities, is introduced.

The variable $s_{k|t}$, which represents the size of the adaptive flexible tube, is calculated by decision variable $w_{k|t}$ to change offline calculated constraints (11). The decision variable $w_{k|t}$ is subject to nonlinear function $\tilde{w}_\delta(s_{k|t})$. Then, optimal control problem (14) becomes:

$$\min_{\bar{u}_{k|t}, \bar{w}_{k|t}, k \in \mathbb{K}_{0:N_T-1}} J \tag{38}$$

$$\begin{matrix} \bar{x}_{0|t} = \bar{x}_0, \bar{u}_{0|t} = \bar{u}_0, \bar{s}_{0|t} = 0 \\ \bar{x}_{k+1|t} = f_d(\bar{x}_{k|t}, \bar{u}_{k|t}, 0) \\ s_{k+1|t} = \rho s_{k|t} + w_{k|t} \\ \text{s.t.} h_j(\bar{x}_{k|t}, \bar{u}_{k|t}) + c_j s_{k|t} \leq 0, j = 1, 2, \cdots, p \\ w_{k|t} \leq \bar{w}, s_{k|t} \leq \bar{s} \\ w_{k|t} \geq \tilde{w}_\delta(s_{k|t}) \\ \bar{x}_{N_T|t} \in X_f \end{matrix}$$

where constant $\rho \in (0,1)$ is a decay factor, and c_j is a positive constant. Constraint upper bound $\bar{w}$ and $\bar{s}$ are all positive constants.

The nonlinear constraint $\tilde{w}_\delta(s_{k|t})$ is given by:

$$\tilde{w}_\delta(s_{k|t}) = \sqrt{c_{\delta,u}} c_\omega + \alpha_w(s_{k|t}) \tag{39}$$

where $\alpha_w(s_{k|t}) := \sum_{i=0}^{l} a_i s_{k|t}^i$ is a polynomial with $a_i \geq 0$. $c_{\delta,u}$ and c_ω are positive constants.

4.3. Implementation of the Proposed Strategy

To conclude, the proposed tube-based event-triggered path-tracking strategy consists of an offline strategy and an online strategy. To achieve good real-time performance, the offline strategy is introduced in Algorithm 1. LMI (31) is used to calculate the tight constraint Ω and the feedback matrix K. Optimal control problem (37) is used to calculate the terminal feasible set X_f. The online strategy is introduced in Algorithm 2. The optimal control problem (29) in the LMPC controller is first solved to obtain the speed control law. Then, the control law of the tube MPC controller (10), which consists of a nominal control law and a feedback control law, is respectively calculated to track the speed control law based on these decoupled models (19), (20), and (22). According to the surge speed step signal Δu_s, optimal control problem (38) or optimal control problem (14) is solved to obtain the nominal control law based on the offline calculated tight constraint and terminal feasible set X_f. Then, the offline calculated feedback control matrix is used to calculate the feedback control law to converge the deviation of the nominal and actual states.

Algorithm 1 Offline strategy

1. Define nominal cost function (7); choose state and control input constraints (2)
2. Choose appropriate parameters λ and L to solve LMI (31)
3. Obtain feedback matrix K and RPI set Ω (32)
4. Calculate invariant rigid tube (11)
5. Choose appropriate weight matrices Q and R to solve optimal control problem (37)
6. Obtain terminal feasible set X_f

Algorithm 2 Online AUV path-tracking algorithm

1. Measure AUV's actual state η_t, ν_t, and nominal state $\bar{\eta}_t, \bar{\nu}_t$
2. Solve optimal control problem (29) to obtain the speed control law $\bar{\nu}_{d_t}$
3. If $\Delta u_{st} \leq \Delta \bar{u}_s$:
4. Based on these decoupling models (19–20,22), separately formulate optimal control problem (38) to obtain nominal control vector $\bar{\tau}_t = (\bar{F}_{xt}, \bar{\delta}_{rt}, \bar{\delta}_{st})^T$
5. Otherwise:
6. Based on these decoupling models (19–20,22), separately formulate optimal control problem (14) to obtain nominal control vector $\bar{\tau}_t = (\bar{F}_{xt}, \bar{\delta}_{rt}, \bar{\delta}_{st})^T$
7. End
8. Calculate the AUV's control vector $\tau_t = \bar{\tau}_t + K(\nu_t - \bar{\nu}_t)$
9. Set $t = t + 1$, and go back to 1

---

5. Numerical Simulation

Numerical simulations are conducted to demonstrate the control performance of the proposed tube-based event-triggered path-tracking strategy. Path-tracking deviation, control input smoothness, and real-time performance are used to evaluate the control performance. In order to show the variation trend of the path-tracking deviation intuitively, the path-tracking integral deviation index is introduced, e.g., Se_x denotes the integral deviation of x in global coordinate system $E - \xi\eta\zeta$:

$$Se_{xt} = \int_0^t |e_{xj}| dj \tag{40}$$

Problem 1 is always solved by the proposed LMPC controller. These contrasting simulations differ in the method for solving Problem 2: "MPC" denotes simulation results using a nominal MPC controller from [15]. "RTMPC" denotes the simulation results using the tube MPC scheme (10)–(14) [23]. "ATMPC" denotes simulation results using the proposed tube-based event-triggered path-tracking strategy. To verify the superiority of the proposed path-tracking strategy, three real-time simulations were carried out, where "ATMPC" is compared with "RTMPC" and "MPC".

Numerical simulations were carried out using Simulink/Matlab, with AMD Ryzen Threadripper PRO 3995WX 64-Cores 2.70Ghz CPU and 256 GB RAM running Windows 10. Following this, optimal control problem (29) can be converted to a standard quadratic programming (QP) problem [22]. Then, the 'quadprog' function in Matlab can be used to solve the QP problem. When formulating the general optimal control problem (38), we refer to the open-source code, gitlab.ethz.ch/ics/RAMPC-CCM.

5.1. Parameters Set

Note that the influence of parameters on control performance in the optimal control problem is significant. To focus on evaluating the control performance of the proposed path-tracking strategy, in the contrasting simulations, these same parameters in different methods are all set to the same value. Then, the parameters in the numerical simulation are given as follows. Following [10], the external sinusoidal disturbance term is set as $\tau_d = [1.25\sin(t); 0.785\sin(t); 0.485\sin(t); 0.0325\sin(t); 0.325\sin(t)]$. The upper bound of the surge speed step signal $\Delta \bar{u}_s$ is set as 0.05. Following [10], the parametric uncertainties are reflected by the percentage of the hydrodynamic term. Then, ΔF_{CD} is set as $\Delta F_{CD} = 0.2(C(v) + D(v))$.

Note that the proposed path-tracking strategy consists of a LMPC controller and a tube MPC controller. The LMPC controller is used to calculate the speed control law to converge the path-tracking deviation, and the tube MPC controller is used to track the speed control law.

For the LMPC controller, these parameters in (24)–(29) are listed in Table 1. In the LMPC controller, weighting matrix Q_η and Q_v are for minimizing the path-tracking deviation e_η. The weight matrix R_v is for the smooth change in AUV's speed. With these weight matrices set appropriately, the speed control law can efficiently converge the path-tracking deviation, avoiding abrupt changes in AUV's speed.

Table 1. Parameter value in the LMPC controller.

Parameter	Value	Parameter	Value	Parameter	Value
Q_η	diag{4.4, 19.2, 5.2, 20.5, 25.5}	P_η	diag{4.4, 19.2, 5.2, 20.5, 25.5}	$\bar{v}_{min}$	$-[0; 0.06; 0.01; 0.03; 0.08]$
Q_v	25.5	P_v	25.5	N_l	4
R_v	diag{2, 0.3, 5, 2, 0.1}	$\bar{u}_{l\,max}$	$[0.2; 0.01; 0.01; 0.03; 0.05]$	$\bar{v}_{max}$	$[1.2; 0.06; 0.01; 0.03; 0.08]$

Note that the tube MPC controller is used for surge speed control, heading control, and depth control, respectively, based on these decoupled models (19), (20), and (22). These corresponding parameters of each controller in (24)–(29) and (38) are listed in Tables 2–4. ΔF_x is the increment of the stern thruster force. $\Delta \delta_r$ is the increment of the vertical plane deflection. $\Delta \delta_s$ is increment of the translation plane deflection. In the tube MPC controller, the weighting matrices play a similar role. With the appropriate Q_T and R_T set, the control input of the AUV can change smoothly to track the nonmail speed control law. The RPI set in Definition 1 is used to obtain the tight constraint in nominal system dynamics to ensure that the deviation z (9) also contained in the RPI set. The feedback matrix is used to converge the deviation. As mentioned in Section 4.2, with appropriate parameters λ, μ and P_R obtained, the tube MPC controller can efficiently track the speed control law.

Table 2. Parameter value in the tube MPC controller for surge speed control.

Parameter	Value	Parameter	Value	Parameter	Value	Parameter	Value	Parameter	Value
Q_T	20.5	R_T	50.5	P_T	138.8	N_T	5
T	0.05	P_v	25.5	λ	2.7	P_R	2.3
$\mathcal{X}$	$\{u_s \mid 0 \leq u_s \leq 1.2\}$	$\mathcal{U}$	$\{(F_x, \Delta F_x) \mid \lvert F_x \rvert \leq 15, \lvert \Delta F_x \rvert \leq 2\}$	c_ω	0.12	μ	1.7
K	-182.38	$\overline{w}$	5	$c_{\delta,u}$	0.01	ρ	0.5
l	3	a_1	0.2	a_2	0.1	a_3	0.05

Table 3. Parameter value in the tube MPC controller for heading control.

Parameter	Value	Parameter	Value	Parameter	Value	Parameter	Value	Parameter	Value
Q_T	$\mathrm{diag}\{190.5, 180.5\}$	R_T	50	P_T	$[6462.1, 215.8; 215.8, 3688.3]$	N_T	9
T	0.05	$\overline{s}$	5°	λ	0.6	P_R	$\mathrm{diag}\{0.91, 0.61\}$
$\mathcal{X}$	$\{(v,r) \mid \lvert v \rvert \leq 0.01, \lvert r \rvert \leq 0.05\}$	$\mathcal{U}$	$\{(\delta_r, \Delta \delta_r) \mid \lvert \delta_r \rvert \leq 20°, \lvert \Delta \delta_r \rvert \leq 5°\}$	c_ω	0.07	μ	2.6
K	$[-28.29; 11.54]$	$\overline{w}$	10°	$c_{\delta,u}$	0.01	ρ	0.5
l	3	a_1	0.2	a_2	0.1	a_3	0.05

Table 4. Parameter value in the tube MPC controller for depth control.

Parameter	Value	Parameter	Value	Parameter	Value	Parameter	Value	Parameter	Value
Q_T	$\mathrm{diag}\{2.5, 5.5\}$	R_T	5	P_T	$[199.5, 25.6; 25.6, 103.9]$	N_T	9
T	0.05	$\overline{s}$	5°	λ	1.8	P_R	$\mathrm{diag}\{0.29, 0.59\}$
$\mathcal{X}$	$\{(w,q) \mid \lvert w \rvert \leq 0.02, \lvert q \rvert \leq 0.07\}$	$\mathcal{U}$	$\{(\delta_s, \Delta \delta_s) \mid \lvert \delta_s \rvert \leq 14°, \lvert \Delta \delta_s \rvert \leq 5°\}$	c_ω	0.05	μ	2.9
K	$[-28.29; 11.54]$	$\overline{w}$	10°	$c_{\delta,u}$	0.01	ρ	0.5
l	3	a_1	0.2	a_2	0.1	a_3	0.05

When the adaptive flexible tube is used, two decision variables are used to dynamically adjust these tight constraints. Parameter $\overline{s}$ represents the upper bound of the tight constraints. Parameters $\overline{w}$, ρ, and nonlinear function $\widetilde{w}_\delta(s_{k|t})$ are used to represent the variation in the tight constraint.

5.2. Analysis and Discussion

The reference path of the AUV is generated by tracking the sinusoidal shape trajectory, and the initial state of the AUV is set as: $\eta_0 = [0; 0; 0; 0; 27 * \pi/180]$, $v_0 = [0.1; 0; 0; 0; 0]$.

To visually compare the control performance of "MPC", "RTMPC", and "ATMPC", AUV's trajectories during path tracking are shown in Figure 3.

Intuitive path-tracking performance can be visualized in the trajectory of AUV during path tracking, which is the position of the AUV given in Section 3.1. Figure 3 shows a three-dimensional view of the AUV's path-tracking control performance of "MPC", "RTMPC", and "ATMPC". The trajectory of "MPC" fails to track the reference trajectory well. Although the path-tracking deviation of "MPC" tends to converge, there are still several obvious position offsets, especially at the beginning of path tracking. Actual trajectories of both "RTMPC" and "ATMPC" can separately track the nominal trajectory. Note that the nominal trajectory of "ATMPC" tracks the reference trajectory better, compared with that of "RTMPC".

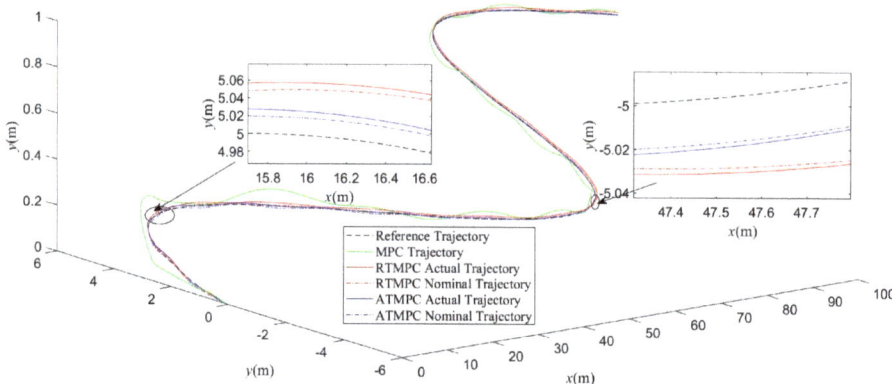

Figure 3. AUV trajectory during path tracking.

$$\begin{cases} x_r = t \\ y_r = 5\sin(0.1t) \\ z_r = 0.01t \\ \theta_r = -\text{atan}(\frac{0.01}{\sqrt{1+0.5\cos^2(0.1t)}}) \\ \psi_r = \text{atan}(0.5\cos(0.1t)) \\ u_{sr} = 1 \end{cases} \qquad (41)$$

where x_r, y_r, and z_r are the reference positions. u_{sr} is the reference surge speed.

To compare the path-tracking deviation in detail, the path-tracking deviation and path-tracking integral deviation are respectively shown in Figures 4 and 5. The maximum deviations in position and attitude angles are given in Table 5. Section 3.3 introduced the path-tracking deviation, whose absolute value is used. The definition of path-tracking integral deviation is given in (41). As shown in Figure 4, under sinusoidal external disturbances and parametric uncertainties, position and attitude angle deviations of three methods all have a bounded and convergent tendency over time. Compared with the position and attitude deviations of "MPC", that of "RTMPC" has been all effectively reduced in every moment. As shown in Figure 5, the growth trend of the integral deviation is also much slower. In addition, the maximum position deviation of "RTMPC" can be reduced from 0.38 m to 0.12 m. That is a reduction of about 68%. The maximum pitch angle deviation of "RTMPC" can be reduced from 3.45° to 0.58°. That is a reduction of about 83%. The maximum yaw angle deviation of "RTMPC" can be reduced from 3.45° to 1.07°. That is a reduction of about 69%. It can be seen the "RTMPC" has good robustness against external disturbances and parametric uncertainties.

Compared with "RTMPC", the proposed tube-based event-triggered path-tracking strategy has a smaller position and attitude angle deviations. The maximum position deviation of "ATMPC" can be reduced from 0.12 m to 0.04 m. That is a reduction of about 67%. As shown in Figure 4, compared with the position deviation in the x direction and of "RTMPC", that of "ATMPC" is almost the same in every moment. However, after about 20 s, the position deviation of y is much smaller in every moment. In addition, the maximum yaw angle deviation can be reduced from 1.07° to 0.45°. That is a reduction of about 58%. As shown in Figure 4, after about 10 s, the yaw angle deviation is almost smaller in every moment. Integral deviations can intuitively show the variation trend of these position and attitude deviations in Figure 5.

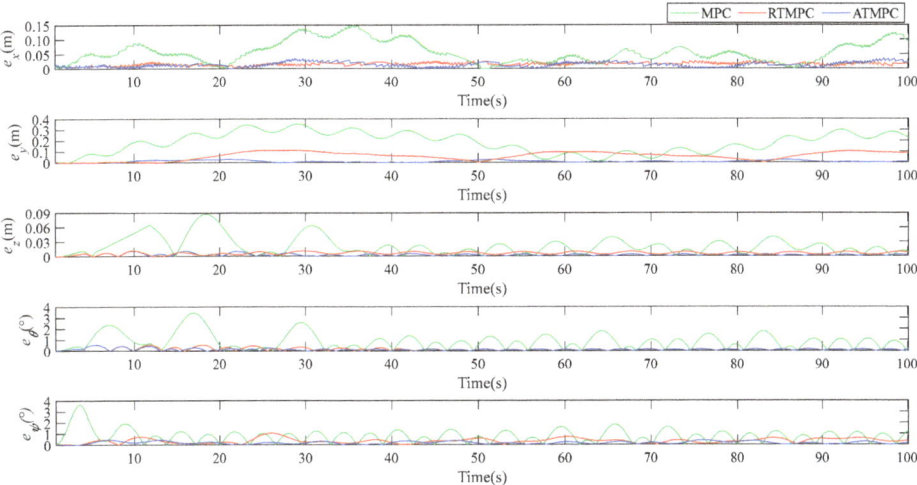

Figure 4. Deviation of position and attitude angle.

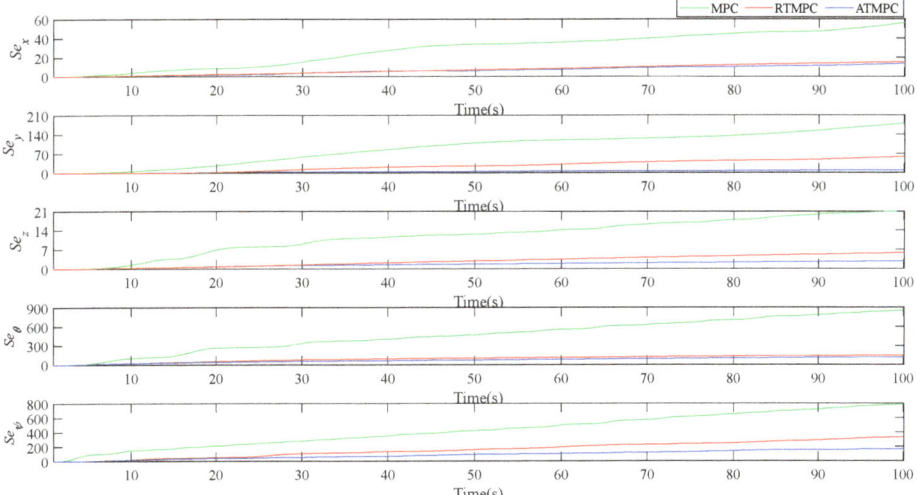

Figure 5. Integral deviation of position and attitude angle.

Table 5. Maximum deviation of position and attitude angles.

Method	Max Position Deviation (m)	Max Pitch Angle Deviation (°)	Max Yaw Angle Deviation (°)
MPC	0.38	3.45	3.45
RTMPC	0.12	0.58	1.07
ATMPC	0.04	0.57	0.45

To compare control input smoothness, the range of the AUV's speed and the control input are respectively shown in Figures 6 and 7. AUV's speed and the control input have been given in Section 3.1. As shown in Figure 6, the surge speed of "ATMPC" and "RTMPC" tracks the desired surge speed well. The sway speed, heave speed, pitch speed, and yaw speed changes in "ATMPC" and "RTMPC" occur more smoothly, compared with those of "MPC".

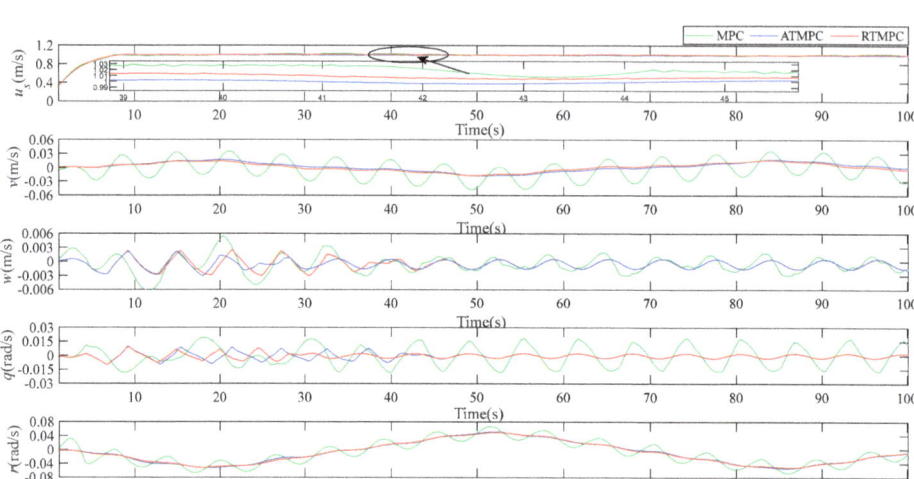

Figure 6. Range of the AUV's speed.

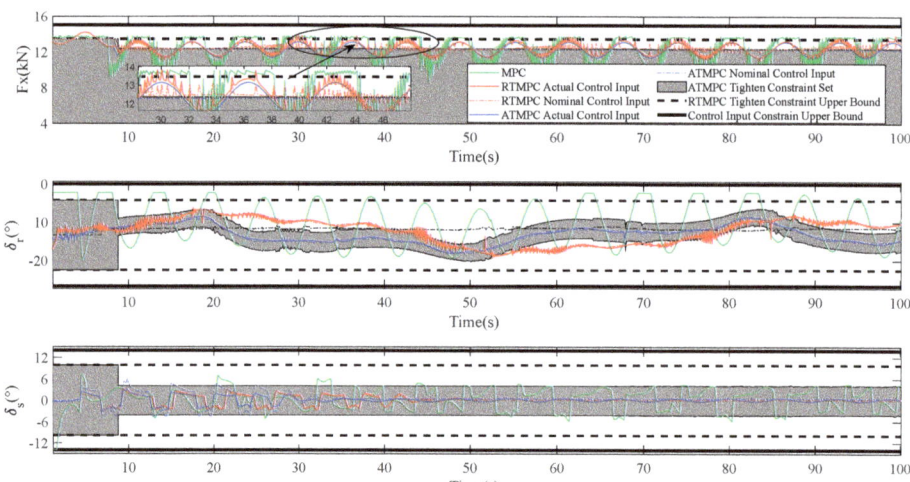

Figure 7. Range of the AUV's control input.

As shown in Figure 7, compared with the stern thruster force of "MPC", that of "ATMPC" and "RTMPC" have better smoothness, avoiding the high-frequency oscillation phenomenon. As shown in the local zoom-in of Figure 7, the smoothness of the stern thruster force of "ATMPC" is enhanced, compared with that of "RTMPC". It can be seen that the nominal control input of the "ATMPC" is within the upper bound of the tight constraint, and the output value of "ATMPC" is almost the lowest, which may be consistent with the purpose of energy conservation in real-world application.

Like the vertical plane deflection shown in Figure 7, those of "RTMPC" and "ATMPC" can all avoid large periodic changes, compared with that of "MPC". At the beginning of the simulation, the range of the vertical plane deflection of "MPC" has a tendency to be unstable. With the adaptive tight constrain introduced, the vertical plane deflection of "ATMPC" changes more smoothly. As the simulation time goes on, there is almost no oscillation phenomenon. As shown in Figure 7, the blue line is contained within the gray area, and the trend of the upper and lower limits in the gray area is consistent with the trend of the blue line.

Like the translational plane deflection shown in Figure 7, that of "RTMPC" and "ATMPC" can also avoid large periodic changes, compared with that of "MPC". With the adaptive tight constraint introduced, the translational plane deflection of "ATMPC" changes more smoothly, and tends to stabilize more quickly.

To analyze the real-time performance, the time consumption of different methods is recorded in Table 6. Note that the tightened constraint set of "RTMPC" is calculated offline. It can be explained that the average time consumption and the maximal time consumption of "RTMPC" are almost the same as those of "MPC". The average time consumption and the maximal time consumption of "ATMPC" will not increase by much: the average time consumption increases by 2.91 ms, and the maximal time consumption increases by 3.47 ms.

Table 6. Time consumption of different methods.

Method	Max Time Consumption (ms)	Average Time Consumption (ms)
MPC	7.25	7.27
RTMPC	7.88	7.34
ATMPC	11.35	10.25

6. Conclusions

In this paper, a novel tube-based event-triggered path-tracking strategy against disturbance is proposed, which consists of an LMPC controller and a tube MPC controller. In the LMPC controller, based on the nominal kinematics model of the AUV, a nominal optimal speed control law is obtained to converge the nominal path-tracking deviation. In the tube MPC controller, AUV's available control inputs are separately calculated based on a decoupled model. Considering the nonlinear hydrodynamic characteristics of the AUV, an LMI is formulated to calculate the feedback matrix and tight constraints offline. The terminal region in the tube MPC controller is obtained offline using linear differential inclusion technology. When the surge speed step signal does not exceed the upper bound, the tight constraints become adaptive. Numerical simulation results show that the feedback matrix is successfully used to match the actual trajectory and the nominal trajectory. With the adaptive constraints introduced, the nominal trajectory tracks the reference better. Note that the online computing time of the tube MPC is acceptable, and these corresponding control inputs are also smooth. Therefore, the proposed tube-based event-triggered path-tracking strategy can enhance the path-tracking performance and ensure good real-time performance.

In the MPC controller, the disturbance upper bound needs to be set appropriately. If the bound is too small, the robustness is weak; otherwise, the tube will be too conservative. The RPI set may not be obtained, or the optimal control problem is easy to be infeasible. In numerical simulation, the disturbance upper bound is still easy to set appropriately. In the application of a real-world system, it may be a challenge. The disturbance bound is different for different real-time scenarios, which may be difficult to accurately set. This may lead to degradation of the control performance. In future research, the work will be extended to predict the model mismatches due to parametric uncertainties and external disturbances to improve the accuracy of the nominal model, based on data-driven technology, such as machine learning. The RPI set is used to address the bounded prediction deviation. If the prediction deviation is convergent and bounded, it can effectively solve the problem of setting the disturbance upper bound in real-world applications.

Author Contributions: Conceptualization, Y.C. and Y.B.; methodology, Y.C. and Y.B.; software, Y.C. and Y.B.; validation, Y.C. and Y.B.; formal analysis, Y.C. and Y.B.; investigation, Y.C. and Y.B.; resources, Y.C. and Y.B.; data curation, Y.C. and Y.B.; writing—original draft preparation, Y.C.; writing—review and editing, Y.C. and Y.B.; visualization, Y.C. and Y.B.; supervision, Y.B.; project administration, Y.B.; funding acquisition, Y.B. All authors have read and agreed to the published version of the manuscript.

Funding: This work was supported by the Hunan Provincial Natural Science Foundation of China (2021JC0010), and Defense Industrial Technology Development Program (JCKY2021110B024, JCKY2022110C072).

Data Availability Statement: Data is unavailable due to privacy or ethical restrictions.

Acknowledgments: Haicheng Zhang (Hunan University) and Weisheng Zou (Hunan University) were acknowledged to provide careful guidance.

Conflicts of Interest: The authors declare no conflict of interest.

Nomenclature

$\ominus$	Pontryagin difference, $A \ominus B = \{x \mid x + y \in A, y \in B\}$	I_n	n-dimensional identity matrix
$\alpha_{\min}(\cdot)(\alpha_{\max}(\cdot))$	the smallest (largest) real part of eigenvalues of a matrix	$R^{m \times n}$	A matrix with m rows and n columns
w	bounded external disturbance	c_ω	disturbance upper bound
Q, R, P	positive weight matrix	$\|\cdot\|_Q^2$	quadratic norm of a vector with positive weight matrix Q
$g(\cdot)$	Lipschitz nonlinear function	L	Lipschitz constant
$\bar{x}, \bar{u}$	nominal state and control input	x, u	actual state and control input
J_N, J_1	cost function	K	feedback matrix
X_f	terminal feasible set	Ω	robust positively invariant (RPI) set
$h(\cdot) < 0$	inequality constraint	M	constraint set
$f_d(\cdot)$	state transition function	N_T, N_l	predictive horizon
$\Theta(\cdot)$	linear differential inclusion function	$\text{Co}\Theta(\cdot)$	minimum convex polytope
$\det(\cdot)$	determinant calculation	$\alpha_w(\cdot)$	polynomial function
$\mathbb{K}_{N_1:N_2}$	set $\{N_1, N_1+1, \cdots, N_2-1, N_2\}$		

References

1. Bibuli, M.; Zereik, E.; De Palma, D.; Ingrosso, R.; Indiveri, G. Analysis of an Unmanned Underwater Vehicle Propulsion Model for Motion Control. *J. Guid. Control Dyn.* **2022**, *45*, 1046–1059. [CrossRef]
2. Guerrero, J.; Torres, J.; Creuze, V.; Chemori, A.; Campos, E. Saturation based nonlinear PID control for underwater vehicles: Design, stability analysis and experiments. *Mechatronics* **2019**, *61*, 96–105. [CrossRef]
3. Petritoli, E.; Bartoletti, C.; Leccese, F. Preliminary Study for AUV: Longitudinal Stabilization Method Based on Takagi-Sugeno Fuzzy Inference System. *Sensors* **2021**, *21*, 1866. [CrossRef] [PubMed]
4. Sedghi, F.; Arefi, M.M.; Abooee, A.; Kaynak, O. Adaptive Robust Finite-Time Nonlinear Control of a Typical Autonomous Underwater Vehicle with Saturated Inputs and Uncertainties. *IEEE/ASME Trans. Mechatron.* **2021**, *26*, 2517–2527. [CrossRef]
5. Yang, X.; Yan, J.; Hua, C.C.; Guan, X.P. Trajectory Tracking Control of Autonomous Underwater Vehicle with Unknown Parameters and External Disturbances. *IEEE Trans. Syst. Man Cybern. Syst.* **2021**, *51*, 1054–1063. [CrossRef]
6. Jia, Z.; Qiao, L.; Zhang, W. Adaptive tracking control of unmanned underwater vehicles with compensation for external perturbations and uncertainties using Port-Hamiltonian theory. *Ocean Eng.* **2020**, *209*, 107402. [CrossRef]
7. Kou, L.; He, S.; Li, Y.; Xiang, J. Constrained Control Allocation of a Quadrotor-Like Autonomous Underwater Vehicle. *J. Guid. Control Dyn.* **2021**, *44*, 659–666. [CrossRef]
8. Shen, C.; Shi, Y.; Buckham, B. Trajectory Tracking Control of an Autonomous Underwater Vehicle Using Lyapunov-Based Model Predictive Control. *IEEE Trans. Ind. Electron.* **2018**, *65*, 5796–5805. [CrossRef]
9. Li, S.; Li, Z.; Yu, Z.; Zhang, B.; Zhang, N. Dynamic Trajectory Planning and Tracking for Autonomous Vehicle with Obstacle Avoidance Based on Model Predictive Control. *IEEE Access* **2019**, *7*, 132074–132086. [CrossRef]
10. Long, C.; Hu, M.; Qin, X.; Bian, Y. Hierarchical trajectory tracking control for ROVs subject to disturbances and parametric uncertainties. *Ocean Eng.* **2022**, *266*, 112733. [CrossRef]
11. Li, H.; Swartz, C.L.E. Robust model predictive control via multi-scenario reference trajectory optimization with closed-loop prediction. *J. Process Control* **2021**, *100*, 80–92. [CrossRef]

12. Wang, W.; Yan, J.; Wang, H.; Ge, H.; Zhu, Z.; Yang, G. Adaptive MPC trajectory tracking for AUV based on Laguerre function. *Ocean Eng.* **2022**, *261*, 111870. [CrossRef]
13. Zhang, Y.; Liu, X.; Luo, M.; Yang, C. MPC-based 3-D trajectory tracking for an autonomous underwater vehicle with constraints in complex ocean environments. *Ocean Eng.* **2019**, *189*, 106309. [CrossRef]
14. Yan, Z.; Gong, P.; Zhang, W.; Wu, W. Model predictive control of autonomous underwater vehicles for trajectory tracking with external disturbances. *Ocean Eng.* **2020**, *217*, 107884. [CrossRef]
15. Gong, P.; Yan, Z.; Zhang, W.; Tang, J. Trajectory tracking control for autonomous underwater vehicles based on dual closed-loop of MPC with uncertain dynamics. *Ocean Eng.* **2022**, *265*, 112697. [CrossRef]
16. Dai, L.; Lu, Y.; Xie, H.; Sun, Z.; Xia, Y. Robust Tracking Model Predictive Control with Quadratic Robustness Constraint for Mobile Robots with Incremental Input Constraints. *IEEE Trans. Ind. Electron.* **2021**, *68*, 9789–9799. [CrossRef]
17. Long, C.Q.; Qin, X.H.; Bian, Y.G.; Hu, M.J. Trajectory tracking control of ROVs considering external disturbances and measurement noises using ESKF-based MPC. *Ocean Eng.* **2021**, *241*, 109991. [CrossRef]
18. Nikou, A.; Verginis, C.K.; Heshmati-alamdari, S.; Dimarogonas, D.V. A robust non-linear MPC framework for control of underwater vehicle manipulator systems under high-level tasks. *Iet Control Theory Appl.* **2021**, *15*, 323–337. [CrossRef]
19. Vu, Q.V.; Dinh, T.A.; Nguyen, T.V.; Tran, H.V.; Le, H.X.; Pham, H.V.; Nguyen, L. An Adaptive Hierarchical Sliding Mode Controller for Autonomous Underwater Vehicles. *Electronics* **2021**, *10*, 2316. [CrossRef]
20. Blanchini, F. Control synthesis for discrete time systems with control and state bounds in the presence of disturbances. *J. Optim. Theory Appl.* **1990**, *65*, 29–40. [CrossRef]
21. Rakovic, S.V.; Kerrigan, E.C.; Kouramas, K.I.; Mayne, D.Q. Invariant approximations of the minimal robust positively Invariant set. *IEEE Trans. Autom. Control* **2005**, *50*, 406–410. [CrossRef]
22. Chryssochoos, I.; Raković, S.; Mayne, D.Q. Robust model predictive control using tubes. *Automatica* **2004**, *40*, 125–133.
23. Yu, S.; Maier, C.; Chen, H.; Allgöwer, F. Tube MPC scheme based on robust control invariant set with application to Lipschitz nonlinear systems. *Syst. Control Lett.* **2013**, *62*, 194–200. [CrossRef]
24. Chen, Y.; Bian, Y.; Cui, Q.; Dong, L.; Xu, B.; Hu, M. LTVMPC for Dynamic Positioning of An Autonomous Underwater Vehicle. In Proceedings of the 2021 5th CAA International Conference on Vehicular Control and Intelligence (CVCI), Tianjin, China, 29–31 October 2021; pp. 1–5.
25. Chen, W.H.; O'Reilly, J.; Ballance, D.J. On the terminal region of model predictive control for non-linear systems with input/state constraints. *Int. J. Adapt. Control Signal Process.* **2003**, *17*, 195–207. [CrossRef]
26. Cox, W.; While, L.; Reynolds, M. A Review of Methods to Compute Minkowski Operations for Geometric Overlap Detection. *IEEE Trans. Vis. Comput. Graph.* **2021**, *27*, 3377–3396. [CrossRef]

Disclaimer/Publisher's Note: The statements, opinions and data contained in all publications are solely those of the individual author(s) and contributor(s) and not of MDPI and/or the editor(s). MDPI and/or the editor(s) disclaim responsibility for any injury to people or property resulting from any ideas, methods, instructions or products referred to in the content.

MDPI
St. Alban-Anlage 66
4052 Basel
Switzerland
www.mdpi.com

Electronics Editorial Office
E-mail: electronics@mdpi.com
www.mdpi.com/journal/electronics

Disclaimer/Publisher's Note: The statements, opinions and data contained in all publications are solely those of the individual author(s) and contributor(s) and not of MDPI and/or the editor(s). MDPI and/or the editor(s) disclaim responsibility for any injury to people or property resulting from any ideas, methods, instructions or products referred to in the content.

www.ingramcontent.com/pod-product-compliance
Lightning Source LLC
LaVergne TN
LVHW070409100526
838202LV00014B/1416